VOLUME ONE HUNDRED AND FORTY SIX

Progress in
MOLECULAR BIOLOGY AND TRANSLATIONAL SCIENCE

Molecular Biology of Aging

VOLUME ONE HUNDRED AND FORTY SIX

Progress in MOLECULAR BIOLOGY AND TRANSLATIONAL SCIENCE

Molecular Biology of Aging

Edited by

P. HEMACHANDRA REDDY
Garrison Institute on Aging,
Texas Tech University Health Sciences Center,
Lubbock, TX, United States

ELSEVIER

ACADEMIC PRESS
An imprint of Elsevier
elsevier.com

Academic Press is an imprint of Elsevier
50 Hampshire Street, 5th Floor, Cambridge, MA 02139, United States
525 B Street, Suite 1800, San Diego, CA 92101-4495, United States
The Boulevard, Langford Lane, Kidlington, Oxford OX5 1GB, United Kingdom
125 London Wall, London EC2Y 5AS, United Kingdom

First edition 2017

ISBN: 978-0-12-811532-9
ISSN: 1877-1173

Publisher: Zoe Kruze
Acquisition Editor: Alex White
Editorial Project Manager: Helene Kabes
Production Project Manager: Magesh Kumar Mahalingam
Cover Designer: Alan Studholme

Typeset by SPi Global, India

CONTENTS

IN MEMORIAM

P. Michael Conn, PhD

P. Michael Conn unexpectedly passed away on Saturday, November 26, 2016 in Lubbock, TX, United States. Dr. Conn was a pioneer in the field and a role model with great dedication to scientific discovery. At Texas Tech University Health Sciences Center (TTUHSC), Dr. Conn was an outstanding and highly respected researcher, educator, leader, director, consultant, and manager of university programs. He elevated the TTUHSC research mission by supporting its scientists across disciplines, departments, and schools.

He received his Bachelor of Science and a teaching certificate from the University of Michigan (1971), a Master of Science from North Carolina State University (1973), and a doctorate from Baylor College of Medicine (1976). Dr. Conn received a postdoctoral fellowship to study the endocrine research methods at the National Institutes of Health in Bethesda, MD (1976–78). He then joined the faculty of the Department of Pharmacology at Duke University Medical Center (1978) as an assistant professor and was promoted to associate professor in 1982. In 1984, Dr. Conn went to the University of Iowa College of Medicine, where he accepted a position as professor and head of the Department of Pharmacology, a position he held for 11 years.

Dr. Conn joined TTUHSC in December 2013 as the Senior Vice President for Research. He was also Associate Provost and the Robert C. Kimbrough Professor of Internal Medicine at TTUHSC, with joint appointment in the Department of Cell Biology and Biochemistry. He was previously the Director of Research Advocacy at TTUHSC. Before coming to TTUHSC, Dr. Conn was a professor in the Departments of Physiology and Pharmacology, Cell Biology and Development, and Obstetrics and Gynecology at the Oregon Health and Science University, and a senior scientist at the Oregon National Primate Research Center. Dr. Conn served for 12 years as a special assistant to the primate center director before becoming Associate Director.

For nearly 40 years, he was a leading scientist in elucidating the function of the G protein-coupled receptor in the gonadotropin-releasing hormone receptor system. Dr. Conn's research led to a better understanding of therapeutic targets to help patients with endocrine disease.

Dr. Conn was the first to report that membrane receptors, when they bound to agonists, but not to antagonists. Dr. Conn's research into membrane receptors changed the way these proteins were viewed by the scientific community.

Another line of research that Dr. Conn pursued was receptor–receptor interactions. This research contributed to our understanding of the function of membrane receptors, and it led to what was then called microaggregation—the massing of receptor dimers, which Dr. Conn distinguished from macroaggregation. Dr. Conn's research into membrane receptors and their interactions led to the development of oligomerization, a chemical process that converts molecular aggregates into molecular complexes. This process is important for our present understanding of how receptors regulate and communicate information to other receptors. Dr. Conn also contributed significantly to the scientific community's understanding of the use of diacylglycerols, which are lipids. Working with Jim Neidel, Dr. Conn revealed how these lipids are involved in hormonal action.

Dr. Conn demonstrated that many receptor mutations result in the misrouting of molecules. With this information, Dr. Conn developed a treatment strategy that restores mutant receptors to function. This strategy appears useful in restoring a range of mutant receptors to normal function, including receptors in cystic fibrosis, nephrogenic diabetes insipidus, hypercholesterolemia, retinitis pigmentosa, and a range of digestive diseases. Dr. Conn also created high-throughput screening, a drug-discovery process widely used in the pharmaceutical industry, to automatically assay the biochemical activity of drug-like compounds, from which chemical libraries are formed. Dr. Conn's research into high-throughput screenings has resulted in the appreciation of pharmacoperone drugs as a new class of drugs to treat abnormal receptors.

Dr. Conn authored or coauthored over 350 publications in receptor research, and he wrote or was the editor of over 200 books, including text books on neuroscience, molecular biology, and endocrinology. Dr. Conn served as the editor of many professional journals and book series, including *Endocrinology*, *Journal of Clinical Endocrinology and Metabolism*, *Endocrine*, *Methods*, *Progress in Molecular Biology and Translational Science*, and

Contemporary Endocrinology. Dr. Conn was also a member of numerous study sections, and advisory committees and groups: 1986–87, *Biochemical Endocrinology*; 1991–95, *Pharmacological Sciences*; 1985–89, American Society for Cell Biology, Council Member; 1992–97, The Endocrine Society, Council Member; 1996, The Endocrine Society, President; 1997–2000, The Hormone Foundation Board of Directors; 1998–2000, National Diabetes Education Program Steering Committee; 1995–2002, Pituitary Tumor Network Association Scientific Advisory Committee; and 2000–02, FASEB Board of Directors.

Dr. Conn served on the National Board of Medical Examiners, including 2 years as the Chair of its Reproduction and Endocrinology Committee, and he was on the Board of Scientific Councilors for the Intramural Program in NICHD at the National Institutes of Health. Dr. Conn was a member of Council for the American Society for Cell Biology and the Endocrine Society, and he was a former president of the Endocrine Society, during which time he founded the Hormone Foundation and worked with political leaders throughout the United States to heighten the public's awareness of diabetes. Dr. Conn was an elected member of the Mexican Institute of Medicine and a fellow of the American Association for the Advancement of Science.

In recognition of Dr. Conn as an extraordinary scientist and educator, he received many awards and honors. Dr. Conn's students and fellows have gone on to become leaders in industry and academia. Dr. Conn was given a MERIT award from the National Institutes of Health; the J.J. Abel Award of the American Society for Pharmacology and Experimental Therapeutics; the Weitzman, Oppenheimer and Ingbar Awards of the Endocrine Society; the National Science Medal of Mexico (the Miguel Aleman Prize); and the Stevenson Award of Canada. He was also the recipient of the Medical Research Foundation Oregon Award for Discovery, the Media Award of the American College of Neuropsychopharmacology, and a distinguished alumnus of Baylor College of Medicine. Dr. Conn's honors included the Dean's Award from TTUHSC, bestowed upon him for outstanding work as a scientist.

P. Michael Conn was our friend, teacher, and mentor, and we will miss him dearly.

P. Hemachandra Reddy, PhD

CONTRIBUTORS

F. Akhter
School of Pharmacy, Higuchi Bioscience Center, University of Kansas, Lawrence, KS, United States

N. Basisty
University of Washington, Seattle, WA, United States

G.K. Bhatti
UGC Centre of Excellence in Nano Applications, Panjab University, Chandigarh, India

J.S. Bhatti
Garrison Institute on Aging, Texas Tech University Health Sciences Center, Lubbock, TX, United States; Department of Biotechnology, Sri Guru Gobind Singh College, Chandigarh, India

D. Chen
School of Pharmacy, Higuchi Bioscience Center, University of Kansas, Lawrence, KS, United States

Y.-A. Chiao
University of Washington, Seattle, WA, United States

J.W. Culberson
Texas Tech University Health Sciences Center, Lubbock, TX, United States

D.-F. Dai
University of Washington, Seattle, WA, United States

C. Hayley
University of Kansas Alzheimer's Disease Center, University of Kansas School of Medicine, Landon Center on Aging, Kansas City, KS, United States

Y. Ji
University of Kansas Alzheimer's Disease Center, University of Kansas School of Medicine, Landon Center on Aging, Kansas City, KS, United States

R. Kandimalla
Garrison Institute on Aging, Texas Tech University Health Sciences Center, Lubbock, TX, United States

S. Koppel
University of Kansas Alzheimer's Disease Center, University of Kansas School of Medicine, Landon Center on Aging, Kansas City, KS, United States

S. Kumar
Garrison Institute on Aging, Texas Tech University Health Sciences Center, Lubbock, TX, United States

C.S. Kuruva
Garrison Institute on Aging, Texas Tech University Health Sciences Center, Lubbock, TX, United States

M. Manczak
Garrison Institute on Aging, Texas Tech University Health Sciences Center, Lubbock, TX, United States

D.J. Marcinek
University of Washington, Seattle, WA, United States

G.M. Martin
University of Washington, Seattle, WA, United States

S.S. Prabhakar
Texas Tech University Health Sciences Center, Lubbock, TX, United States

S. Pugazhenthi
University of Colorado, Aurora; Eastern Colorado Health Care System, Denver, CO, United States

E.K. Quarles
University of Washington, Seattle, WA, United States

P.S. Rabinovitch
University of Washington, Seattle, WA, United States

A.P. Reddy
Texas Tech University Health Sciences Center, Lubbock, TX, United States

P.H. Reddy
Garrison Institute on Aging, Texas Tech University Health Sciences Center; Texas Tech University Health Sciences Center, Lubbock, TX, United States

F. Smith
Garrison Institute on Aging, Texas Tech University Health Sciences Center, Lubbock, TX, United States

H. Sobamowo
Texas Tech University Health Sciences Center, Lubbock, TX, United States

R.H. Swerdlow
University of Kansas Alzheimer's Disease Center, University of Kansas School of Medicine, Landon Center on Aging, Kansas City, KS, United States

M. Vijayan
Garrison Institute on Aging, Texas Tech University Health Sciences Center, Lubbock, TX, United States

R. Wang
Garrison Institute on Aging, Texas Tech University Health Sciences Center, Lubbock, TX, United States

I. Weidling
University of Kansas Alzheimer's Disease Center, University of Kansas School of Medicine, Landon Center on Aging, Kansas City, KS, United States

H.M. Wilkins
University of Kansas Alzheimer's Disease Center, University of Kansas School of Medicine, Landon Center on Aging, Kansas City, KS, United States

J. Williams
Garrison Institute on Aging, Texas Tech University Health Sciences Center, Lubbock, TX, United States

S.F. Yan
School of Pharmacy, Higuchi Bioscience Center, University of Kansas, Lawrence, KS, United States

S.S. Yan
School of Pharmacy, Higuchi Bioscience Center, University of Kansas, Lawrence, KS, United States

X. Yin
Garrison Institute on Aging, Texas Tech University Health Sciences Center, Lubbock, TX, United States

PREFACE

The biology of aging is a topic that has long interested me. When I was child, I witnessed my grandfather dying at the relatively young age of 55 years. And my father died at 61 years of age. These losses have led me to try to answer why some people die so early and others live so long, 90 years and beyond. What are factors that cause early death? Are they genetic or environmental, or both? Is lifestyle the main factor that may shorten a life span?

In this modern era, with medical and technological advancements and with increased social media about health care, I am appreciative that longevity is increasing, but not without costs. Dementia rates in persons older than 80 years are alarmingly increasing in many populations of the world. It is important to better understand the biology behind aging and the aging brain. And it is also important to identify biomarkers of aging in order to develop effective therapeutic strategies.

There has been much research on aging and age-related diseases, and important contributions to better understanding the molecular biology of aging. This area is too broad to cover fully in this book or in a few books. I have narrowed the scope of this book to four topics under the rubric of molecular biology of aging: (1) molecular, cellular, and physiological bases of metabolic syndromes, including diabetes, obesity, and Alzheimer's disease; (2) the role of mitochondria in aging and Alzheimer's disease; (3) the role of microRNAs in aging and age-related human neurological diseases; and (4) the aging kidney and its physiological and pathological implications for diabetes.

Chapters 1, 2, and 8 primarily cover basic biology, and cellular and therapeutic aspects of metabolic syndromes, including diabetes, obesity, and Alzheimer's disease. In the first chapter, John Culberson covers morbidity in chronic diseases, with a focus on the role of aging and age-related chronic diseases, such as sarcopenia. Sarcopenia is an age-related loss of skeletal muscle mass, which is accelerated by chronic inflammation. Sarcopenia results in a cascade of cytokines, insulin resistance, hyperglycemia, and altered mitochondrial glucose signaling pathways. Dr. Culberson also covers neurogenesis and defective neuronal plasticity in the diabetic brain and advanced glycation end-products generated by chronic hyperglycemia identified in postmortem brains of persons with Alzheimer's disease.

In the second chapter, Jasvinder Singh Bhatti and colleagues discuss several features of metabolic disorders, particularly the involvement of mitochondrial dysfunction and oxidative damage in aging and age-related metabolic and neurodegenerative disorders. They focus on the structure, function, and physiology of mitochondria in such disorders as diabetes, obesity, cardiovascular diseases, and stroke. They also cover therapeutic strategies for mitochondrial dysfunction and oxidative stress in different age-related metabolic disorders, including such strategies as lifestyle intervention, and pharmacological and mitochondria-targeted therapeutic approaches.

Subbiah Pugazhenthi's chapter focuses on cellular changes in obesity, diabetes, hypertension, and cardiovascular disease. He describes risk factors for comorbidities, collectively referred to as the *metabolic syndrome.* This syndrome can play a critical role in driving neuroinflammation, an important factor of Alzheimer's disease pathogenesis. His research suggests a role for microglia, the resident immune cells of the brain, in Alzheimer's disease pathogenesis. Metabolic syndrome could reactivate microglia through the interface of blood–brain barrier. As Dr. Pugazhenthi notes, an age-dependent breakdown of the blood–brain barrier has been found in humans with neurological diseases, including those with Alzheimer's disease.

Chapters 6, 7, 9, and 11 focus on mitochondrial abnormalities and mitochondrial dysfunction, and protective effects of mitochondria-targeted antioxidants. In Chapter 6, Arubala P. Reddy and P. Hemachandra Reddy present a systematic review of mitochondria-targeted antioxidants and a summary of antioxidants that researchers have used in studying mouse models of Alzheimer's disease, elderly populations, and clinical trials involving patients with Alzheimer's disease. They also discuss recent progress in the development and testing of mitochondria-targeted molecules, using cell cultures and mouse models of Alzheimer's disease. They cover mitochondria-targeted molecules as potential therapeutic targets to delay or prevent the progression of Alzheimer's disease.

In Chapter 7, Peter Rabinovitch and colleagues discuss catalase mouse models that they have developed in order to understand the role of catalase in delaying aging. They describe three lines of mice—mice overexpressing catalase targeted to mitochondria (mCAT), peroxisomes (pCAT), and the nucleus (nCAT) that they have developed to investigate the role of hydrogen peroxide in aging. They review features of all three mouse models, noting that the mCAT mice have the longest and healthiest life span. Dr. Rabinovitch's group extensively studied mCAT mice and reviewed well in their chapter.

In Chapter 9, Russell Swerdlow and colleagues describe mitochondrial and bioenergetic functional changes in aging and Alzheimer's disease. They link mitochondrial and bioenergetic impairments to the aging brain, and they discuss a new avenue that involves transferring mitochondria from patients with Alzheimer's disease to cell lines depleted of endogenous mitochondrial DNA, in order to develop cytoplasmic hybrid cell lines of mice that exhibit specific biochemical, molecular, and histologic features of Alzheimer's disease. They also discuss their proposed mitochondrial cascade hypothesis that places mitochondrial dysfunction at the apex of the pathology pyramid for Alzheimer's disease.

In Chapter 11, Shirley ShiDu Yan and colleagues provide a review of major recent findings on mitochondrial abnormalities and synaptic dysfunction relevant to aging, neurodegeneration, and cognitive decline in persons with Alzheimer's disease and diabetes. Dr. Yan argues that elucidation of the role of mitochondrial perturbation can inform the development of specific small molecules capable of targeting aberrant mitochondrial function as a therapeutic delivery system for combating aging-related dementia and neurodegenerative diseases.

Chapters 3–5 focus on the role of microRNAs in aging and age-related diseases. In Chapter 3, Murali Vijayan and colleagues focused on ischemic stroke in aging and Alzheimer's disease, explaining that stroke and vascular dementia increase with an increase in a number of modifiable factors. They suggest that most strokes can be prevented or controlled through pharmacological and surgical interventions, and lifestyle changes. They also identify cellular changes that are implicated in ischemic stroke, including inflammatory responses, microRNA alterations, and marked changes in brain proteins. They review the latest developments of research that identifies protein biomarkers in peripheral and central nervous system tissues from aged persons.

In Chapter 4, Subodh Kumar and colleagues review research on the biogenesis of microRNAs and the role of miRNAs, particularly circulatory mRNAs, in detecting aging and neurodegenerative diseases, particularly Alzheimer's, Parkinson's, and Huntington's diseases. They hypothesize that, at a pathological level, changes in cellular homeostasis lead to the modulation of molecular function in cells, resulting in the deregulation of miRNA expression. They suggest that identification of these changes may open a new avenue for developing biomarkers capable of detecting aging and cellular senescence.

In Chapter 5, P. Hemachandra Reddy and colleagues discuss several aspects of aging, including oxidative damage, mitochondrial dysfunction, telomere shortening, and inflammation, all of which leads to cellular senescence. Reddy and colleagues hypothesize that cellular senescence may induce age-related human diseases, including Alzheimer's, Parkinson's, multiple sclerosis, amyotrophic lateral sclerosis, cardiovascular, cancer, and skin diseases. They also discuss microRNAs in aging persons and persons with Alzheimer's disease, as possible blood-based peripheral biomarkers of Alzheimer's disease.

In Chapter 10, Hezekiah Sobamowo and Sharma Prabhakar cover the physiology and pathology of the aging kidney, noting that aging is linked to a progressive decline in renal function along with concurrent morphological changes in kidney, ultimately leading to glomerulosclerosis. They also discuss cellular changes in the aging kidney.

I sincerely thank all the contributors for their outstanding chapters. I also thank Magesh Mahalingham, Helene Kabes, and Alex White at Elsevier, for their support and help in assembling this volume. I also recognize and thank P. Michael Conn, PhD, posthumously for introducing me to the first volume in the series *Molecular Biology of Aging in the Progress in Molecular Biology and Translational Science.*

P. Hemachandra Reddy, PhD

CHAPTER ONE

Clinical Aspects of Glucose Metabolism and Chronic Disease

J.W. Culberson[1]
Texas Tech University Health Sciences Center, Lubbock, TX, United States
[1]Corresponding author: e-mail address: john.culberson@ttuhsc.edu

Contents

Abstract

The burden of chronic disease is an emerging world health problem. Advances made in the treatment of individual disease states often fail to consider multimorbidity patterns in clinical research models. Adjusting for age as a confounder ignores its contribution as a powerful risk factor for most chronic diseases. Sarcopenia is an age-related loss of skeletal muscle mass, which is accelerated by chronic inflammation and its resulting cascade of cytokines. Skeletal muscle loss results in insulin resistance, hyperglycemia, and altered mitochondrial glucose signaling pathways. Vascular disease in the brain may alter blood–brain barrier function, allowing transport of substances into the brain which adversely affect the "astrocyte-centric" subunit. Neurogenesis that provides neuronal plasticity is impaired in the diabetic brain, while insulin resistance markers such as insulin-like growth factor (IGF-1) and insulin receptor substrate (IRS-1) are associated with poor cognitive performance. Advanced glycation end products generated by chronic hyperglycemia are found in postmortem AD brain. Intranasal insulin administration, a preferential route for CNS delivery, improved cognitive function in healthy adults, without affecting circulating levels of insulin or glucose. Exercise has demonstrated a neuroprotective effect through induction of antioxidative enzymes, neurotrophic, and vascular endothelial

Progress in Molecular Biology and Translational Science, Volume 146
ISSN 1877-1173
http://dx.doi.org/10.1016/bs.pmbts.2016.12.011

growth factors. Sarcopenia appears to be a dynamic process and is potentially reversible with attention to nutrition and cardiovascular fitness. Early detection and intervention may slow the progression of multimortality disease states and should be a focus of worldwide health systems.

1. INTRODUCTION

The burden of chronic disease is an emerging world health problem. Improved sanitation and medical care is dramatically decreasing the mortality of communicable disease, and life expectancy has increased sharply in many developing countries.[1] Advancing agricultural infrastructure, technology, and cultural shifts are resulting in behavioral and lifestyle changes across a large portion of the world population. Many of these changes significantly increase the risk of chronic disease, including diabetes mellitus, cardiovascular disease (CVD), chronic renal disease, and dementia.[2] The aging population demographic and a gradual shift toward multimorbidity patterns are challenging all healthcare systems. In recognition of the increasing importance of chronic diseases, the 2008 World Health Assembly endorsed a Global Non-Communicable Disease (NCD) Action Plan for NCD prevention and control.[3] While more highly developed health care systems have made great advances toward treating individual disease states, multimorbidity management systems, and prevention programs have not been a priority.

2. DIABETES MELLITUS

The number of people with diabetes mellitus worldwide has more than doubled over the past 3 decades and is projected to affect 7.7% of the total adult population of the world by 2030.[4,5] The most common form of diabetes, Type 2 Diabetes (T2DM), is characterized by insulin resistance (IR) and is considered a metabolic disorder closely tied to overweight (BMI > 25%) or obesity (BMI > 30%).[6] The prevalence of overweight or obesity in the world's population is predicted to rise from 33% in 2005 to 58% in 2030.[2] Ongoing research has demonstrated that the diabetes epidemic is the result of a complex interaction between genetic and epigenetic predispositions and societal factors that, in combination, determine behavioral and environmental risks.[7]

3. CARDIOVASCULAR DISEASE

CVD remains the most significant global health burden, and its prevalence in developing countries is expected to increase significantly due, in part, to the impending worldwide epidemic of DMT2.[8] While the risk of CVD is known to increase with obesity, recent work has found that metabolic status is more important than measures of adipose tissue quantity in estimating cardiac risk in individuals with T2DM.[9] Additionally, excessive dietary salt and caloric intake are linked not only to increased risk due to elevated blood pressure, but also to insulin resistance and impaired glucose metabolism. Insulin resistance, in turn, affects not only skeletal muscle but also the cardiovascular system, where it increases the risk of both CVD and chronic kidney disease (CKD).[10]

4. CHRONIC KIDNEY DISEASE

CKD is another common comorbidity in patients with T2DM. The presence and severity of T2DM, and associated cardiovascular complications, markedly influence the progression of CKD.[3] CKD is more common in certain patient populations including the elderly, obese, and certain ethnic groups. Similar to CVD, the association between obesity and CKD may have a metabolic component, and a favorable type of body fat, with low insulin resistance and low subclinical inflammation has been identified.[11] Both CKD and T2DM are increasing in prevalence in low to middle income countries. The increasing prevalence of younger individuals with T2DM, along with improvements in cardiovascular survival, may contribute to increase the prevalence and burden of CKD.[12] CKD is a key determinant of the poor health outcomes of diabetes and CVD.[3] Progression to end-stage renal disease (ESRD) is accompanied by increasing IR and loss of lean muscle mass.[10] Furthermore, aging populations will exert pressure to increase the absolute number of people with ESRD who are dependent on renal replacement therapy (RRT). As developing nations realize economic prosperity, access to RRT will further increase the total economic burden of CKD.[13]

5. SARCOPENIA

A primary mechanism for the increased morbidity and mortality associated with CKD involves a loss of muscle mass, strength, and function known as sarcopenia. The reported prevalence of sarcopenia in persons over 60 years-of-age varies from 8% to 40% depending on diagnostic criteria and method of assessment. Muscle mass is reported to decline in disease-free individuals at an annual rate of approximately 1.5%–3% per year after age 60 and becomes more rapid after age 75.[14] Complex signaling pathways have been identified, which may regulate sarcopenia and its relationship with T2DM, CVD, and advanced CKD.[15] T2DM will accelerate the reduction of muscle mass and strength because of hyperglycemia, diabetic complications, and insulin resistance.[14] Insulin resistance in skeletal muscle is particularly important because skeletal muscle mass is responsible for more than 75% of all insulin mediated the glucose disposal.[15] Accumulating evidence indicates that inflammatory cytokines such as interleukin 6 (IL6) and tumor necrosis factor (TNF) contribute to the link between elevated levels of inflammation and oxidative stress common in T2DM, CVD, and CKD, and the development of skeletal muscle insulin resistance, and ultimately, sarcopenia.[16] Other studies have provided evidence that variation in apoptosis and transcription regulation-related genes related to inflammation and muscle maintenance appears to be associated with frailty.[17]

Other chronic inflammatory diseases may contribute to the overall systemic burden of inflammatory factors. Osteoarthritis is a very common chronic joint disease most commonly affecting overweight older adults and associated with increased peripheral inflammatory markers.[18] Periodontitis, a common chronic oral infection, has long been associated with CVD and T2DM, and more recently, with an increased risk of Alzheimer's disease.[19,20]

6. THE FRAILTY SYNDROME

Sarcopenia is a significant risk factor for frailty, a common syndrome in older adults that carries an increased risk of poor health outcomes including falls, incident disability, hospitalization, and mortality due to decreased physiological reserves.[14] Frailty has been associated with a measurable loss of reserve in the respiratory, cardiovascular, renal, hematopoietic, and clotting systems.[21] Nutritional status can also be a mediating factor. Decline in skeletal muscle mass results in decreased basal metabolic rate and subsequent appetite and nutritional deficits.[22] Studies have found that variation in apoptosis and transcription regulation-related genes related to inflammation and muscle maintenance appears to be associated with frailty.[17] Recently, the concept of sarcopenic obesity (SO) has described a syndrome present in a group of older adults in whom obesity is accompanied by sarcopenia and insulin resistance. The prevalence of SO in the United States was estimated to be 80% of women and 42% of men at age 70 years. These individuals demonstrate two to three times the normal risk of developing disability associated with reduced activities of daily living.[23] Sarcopenia and fluctuations in systemic blood sugar levels have been associated with impaired grip strength, exhaustion, slow gait speed, weight loss, and reduction in activities.[21] At the cellular level, mitochondria contribute to the dynamics of cellular metabolism, the production of reactive oxygen species, and apoptotic pathways. Consequently, mitochondrial genetic variation may be related to vulnerability to disease and contribute to altered susceptibility to the frailty syndrome in older adults.[24]

The presence of frailty also increases the risk of dementia. Clinical studies of older persons without dementia at baseline have found that greater muscle strength was associated with a decreased risk of developing AD.[25] These findings have motivated work to determine whether the frailty syndrome should be expanded to include aspects of cognition and affect.[26]

7. DEMENTIA

Dementia rates are growing at an alarming proportion in all regions of the world and are related to population aging. The Global Burden of Disease 2010 study identified dementia as the third leading cause of "years lived with disability" at the global level. In 2010, there were an estimated 35.6 million people with Alzheimer's disease and other dementias worldwide. This number will increase with an aging population, and will reach 66 million by the year 2030, and 115 million by 2050.[27] The main increase will take place in low and middle income countries, where more than 70% of the people with dementia will live by 2050.[28] Loss of lean muscle mass has been found to accelerate the progression of Alzheimer's disease (AD) and is associated with brain atrophy and lower cognitive performance. This may be a direct or indirect consequence of the pathophysiology of AD, or a shared mechanism.[29]

Most dementia in older individuals is due to a combination of Alzheimer's disease, neurodegeneration, and vascular pathology.[30] Prospective evaluation using MRI found that 44% of the incident dementia cases in older individuals had vascular disease, either as the sole cause or a contributory factor, usually with Alzheimer's disease.[31] Vascular disease in the brain may alter the blood–brain barrier (BBB) function allowing transport of substances into the brain and adversely affect the perivascular clearance of amyloid from brain to periphery.[32] Additionally, ischemic injury within the "astrocyte-centric" subunit may result in an inflammatory response and increased intracellular phosphorylation of tau protein and resulting neurodegeneration.[33]

Chronic inflammation is a characteristic of metabolic disorders, frailty, and AD. A metaanalysis of 40 studies found that AD is accompanied by higher peripheral concentrations of a number of inflammatory markers, including IL6 and TNF.[34] Damage to the BBB can lead to infiltration of immune cells into the brain, potentially contributing to central inflammation. Inflammatory mediators have adverse effects on beta amyloid and glucose metabolism. Impaired metabolism of brain glucose and lower hippocampal volume, hallmarks of AD, are strongly associated with peripheral insulin resistance.[35]

Neuroimaging has supported the hypothesis that T2DM is associated with accelerated cognitive decline and dementia. The structural basis for these cognitive deficits includes both vascular lesions and global cerebral atrophy.[36] Vascular complications associated with chronic T2DM have been shown to cause BBB breakdown that proceeds and drives the pathological changes within the white matter progressing to symptomatic AD.[37]

Impaired brain insulin signaling contributes to Alzheimer's disease pathogenesis as first proposed by Hoyer.[38] Increased levels of the insulin resistance markers, insulin growth factor (IGF-1), and insulin receptor substrate (IRS-1) are associated with poor performance on tests of working an episodic memory.[39] Increased amounts of advanced glycation end products generated by chronic hyperglycemia are found in postmortem AD brain. Adult neurogenesis that provides neuronal plasticity is also impaired in the diabetic brain.[40]

8. EXERCISE AND BRAIN METABOLISM

Exercise is one of the most effective strategies to promote brain plasticity, increase cognition, and reduce risk of cognitive decline in later life.[41] There is evidence that an evolutionary requirement for exercise promotes beneficial adaptive responses that may inhibit, or even reverse, the effects of a sedentary, overindulgent lifestyle.[42] The beneficial effects resulting from increased physical activity occur at different levels of cellular organization, with mitochondria being preferential target organelles. Mitochondrial adaptations to exercise include an improvement of redox modulation bioenergetics, decreased apoptotic signaling, activation of mitochondrial biogenesis, and modulation of autophagy.[43]

Moderate and high intensity exercise have demonstrated a neuroprotective effect through the induction of antioxidative enzymes, neurotrophic factors, IGF-1, and vascular endothelial growth factor. Expression of these cellular products has been shown to reduce AB amyloid plaques and tau phosphorylation in cognitive regions, improve cerebral blood flow, and stimulate formation of synaptic connections.[44]

Diffusion-tensor magnetic resonance imaging has shown that greater cardiorespiratory fitness is positively associated with more brain volume and greater neuronal white matter integrity.[45] Exercise involving power and balance improves several aspects of cognitive function, including attentional capacity, processing speed, executive function, episodic memory, and procedural memory.[46]

Skeletal muscle activation by exercise appears to play a role in the cognitive effects of aerobic activity through transcriptional factors regulating muscle fiber contraction and metabolic genes.[47] Findings suggest that brain plasticity is maintained throughout lifespan and that it can be enhanced by exercise and other interventions that activate AMP-activated protein kinase (AMPK). AMPK acts as a mediator for metabolic hormones and cytokines such as leptin, adiponectin, and ghrelin, which regulate glucose homeostasis, appetite, and exercise physiology.[48] Regular exercise promotes an energy demanding, stress response efficiency, which involves a host of evolved neuroendocrine responses.[47] The brain plays fundamental roles in regulating peripheral glucose metabolism by pathways and signaling mechanisms that remain incompletely understood.[49] Overall, physical activity produces numerous changes in the brain and the periphery that converge to promote stress robustness.[42] Although challenging the brain and body intermittently through physical exercise is beneficial, a society-wide effort will be required to implement brain and body health programs in the educational and healthcare systems, communities, and work places.[47]

9. PHARMACOLOGICAL TREATMENTS FOR DEMENTIA

Shared mechanisms and associations between T2DM, CVD, CKD, and AD provide an opportunity to prevent, and significantly reduce, chronic disease burden. Unfortunately, few clinical trials have demonstrated efficacy of lowering blood pressure, cholesterol, or treating diabetes to reduce the risk of dementia.[32] A search for interventions to prevent, treat, reduce the progression, or cure AD continues. To date, most proposed treatments for AD have disappointingly failed in clinical trials.[50] Restoring insulin signaling might be beneficial to AD patients. Intranasal insulin administration, a preferential route for CNS delivery, improved memory in healthy adults, without affecting circulating levels of insulin or glucose. Intranasal insulin also received enhanced verbal memory in memory-impaired subjects and improved cognitive performance in early AD patients.[51] Insulin was found to protect neurons against AD causing neurotoxins in cellular and animal models of AD.[52] Brain insulin resistance may have a role in prenatal development as well. It is conceivable that brain insulin resistance is a cause rather than a consequence of obesity and T2DM, and perhaps even a precursor to AD.[35]

10. REDUCING CHRONIC DISEASE BURDEN

Aging is typically considered an important confounder in clinical research. Adjusting for age ignores the contribution of age as the most powerful risk factor for many chronic diseases, and the clinical fact that chronological age is a poor approximation of physiological aging. The phenotypes of aging and frailty may both result from a core set of mechanisms that contribute to a multimorbidity that is modifiable with appropriate interventions. This leads to the possibility that chronic diseases in older age and frailty both originate from accelerated aging and may precipitate or exacerbate one another.[53] The ability to identify and modify the course of most chronic medical conditions is well established. The U.S. Preventive Services Task Force (USPSTF), World Health Organization (WHO), and other medical specialty organizations have issued evidence-based guidelines for the primary, secondary, and tertiary prevention of T2DM, CVD, CKD, and dementia.[54–57] Sarcopenia may be an intermediate step in the development of frailty in people with chronic illness, including AD.[21] It appears to be a dynamic process and potentially reversible. Therefore, early detection and interventions which specifically target the molecular mechanisms of sarcopenia should be a focus of worldwide health systems.[14]

REFERENCES

1. Garin N, Koyanagi A, Chatterji S, et al. Global multimorbidity patterns: a cross-sectional, population-based, multi-country study. *J Gerontol Ser A Biol Sci Med Sci*. 2016;71(2): 205–214.
2. Kelly T, Yang W, Chen C, Reynolds K, He J. Global burden of obesity in 2005 and projections to 2030. *Int J Obes*. 2008;32(9):1431–1437.

3. Couser WG, Remuzzi G, Mendis S, Tonelli M. The contribution of chronic kidney disease to the global burden of major noncommunicable diseases. *Kidney Int.* 2011;80(12):1258–1270.
4. Danaei G, Finucane MM, Lu Y, et al. National, regional, and global trends in fasting plasma glucose and diabetes prevalence since 1980: systematic analysis of health examination surveys and epidemiological studies with 370 country-years and 27 million participants. *Lancet.* 2011;378(9785):31–40.
5. Shaw JE, Sicree RA, Zimmet PZ. Global estimates of the prevalence of diabetes for 2010 and 2030. *Diabetes Res Clin Pract.* 2010;87(1):4–14.
6. Hu FB, Manson JE, Stampfer MJ, et al. Diet, lifestyle, and the risk of type 2 diabetes mellitus in women. *N Engl J Med.* 2001;345(11):790–797.
7. Chen L, Magliano DJ, Zimmet PZ. The worldwide epidemiology of type 2 diabetes mellitus—present and future perspectives. *Nat Rev Endocrionol.* 2012;8(4): 228–236.
8. McAloon CJ, Boylan LM, Hamborg T, et al. The changing face of cardiovascular disease 2000–2012: an analysis of the world health organization global health estimates data. *Int J Cardiol.* 2016;224:256–264.
9. Franssens BT, Westerink J, van der Graaf Y, Nathoe HM, Visseren FLJ. Metabolic consequences of adipose tissue dysfunction and not adiposity per se increase the risk of cardiovascular events and mortality in patients with type 2 diabetes. *Int J Cardiol.* 2016;222:72–77.
10. Lastra G, Dhuper S, Johnson MS, Sowers JR. Salt, aldosterone, and insulin resistance: impact on the cardiovascular system. *Nat Rev Cardiol.* 2010;7(10):577–584.
11. Stefan N, Häring HU, Hu FB, Schulze MB. Metabolically healthy obesity: epidemiology, mechanisms, and clinical implications. *Lancet Diabetes Endocrinol.* 2013;1(2): 152–162.
12. Thomas MC, Cooper ME, Zimmet P. Changing epidemiology of type 2 diabetes mellitus and associated chronic kidney disease. *Nat Rev Nephrol.* 2016;12(2):73–81.
13. Eggers PW. Has the incidence of end-stage renal disease in the USA and other countries stabilized? *Curr Opin Nephrol Hypertens.* 2011;20(3):241–245.
14. Jang HC. Sarcopenia, frailty, and diabetes in older adults. *Diabetes Metab J.* 2016;40(3):182–189.
15. Stefan N, Artunc F, Heyne N, Machann J, Schleicher ED, Häring HU. Obesity and renal disease: not all fat is created equal and not all obesity is harmful to the kidneys. *Nephrol Dial Transplant.* 2016;31(5):726–730.
16. Wei Y, Chen K, Whaley-Connell AT, Stump CS, Ibdah JA, Sowers JR. Skeletal muscle insulin resistance: role of inflammatory cytokines and reactive oxygen species. *Am J Physiol Regul Integr Comp Physiol.* 2008;294(3):R680.
17. Ho Y, Matteini AM, Beamer B, et al. Exploring biologically relevant pathways in frailty. *J Gerontol Ser A Biol Sci Med Sci.* 2011;66(9):975–979.
18. Fernandes GS, Valdes AM. Cardiovascular disease and osteoarthritis: common pathways and patient outcomes. *Eur J Clin Invest.* 2015;45(4):405–414.
19. Otomo-Corgel J, Pucher JJ, Rethman MP, Reynolds MA. State of the science: chronic periodontitis and systemic health. *J Evid Based Dent Pract.* 2012;12(suppl 3):20–28.
20. Abbayya K, Puthanakar NY, Naduwinmani S, Chidambar YS. Association between periodontitis and Alzheimer's disease. *N Am J Med Sci.* 2015;7(6):241–246.
21. Clegg A, Young J, Iliffe S, Rikkert MO, Rockwood K. Frailty in elderly people. *Lancet.* 2013;381(9868):752–762.
22. Fried LP, Tangen CM, Walston J, et al. Frailty in older adults: evidence for a phenotype. *J Gerontol Ser A Biol Sci Med Sci.* 2001;56(3):M156.
23. Cleasby ME, Jamieson PM, Atherton PJ. Insulin resistance and sarcopenia: mechanistic links between common co-morbidities. *J Endocrinol.* 2016;229(2):R81.

24. Moore AZ, Biggs ML, Matteini A, et al. Polymorphisms in the mitochondrial DNA control region and frailty in older adults. *PLoS One*. 2010;5(6):e11069.
25. Boyle PA, Buchman AS, Wilson RS, Leurgans SE, Bennett DA. Association of muscle strength with the risk of Alzheimer disease and the rate of cognitive decline in community-dwelling older persons. *Arch Neurol*. 2009;66(11):1339–1344.
26. Searle SD, Rockwood K. Frailty and the risk of cognitive impairment. *Alzheimers Res Ther*. 2015;7(1):54.
27. Rizzi L, Rosset I, Roriz-Cruz M. Global epidemiology of dementia: Alzheimer's and vascular types. *Biomed Res Int*. 2014;2014:1–8.
28. Wortmann M. Dementia: a global health priority—highlights from an ADI and world health organization report. *Alzheimers Res Ther*. 2012;4(5):40–42.
29. Burns JM, Johnson DK, Watts A, Swerdlow RH, Brooks WM. Reduced lean mass in early Alzheimer disease and its association with brain atrophy. *Arch Neurol*. 2010;67(4):428–433.
30. Lopez OL, Klunk WE, Mathis C, et al. Amyloid, neurodegeneration, and small vessel disease as predictors of dementia in the oldest-old. *Neurology*. 2014;83(20):1804–1811.
31. Kuller LH, Lopez OL, Jagust WJ, et al. Determinants of vascular dementia in the cardiovascular health cognition study. *Neurology*. 2005;64(9):1548–1552.
32. Kuller LH, Lopez OL. Cardiovascular disease and dementia risk: an ever growing problem in an aging population. *Expert Rev Cardiovasc Ther*. 2016;14(7):771–773.
33. Jo WK, Law ACK, Chung SK. The neglected co-star in the dementia drama: the putative roles of astrocytes in the pathogeneses of major neurocognitive disorders. *Mol Psychiatry*. 2014;19(2):159–167.
34. Swardfager W, Lanctt K, Rothenburg L, Wong A, Cappell J, Herrmann N. A meta-analysis of cytokines in Alzheimer's disease. *Biol Psychiatry*. 2010;68(10):930–941.
35. Kullmann S, Heni M, Hallschmid M, Fritsche A, Preissl H, Häring HU. Brain insulin resistance at the crossroads of metabolic and cognitive disorders in humans. *Physiol Rev*. 2016;96(4):1169–1209.
36. Tiehuis AM, van den Berg E, Kappelle LJ, Biessels GJ. Cognition and dementia in type 2 diabetes: brain imaging correlates and metabolic and vascular risk factors. *Aging Health*. 2007;3(3):361–373.
37. Goldwaser EL, Acharya NK, Sarkar A, Godsey G, Nagele RG. Breakdown of the cerebrovasculature and blood–brain barrier: a mechanistic link between diabetes mellitus and Alzheimer's disease. *J Alzheimers Dis*. 2016;54(2):445–456.
38. Hoyer S, Nitsch R. Cerebral excess release of neurotransmitter amino acids subsequent to reduced cerebral glucose metabolism in early-onset dementia of Alzheimer type. *J Neural Transm*. 1989;75(3):227–232.
39. Talbot K, Wang HY, Kazi H, et al. Demonstrated brain insulin resistance in Alzheimer's disease patients is associated with IGF-1 resistance, IRS-1 dysregulation, and cognitive decline. *J Clin Invest*. 2012;122(4):1316–1338.
40. Pugazhenthi S, Qin L, Reddy PH. Common neurodegenerative pathways in obesity, diabetes, and Alzheimer's disease. *Biochem Biophys Acta*. 2016. http://dx.doi.org/10.1016/j.bbadis.2016.04.017, pii: S0925-4439(16)30097-7. [Epub ahead of print].
41. Chen Z, Zhong C. Decoding Alzheimer's disease from perturbed cerebral glucose metabolism: implications for diagnostic and therapeutic strategies. *Prog Neurobiol*. 2013; 108:21–43.
42. Mattson MP. Energy intake and exercise as determinants of brain health and vulnerability to injury and disease. *Cell Metab*. 2012;16(6):706–722.
43. Bernardo TC, Marques-Aleixo I, Beleza J, Oliveira PJ, Ascenso A, Magalhes J. Physical exercise and brain mitochondrial fitness: the possible role against Alzheimer's disease. *Brain Pathol*. 2016;26(5):648–663.

44. Radak Z, Hart N, Sarga L, et al. Exercise plays a preventive role against Alzheimer's disease. *J Alzheimers Dis*. 2010;20(3):777–783.
45. Zhu N, Jacobs DR, Schreiner PJ, et al. Cardiorespiratory fitness and brain volume and white matter integrity: the CARDIA study. *Neurology*. 2015;84(23):2347–2353.
46. Smith PJ, Blumenthal JA, Hoffman BM, et al. Aerobic exercise and neurocognitive performance: a meta-analytic review of randomized controlled trials. *Psychosom Med*. 2010;72(3):239–252.
47. van Praag X, Fleshner M, Schwartz MW, Mattson MP. Exercise, energy intake, glucose homeostasis, and the brain. *J Neurosci*. 2014;34(46):15139–15149.
48. Hardie DG. The AMP-activated protein kinase pathway—new players upstream and downstream. *J Cell Sci*. 2004;117(23):5479–5487.
49. Schwartz MW, Seeley RJ, Tschp MH, et al. Cooperation between brain and islet in glucose homeostasis and diabetes. *Nature*. 2013;503(7474):59–66.
50. Selkoe DJ. Resolving controversies on the path to Alzheimer's therapeutics. *Nat Med*. 2011;17(9):1060–1065.
51. Freiherr J, Hallschmid M, Frey II WH, et al. Intranasal insulin as a treatment for Alzheimer's disease: a review of basic research and clinical evidence. *CNS Drugs*. 2013;27(7):505–514.
52. De Felice FG. Connecting type 2 diabetes to Alzheimer's disease. *Expert Rev Neurother*. 2013;13(12):1297–1299.
53. Ferrucci L, Fried LP. Etiological role of aging in chronic diseases: from epidemiological evidence to the new geroscience. In: Sierra F, Kohanski R, eds. *Advances in Geroscience*. Switzerland: Springer International Publishing; 2015:37–51.
54. Siu AL, Bibbins-Domingo K, Grossman D, et al. Screening for high blood pressure in adults: U.S. preventive services task force recommendation statement. *Ann Intern Med*. 2015;163(10):778–786.
55. Selph S, Dana T, Blazina I, Bougatsos C, Patel H, Chou R. Screening for type 2 diabetes mellitus: a systematic review for the U.S. preventive services task force. *Ann Intern Med*. 2015;162(11):765–776.
56. Moyer VA. Screening for chronic kidney disease: U.S. preventive services task force recommendation statement. *Ann Intern Med*. 2012;157(8):567–570.
57. Lin JS, O'Connor E, Rossom C, Perdue LA, Eckstrom E. Screening for cognitive impairment in older adults: a systematic review for the U.S. preventive services task force. *Ann Intern Med*. 2013;159(9):601–612.

CHAPTER TWO

Therapeutic Strategies for Mitochondrial Dysfunction and Oxidative Stress in Age-Related Metabolic Disorders

J.S. Bhatti*[,†,1], S. Kumar*, M. Vijayan*, G.K. Bhatti[‡], P.H. Reddy*[,§]
*Garrison Institute on Aging, Texas Tech University Health Sciences Center, Lubbock, TX, United States
[†]Department of Biotechnology, Sri Guru Gobind Singh College, Chandigarh, India
[‡]UGC Centre of Excellence in Nano Applications, Panjab University, Chandigarh, India
[§]Texas Tech University Health Sciences Center, Lubbock, TX, United States
[1]Corresponding author: e-mail address: jasvinder.bhatti@ttuhsc.edu

Contents

Abstract

Mitochondria are complex, intercellular organelles present in the cells and are involved in multiple roles including ATP formation, free radicals generation and scavenging, calcium homeostasis, cellular differentiation, and cell death. Many studies depicted the involvement of mitochondrial dysfunction and oxidative damage in aging and pathogenesis of age-related metabolic disorders and neurodegenerative diseases. Remarkable advancements have been made in understanding the structure, function, and physiology of mitochondria in metabolic disorders such as diabetes, obesity, cardiovascular diseases,

Progress in Molecular Biology and Translational Science, Volume 146
ISSN 1877-1173
http://dx.doi.org/10.1016/bs.pmbts.2016.12.012

and stroke. Further, much progress has been done in the improvement of therapeutic strategies, including lifestyle interventions, pharmacological, and mitochondria-targeted therapeutic approaches. These strategies were mainly focused to reduce the mitochondrial dysfunction caused by oxidative stress and to retain the mitochondrial health in various diseases. In this chapter, we have highlighted the involvement of mitochondrial dysfunction in the pathophysiology of various disorders and recent progress in the development of mitochondria-targeted molecules as therapeutic measures for metabolic disorders.

ABBREVIATIONS

ATP adenosine triphosphate
CAT catalase
ERR estrogen-related receptors
ETC electron transport chain
GPx glutathione peroxidase
GSH glutathione
MetS metabolic syndrome
MitoQ mitochondria-targeted quinone
MtDNA mitochondrial DNA
NAC *N*-acetylcysteine
OXPHOS oxidative phosphorylation
PGC-1α peroxisome proliferator-activated receptor gamma coactivator 1-alpha
RNS reactive nitrogen species
ROS reactive oxygen species
SOD superoxide dismutase
T2DM type 2 diabetes mellitus
TCA tricarboxylic acid
TNF-α tumor necrosis factor-α

1. INTRODUCTION

Aging is basically a degenerative process associated with impaired metabolism and cell damage that leads to decline in all physiological functions. There are several underlying rationales behind the "Free Radical Theory" of aging proposed in 1956 by Harman but the exact reason of aging is still poorly understood. However, the fundamental role of mitochondria in aging has been established several decades ago.[1,2] Mitochondria are self-autonomous intracellular organelles responsible for producing energy in the form of adenosine triphosphate (ATP) by metabolizing nutrients via oxidative phosphorylation (OXPHOS) in concurrence with the oxidation of metabolites by Krebs's cycle and β-oxidation of fatty acids. They are also accountable for many other metabolic processes such as energy metabolism, generation of free radicals and calcium homeostasis, and cell survival and death.[3,4] Currently, impaired mitochondrial functions such as diminished oxidative capacity and antioxidant defense by enhanced production of free radicals reduced OXPHOS and ATP production is substantially linked with biological aging and many other metabolic diseases. There is a decline in mitochondrial biogenesis in aging due to alterations in mitochondrial fission and fusion and the inhibition of mitophagy, a process which eliminates dysfunctional mitochondria.[5] Previous studies demonstrated that aging is one of the major risk factors associated with life-threatening conditions, including cancer, diabetes, obesity, cardiovascular, and neurodegenerative diseases.[6]

The reactive oxygen species (ROS) are a family of free radicals includes superoxide anions, hydroxyl, peroxyl radicals, and other nonradicals capable of generating free radicals.[7,8] Although the intracellular generation of ROS per se is an inevitable process, but cells possess numerous defense systems to counter it. The overproduction of ROS subsequently leads to exponentially increased oxidative damage inflicted on lipids, DNA, and proteins.[4,9] Oxidative stress is an imbalance between the generation of ROS and the antioxidant defense system, wherein the damaging effects of ROS are more powerful compared to the compensatory effect of antioxidants in the cells.[10,11] It is evident from the previous studies that oxidative stress is in associated with various pathophysiological conditions involving aging, cancer and age-related metabolic disorders, and neurodegenerative diseases.[12–22]

The escalating prevalence of age-related metabolic disorders posed major public health problems in the modern society, associated with enormous personal, social, and economic burden across the world.[23–28] Earlier studies demonstrated the interaction of genetic variants and environmental factors contributing the present alarming situation of age-related metabolic disorders.[29–32] However, oxidative stress and mitochondrial dysfunction are implicated in a number of aging pathologies such as cancer, diabetes, obesity, and neurodegenerative diseases.[9,18,19,22,33–46] This chapter highlights the escalating situation of metabolic syndrome (MetS), possible mechanisms linking mitochondrial dysfunction with aging pathologies and promising therapeutic strategies for prevention and treatment of age-related metabolic disorders. We specifically focused on diabetes, obesity, stroke, and heart diseases which are intimately related to mitochondrial dysfunction induced by overproduction of ROS in the cell. Then, pharmacologic strategies translated from the bench to bedside will be provided to target mitochondrial dysfunction for the prevention of risk associated with age-related metabolic diseases.

2. GLOBAL PREVALENCE OF METABOLIC DISORDERS

MetS represents a constellation of several risk factors such as central obesity, elevated blood pressure, abnormal levels of triglycerides and high-density lipoproteins-cholesterol (HDL-C), and impaired glucose tolerance. Previous studies demonstrated the high prevalence of MetS worldwide which leads to the development of type 2 diabetes and cardiovascular diseases (CVD).[47–49] In the past decades, many definitions of MetS have been recommended by various international organizations (Table 1) including the World Health Organization (WHO), the European Group for the Study of Insulin Resistance (EGIR), the National Cholesterol Education Program-Third Adult Treatment Panel (NCEP-ATPIII), the American Association of Clinical Endocrinology (AACE), and the International Diabetes Federation (IDF); however, the most widely accepted and clinically used criteria of MetS are those acclaimed by WHO (1998), modified NCEP-ATPIII (2005), and IDF (2005). As shown in Table 1, each organization has established their own criteria for defining MetS, having thresholds for each component of MetS. Furthermore, in 2009, a joint interim statement for the new definition of MetS was published by the International Diabetes Federation Task Force on Epidemiology and Prevention; National Heart, Lung, and Blood Institute; American Heart Association; World Heart Federation; International Atherosclerosis Society; and International Association for the Study of Obesity.[50] This joint definition stated that obesity and IR are not prerequisites for MetS but that three of the five components would suffice for a diagnosis of MetS, with the thresholds for measuring waist circumference (WC) requiring ethnic and nation specificity.[50,51] Indeed, the varying diagnostic criteria of MetS impede the actual estimates of MetS worldwide, as well as within specific countries, genders, and ethnicities.

Despite ambiguity in the precise definition of MetS, several studies have reported worldwide prevalence of MetS varying between 10% and 84% depending on the ethnicity, age, gender, and race of the population, whereas IDF estimates that one-quarter of the world's population with MetS.[49] Recent studies have shown that approximately one-fifth of the adult US

Table 1 Various Definitions Proposed for the Clinical Diagnosis of Metabolic Syndrome

Parameters	WHO (1998)[52]	EGIR (1999)[53]	ATPIII (2001)[54]	AACE (2003)[55]	Modified NCEP-ATPIII (2005)[56]	IDF (2005)[57]	Harmonizing Criteria (2009)[50]
Insulin resistance	GT, IFG, T2DM, or lowered insulin Sensitivity plus any two of the following	Plasma insulin >75th percentile plus any two of the following	None, but any three of the following five features	IGT or IFG plus any of the following based on the clinical judgment	None	None	None
Abdominal obesity	Men: waist-to-hip ratio >0.90; women: waist-to-hip ratio >0.85 and/or BMI > 30 kg/m	WC ≥94 cm in men or ≥80 cm in women	WC ≥102 cm in men or ≥88 cm in women	BMI ≥ 25 kg/m^2	Increased WC (ethnic-specific) plus any two of the following	Increased WC (population specific) plus any two of the following	Increased WC (population- and country-specific definitions)
Lipids abnormality	TGs ≥150 mg/dL and/or HDL-C <35 mg/dL in men or <39 mg/dL in women	TGs ≥150 mg/dL and/or HDL-C <39 mg/dL in men or women	TGs ≥150 mg/dL HDL-C <40 mg/dL in men or <50 mg/dL in women	TGs ≥150 mg/dL and HDL-C <40 mg/dL in men or <50 mg/dL in women	TGs ≥150 mg/dL HDL-C <40 mg/dL in men or <50 mg/dL in women	TGs ≥150 mg/dL or HDL-C ≤40 mg/dL in males; 50 mg/dL in females (or drug treatment for elevated triglycerides or reduced HDL-C)	TGs ≥150 mg/dL or HDL-C ≤40 mg/dL in males; 50 mg/dL in females (or drug treatment for elevated triglycerides or reduced HDL-C)

Continued

Table 1 Various Definitions Proposed for the Clinical Diagnosis of Metabolic Syndrome—cont'd

Parameters	WHO (1998)[52]	EGIR (1999)[53]	ATPIII (2001)[54]	AACE (2003)[55]	Modified NCEP-ATPIII (2005)[56]	IDF (2005)[57]	Harmonizing Criteria (2009)[50]
Hypertension	≥140/90 mm Hg	≥140/90 mm Hg or on treatment for hypertension	≥130/85 mm Hg	≥130/85 mm Hg	≥130/85 mm Hg or on treatment for hypertension	130/85 mm Hg or on treatment for hypertension	130/85 mm Hg or on treatment for hypertension
Fasting glucose levels	IGT, IFG, or T2DM	IGT or IFG (but not diabetes)	>110 mg/dL (includes diabetes)	IGT or IFG (but not diabetes)	>110 mg/dL (includes diabetes)	≥100 mg/dL (includes diabetes)	≥100 mg/dL or drug treatment for elevated glucose
Other	Microalbuminuria: urinary excretion rate of >20 mg/min or albumin: creatinine ratio of >30 mg/g			Other features of insulin resistance			

population is suffering from MetS.[58] A number of previous studies estimated an increase in the prevalence of MetS (20%–25%) in Asian Indians.[32,59–63] Also, a consistent increase in the prevalence of the MetS associated with abdominal adiposity leads to higher risk of morbidity and mortality in East and Southeast Asian populations.[61] These studies demonstrated potential culprits involved in the development of MetS including rapid urbanization, higher economic growth, increasing age, sedentary lifestyle, bad dietary habits, and Westernization.[64,65] It has been established that subjects with MetS have a five times more risk of developing type 2 diabetes mellitus (T2DM) and three times more risk of developing CVD.[32,66] Recent advances in pharmacological treatment, lifestyle interventions, and disease awareness have slightly limited the progression of MetS all over the world. Furthermore, increasing trends of some components of MetS including hyperglycemia and abdominal obesity still increased risk of MetS.[58] Previous studies implicated the association of oxidative stress with the epidemics of MetS.[22,67]

3. STRUCTURE AND FUNCTIONS OF MITOCHONDRIA

Mitochondria are complex ubiquitous intracellular, rod-shaped organelles ranging from 0.5 to 1.0 μm in diameter, which present in most of the eukaryotic cells. These organelles use carbohydrates, fats, and proteins as fuel and produce energy in the form of ATP and that is why they are called as the "powerhouse of the cell." Mitochondria also perform many other key functions including cell signaling, cell differentiation, and senescence and also regulate cell growth. Each mitochondrion is composed of double membrane with intermembrane space between outer and inner mitochondrial membranes. Outer mitochondrial membrane is smooth and having a number of integral proteins called phorins making it relatively more permeable to nutrients and other molecules. In contrast, the inner membrane is complex, strictly permeable and having many folded structures called cristae and contains enzymes of OXPHOS. This membrane surrounds the mitochondrial matrix, wherein the electrons produced by tricarboxylic acid (TCA) cycle are taken in by electron transport chain (ETC) for the production of ATP. Each mitochondrion contains between 800 and 1000 copies of self-replicating, mitochondrial DNA (mtDNA), which are maternally inherited and packaged in high-ordered nucleoprotein structures called nucleoids.[68] Although nucleoids are distributed throughout

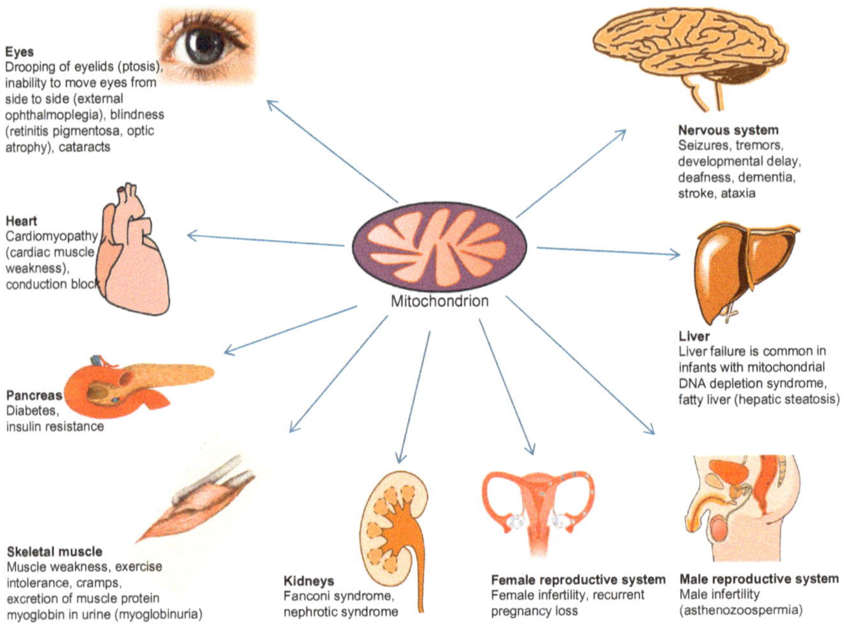

Fig. 1 Mitochondria-related diseases and affected body parts. Mitochondrial dysfunction is involved in the pathophysiology of a variety of metabolic and neurodegenerative disorders affecting important body organs including brain, muscles, eyes, heart, liver, and pancreas in varying levels of severity.

the mitochondrial matrix, they are often located in proximity of the cristae, which carry the OXPHOS system. When the mitochondria do not perform their function properly, they may cause diseases affecting a number of organs including brain, muscles, eyes, heart, liver, and pancreas (Fig. 1). An electrochemical gradient generated across the inner membrane drives the process of OXPHOS.[69] Most of the body's cellular energy (>90%) is generated by mitochondria in the form of ATP via TCA cycle.

3.1 Mitochondrial Dynamics

Mitochondria are highly complex and dynamic organelles that maintain their shape, structure, and functions by constantly undergoing the process of fission and fusion.[42,70–72] The balance between the fusion and fission process determines the morphology of mitochondria (Fig. 2A and B). These mitochondrial events were first described in budding yeast.[73] Disturbed mitochondrial dynamics leads to cell damage and many pathogenic conditions. Over the past several years, many cellular components have been

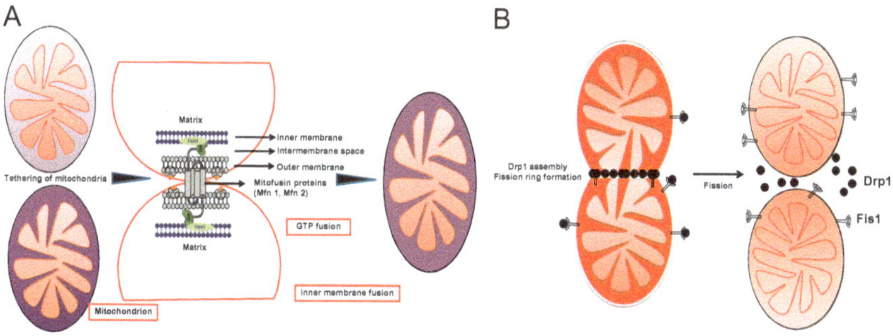

Fig. 2 Mitochondrial fusion and fission processes. (A) There are three GTPase genes that regulate the process of mitochondrial fusion viz mitofusin 1 and 2 (Mfn1/2) and optic atrophy1 (Opa1); Mfn 1 and 2 are located on the outer mitochondrial membrane, whereas Opa 1 is localized on the inner mitochondrial membrane. (B) Mitochondrial fission, on the other hand, is regulated by highly conserved two GTPase genes, Fis1 and Drp1, located on outer membrane and in the cytosol, respectively. Several other Drp1 receptors, including mitochondrial fission factor (Mff), MiD49, and MiD51, reported to involve in the fission process.

identified as the key mediators of mitochondrial fusion and fission processes. There are three GTPase genes that regulate the process of mitochondrial fusion, viz., mitofusin 1 and 2 (Mfn1/2) and optic atrophy1 (Opa1), localized in the outer and inner mitochondrial membranes, respectively.[74,75] Mfn 1 and 2 are located on the outer mitochondrial membrane, whereas Opa1 is localized on the inner mitochondrial membrane. Mitochondrial fission on the other hand is regulated by highly conserved two GTPase genes, Fis1 and Drp1, located on outer membrane and in the cytosol, respectively. In fusion process, mitochondrial content is intermixed and electrical conductivity maintained throughout the mitochondria.[76–78] The mitochondrial dynamics is essential for the maintenance of number of mitochondria in the growing cells, regulation of cell death pathway, and removal of damaged mitochondria.[79] Abnormal and/or impaired fission/fusion directly impacts mitochondrial function, resulting in excessive generation of ROS, altered mitochondrial enzymatic activities, impaired calcium homeostasis, diminished ATP production, and overall reduced energy metabolism in mammalian cells.

3.2 Mitochondrial Biogenesis

Mitochondrial biogenesis is a process by which mitochondria grow in their number and size. Both nuclear and mitochondrial genomes are involved in the biosynthesis of mitochondria. The mitochondrial biogenesis is mediated by physiologic stimuli including physical exercise, dietary restrictions, temperature, and muscle myogenesis. Mitochondrial ETC consists of a group of five multisubunit enzyme complexes, viz., I, II, III, IV, and V located on the inner mitochondrial membrane.[80] The electrons donated by coenzymes, NADH and $FADH_2$ in TCA cycle are accepted and transferred to components of ETC at complex I or II, and then consecutively to complex III, IV, and finally to oxygen. This transfer of electrons along the ETC is coupled with the transport of protons across the inner membrane, establishing the electrochemical gradient that drives complex V to generate ATP.[81] In presence of uncoupling agent, the contributive force will divert to uncoupling proteins (UCPs) and a proton leak occurs. Thus the energy will disburse as heat without producing any ATP. Oxygen free radicals ($^{\bullet}O_2^{-}$) are generated from complex I and III due to incomplete reduction of the oxygen molecule. In normal physiological conditions, these free radicals produced during the process of mitochondrial ETC are scavenged by antioxidant enzymes such as superoxide dismutase (SOD) and catalase (CAT). Several evidences suggest that type 2 diabetes and metabolic disorder are accountable for the declined antioxidant defenses and reduced the mitochondrial functions that lead to lipid accumulation in adipocyte.[82,83] The process of mitochondrial biogenesis is modulated by many transcriptional regulators present in the cell. Peroxisome proliferator-activated receptor (PPAR) γ coactivator 1-alpha (PGC-1α), a cotranscriptional regulation factor is the key regulator of mitochondrial biogenesis which interacts with many transcription factors/proteins such as nuclear respiratory factors (NRF-1 and NRF-2), mitochondrial transcription factor A (Tfam), uncoupling proteins (UCP2), PPARs, thyroid hormone, glucocorticoid, and estrogen and estrogen-related receptors (ERR) α and γ. Fig. 3 shows the role of PGC-1α and other transcriptional factors involved in mitochondrial biogenesis. NRF-1, NRF-2, and Tfam regulate the transcription of key mitochondrial enzymes and mtDNA synthesis.[84–87] Besides these transcription factors, AMP-activated protein kinase (AMPK) also regulate intracellular energy metabolism in energy deprivation.[88] Reduced AMPK activity has been implicated in aging-induced insulin resistance in animals.[89,90] Many experimental and clinical studies revealed the alterations in the morphology and number of mitochondria in heart, skeletal muscles, and liver tissues in pathogenic conditions.[3,91–95]

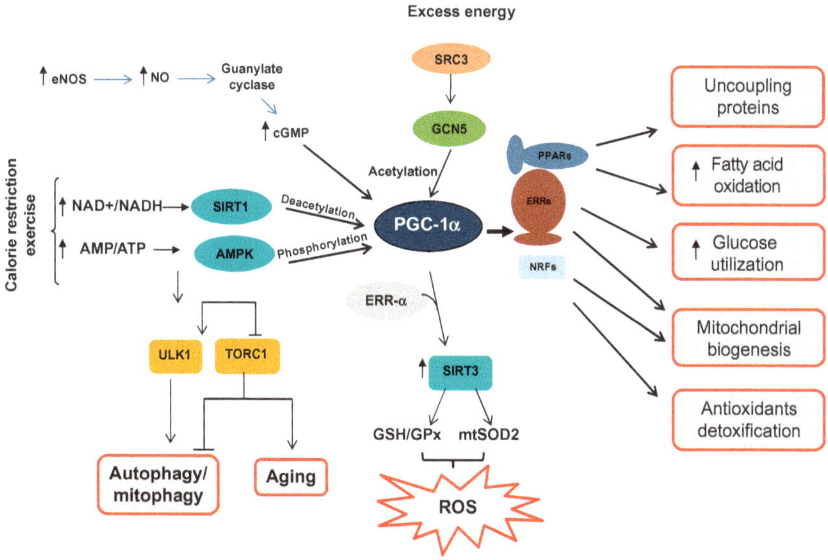

Fig. 3 Regulatory pathways involved in mitochondrial biogenesis. Mitochondrial biogenesis is activated via cellular stress or in response to environmental stimuli. PGC-1α is the main regulator of mitochondrial biogenesis and activated via AMPK, SIRT1, eNOS, SIRTs, TORCs, and AMPK increase the PGC-1α gene transcription, which resulting enhanced NRFs. Endothelial NO synthase (eNOS) display an increase in mtDNA content, cytochrome *c*, as well as PGC-1α, NRF-1, and Tfam mRNA expression. Activation of PGC-1α leads to increased fatty acid oxidation, uncoupling proteins, mitochondrial biogenesis, glucose utilization, and ROS detoxification through PPARs, ERRs, and NRFs.

4. MITOCHONDRIAL DYSFUNCTION IN AGE-RELATED METABOLIC DISORDERS

Mitochondria perform key biochemical functions essential for metabolic homeostasis and are arbiters of cell death and survival. Mitochondrial dysfunction is an assembly of mitochondrial defects that include declined mtDNA copy numbers, reduced mRNA concentration of genes encoding mitochondrial proteins, and decreased antioxidant capacity. It is primarily triggered by toxic by-products such as ROS generated in mitochondria during OXPHOS process.[96] The ROS generated during OXPHOS could cause the diminished mitochondrial biogenesis, altered membrane potential, decreased in number of mitochondria, and altered activities of oxidative proteins.[97] The alterations in the mitochondrial structure and function affect many tissues including brain, skeletal muscles, liver, kidney, respiratory, and endocrine systems and lead to many age-related pathogenic conditions such as cancer, diabetes, stroke, and neurodegenerative diseases.[42,98,99] In a normal cell, mitochondria continuously function to metabolize oxygen and generate ROS. However, either by accident or for a purpose, the flow of electrons through the ETC is an imperfect process in which 0.4%–4% of oxygen consumed by mitochondria is incompletely reduced and leads to production of ROS such as superoxide anion ($^{\bullet}O_2^-$) designated as "primary" ROS.[4,100] Excessive generation of superoxide anion further interacts with many other compounds and generates "secondary" ROS.[40,101] It is earlier established that the interactions of hydroxyl radical ($^{\bullet}OH$) with DNA molecule damages the nitrogenous bases, purine and pyrimidine, and deoxyribose backbone of DNA.[4] Also, the over production of these ROS damage the mitochondrial proteins/enzymes, membranes, and DNA, which leads to interruptions in the process of biosynthesis of ATP and other essential functions in mitochondria.[40,100] Fig. 4 shows the generation of ROS during the process of ATP synthesis in mitochondria. Besides superoxide anion and hydroxyl radicals, the ETC also generates other reactive species such as nitric oxide (NO) and reactive nitrogen species (RNS). Most of the cellular proteins and glutathione (GSH) are affected through nitration induced by RNS. Free radicals are fundamental to any biochemical process and are continuously produced in the body. It is well established that oxidative stress and mitochondrial dysfunction have been involved in the pathophysiology of many metabolic disease states such as insulin resistance, obesity diabetes, and many cardiovascular and neurodegenerative diseases.[102–110] Earlier studies demonstrated the role of impaired mitochondrial dynamics in pathophysiology of several diseases.[42,44,99,111–122] The disturbed balance between mitochondrial fission and fusion processes may also play a pivotal role in the age-dependent decline in mitochondrial biogenesis. Defects in mitochondria lead to impaired oxidative capacity. The defective mitochondria are selectively eliminated by the process of mitophagy. Mitophagy is an autophagy–lysosome system that removes dysfunctional mitochondria through fusion with lysosomes.[123] Advancing age leads to declined mitophagy resulting in accumulation of defective mitochondria, increased oxidative damage, and apoptosis.[5]

The cells have many ways to counter the effects of oxidative damage induced by ROS, either by directly diminishing the generation of free radicals or by scavenging the free radicals by an array of antioxidants, both enzymatic and nonenzymatic mechanisms. Enzymatic defense system includes

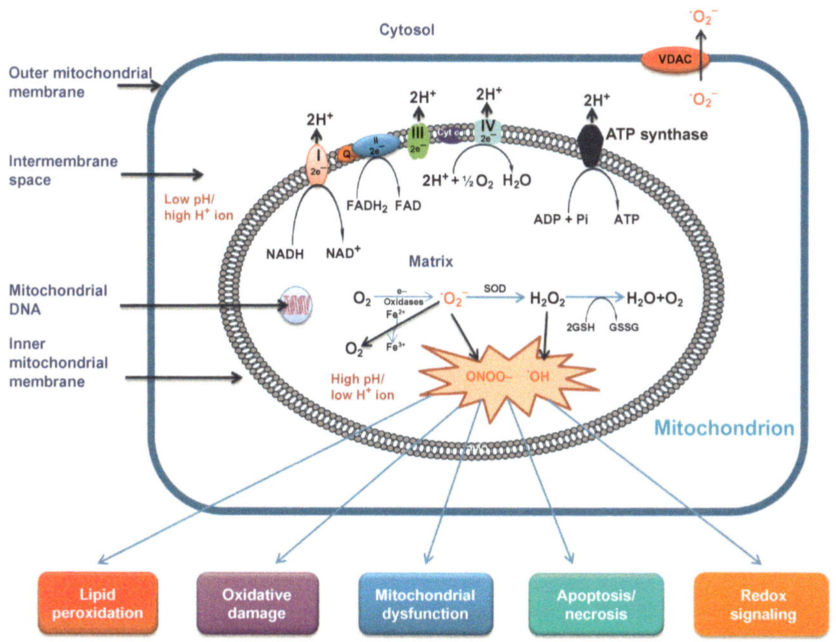

Fig. 4 Generation of ROS during the process of ATP synthesis. Each mitochondrion is composed of a double membrane with intermembrane space between outer and inner mitochondrial membranes. The outer mitochondrial membrane is smooth and permeable to nutrients and other molecules. In contrast, the inner membrane is complex, strictly permeable, and having many folded structures called cristae and contains enzymes of oxidative phosphorylation. This membrane surrounds the mitochondrial matrix, wherein the electrons produced by TCA cycle are taken in by ETC for the production of ATP. ETC is composed of five multisubunit enzyme complexes I, II, III, IV, and V located in the inner mitochondrial membrane. The electrons donated by coenzymes, NADH and FADH2 in TCA cycle are accepted and transferred to components of ETC at complex I or II, and then consecutively to complex III, IV and finally to oxygen through complex V. This transfer of electrons along the ETC is coupled with the transport of protons across the inner membrane, establishing the electrochemical gradient that generated ATP. Normally, mitochondria continuously function to metabolize oxygen and generate ROS. Transfer of electrons to O_2 generates superoxide ($\cdot O_2^-$) which is then converted into hydrogen peroxide (H_2O_2) by the enzyme, superoxide dismutases (SOD) in mitochondria. H_2O_2 is then converted into water by glutathione peroxidase (GPx) or catalase (CAT). Excessive generation of ROS can oxidize proteins, lipids, or mitochondrial DNA (mtDNA).

the ameliorative action of various antioxidant enzymes such as SOD, CAT, glutathione reductase (GR), and glutathione peroxidase (GPx). The other nonenzymatic defenses are the antioxidant compounds which protect the cells against oxidative stress. It includes vitamin E and C, GSH, various carotenoids, and flavonoids. Under normal conditions, the overproduction of ROS is restricted in mitochondria to protect this cellular organelle from oxidative damage via enzymatic and nonenzymatic defense systems. On the other hand, when the antioxidant defenses are overwhelmed, there is overproduction of ROS which then leads to oxidative damage to the proteins, DNA, and lipids in mitochondria.[124] This impaired the enzyme functions in respiratory chain and ultimately leading to mitochondrial dysfunction, reduced mitochondrial biogenesis and a wide range of pathologic conditions such as aging, various metabolic diseases, and neurodegenerative disorders.[9,18,101,125–131] Reduced mitochondrial number and capacity for OXPHOS in diabetes resulted in impaired mitochondrial biogenesis.[132,133] We have discussed the mechanisms of mitochondrial dysfunction in the following metabolic conditions.

4.1 Type 2 Diabetes Mellitus

Insulin resistance has been implicated as a major factor that contributes to T2DM and related complications. It is well established that T2DM is a multifactorial diseases in which both environmental and genetic factors contribute to the pathologic condition of this disease.[28,134–136] Over the past many years, the association of mitochondrial dysfunction with IR and T2DM received considerable attention.[95,137–141] However, it is still not clear whether insulin resistance is the primary cause of mitochondrial dysfunction or vice versa. T2DM results from a combination of reduced tissue sensitivity to insulin and inadequate insulin secretion. Furthermore, it is well established that mitochondrial dysfunction plays a central role in escalating situation of type 2 diabetes, obesity, and dyslipidemia. Previous studies established the role of mitochondria in the normal functioning of β-pancreatic cells involved in the secretion of insulin in response to increasing the glucose levels in the body. Insulin resistance due to altered mitochondrial functions may have contributed the etiology of metabolic and CVD.[142,143] A study done on diabetic and obese patients established the impaired glucose and lipid homeostasis in skeletal muscle.[144] Increased fat mass leads to several factors that inhibit insulin action including decreased glucose transporter type 4 (GLUT4), increased free fatty acids (FFA), and other circulating molecules.[145,146] Due to decreased insulin response in sensitive tissues, excess glucose accumulates leading to chronic hyperglycemia.[147] In humans and animal models, oxidative stress induced by excess of ROS generation or impaired antioxidant defenses in mitochondria plays a crucial role in the pathophysiology of T2DM and its complications such as CVD, neuropathy, and nephropathy.[3,148] Nonetheless, many studies established the association of mitochondrial dysfunction with insulin resistance in various tissues.[93,140,144,149–151] Mitochondrial dysfunction is also associated with insulin resistance in elderly population.[152] In skeletal muscle, altered mitochondrial functions, reduced ATP synthesis, and increased ROS generation lead to insulin resistance and obesity/diabetes.[95,153,154] Recent studies have indicated abnormal mitochondrial dynamics along with overproduction of ROS in diabetic patients.[155–157] Decrease in mitochondrial respiration, ATP production, mitochondrial density, and mRNA concentration has been reported in the insulin resistance and type 2 diabetic patients.[133,158–162] Also, overproduction of ROS has been linked with the pathogenesis of IR.[127] A short-term high-calorie diet resulted in increased markers of oxidative stress and a transient increase in OXPHOS enzyme protein expression. The mtDNA is more predominantly susceptible to oxidative damage induced by excess of ROS during OXPHOS process in mitochondria of brain.[163] Mitochondrial dysfunction inhibits insulin signaling pathway through the overproduction of ROS and interfering with oxidation of acetyl CoA, consequently resulting in increased lipid and diacylglycerol.[3,95,127,164] Disturbed mitochondrial biogenesis may be a reason for decreased number and oxidative capacity in diabetic condition. PGC-1α also regulates the process of mitochondrial biogenesis.[165,166] Furthermore, mitochondrial dysfunction seems to play a key role in the pathophysiology of T2DM and may be considered as a target for therapeutic measures in metabolic diseases.

4.2 Obesity

Obesity is one of the principal components of MetS and known to be a major risk factor in the development of many metabolic disorders.[32] There is overproduction of ROS in adipose tissues with altered activities of NADPH oxidase and antioxidative enzymes in obese mice.[167] Intriguingly, abdominal obesity has been associated with defective mitochondrial biogenesis manifested by impaired mitochondrial dysfunction, oxidative metabolism, low mitochondrial gene expression, and reduced ATP generation in rodents and humans.[83,92,168,169] Also, mtDNA, respiratory protein, and mtDNA transcription factor A (Tfam) gene expressions were markedly reduced in obese mice. Also altered mitochondrial dynamics plays a pivotal role in mitochondrial dysfunction linked to obesity as evident from reduced expression of mitofusin 2 gene in skeletal muscle.[170] The decrease in fatty acid oxidation causes the inhibition of insulin signaling, consequently leading to accumulation of FFA and insulin resistance which further reduces the mitochondrial oxidative capacity and ATP synthesis in obese and insulin resistance models.[171] Increased glucose levels enhance overproduction of ROS which may lead to morphological changes in mitochondria.[155] Inhibition of insulin signaling pathway caused the accumulation of lipids and FFA contributes to insulin resistance and many associated metabolic disorders.[160,172–176] It is evident from these studies that aging, altered mitochondrial biogenesis, and decreased antioxidant defense capacity along with genetic factors caused insulin resistance which is the major cause of many metabolic diseases. Inflammatory responses such as increased concentration and expression of tumor necrosis factor-α (TNF-α), interleukin-6 (IL-6), and monocyte chemoattractant protein-1 (MCP-1) are also associated with adipocyte dysfunction and insulin resistance in obesity and MetS.[177,178] Furthermore, inflammation in adipose tissues could be a causative factor diminishing mitochondrial biogenesis and energy homoeostasis.[92,179,180]

4.3 Cardiovascular Diseases

CVD include atherosclerosis, ischemic heart disease, cardiomyopathy, cardiac hypertrophy, and heart failure. In vivo and ex vivo studies have established the role of oxidative stress induced by the excess of ROS in a wide range of CVD.[12,19,33,40,181–185] It is evident from previous studies that ROS are generated in cardiac myocytes, endothelial cells, and neutrophils. The majority of ROS in the heart are generated by uncoupling of mitochondrial ETC complexes, I and III.[128,186] However, there are other mechanisms such as NADPH oxidase, xanthine oxidoreductase, or NOS by which ROS are generated and induce oxidative damage in heart tissue. The excess of ROS leads to cellular injury and declined antioxidant capacity, which seems to be due to defect in mitochondrial functions and mtDNA damage, endothelial dysfunction, and altered gene expression.[12] The reduced mitochondrial oxidative capacity contributes to cardiac dysfunction. The ROS are significantly enhanced in failing myocardium.[130,185,187–189] Various other studies have established the role of ROS, proinflammatory cytokines, including TNF-α, altered mitochondrial biogenesis and mtDNA damage, structural and morphological changes in mitochondria contributing to the development, and progression of heart diseases such as heart failure and cardiac dysfunction.[92,184,190,191] Moreover,

ROS stimulate contractile function, activate a variety of enzymes, transcription factors, and induce apoptosis. Thus ROS generated in mitochondria play a central role in the pathophysiology of CVD.

4.4 Stroke

Mitochondrial dysfunctions have been demonstrated as a key player in the development of brain stroke as evident by reduced ATP production, the starvation of glucose and oxygen to the tissues and influence on cell death pathways. Oxidative stress is one of the contributing factors leading to cellular damage during ischemic brain injury.[192,193] In experimental models of stroke, the diminished supply of glucose and oxygen leads to impaired oxidative metabolism in brain tissue.[194] The resultant oxygen–glucose deprivation in the brain tissue causes the accumulation of reducing intermediates and leads to enhanced ROS formation.[192] During oxidative stress, the excess of ROS alters antioxidant defense mechanism by reducing the scavenging capacity of antioxidant enzymes that could lead to altered mitochondrial functions by interacting with mitochondrial and cellular components such as DNA, proteins, and lipids.[148,195] In focal ischemia, the role of oxidative stress in necrosis and apoptosis has been explained in the previous studies.[196,197] Peroxynitrite radical also plays an important role in the pathogenesis of brain stroke. Thus overproduction of ROS in mitochondria significantly induces oxidative damage in ischemic and postischemic brain.[198,199]

5. STRATEGIES DIRECTED TO TARGET MITOCHONDRIAL DYSFUNCTION

Mitochondrial stress pathways have been implicated in disease manifestations of mitochondrial dysfunction and could highlight promising therapeutic targets.[200–202] Several shreds of evidence imply that mitochondrial dysfunction plays a central role in the pathophysiology of metabolic disorders, age, and age-related neurodegenerative diseases. So targeting mitochondria might be a promising strategy for potential therapeutic purposes in metabolic disorders and age-related neurodegenerative diseases. Following strategies might be employed to diminish the mitochondrial dysfunction caused by excessive generation of ROS that induced oxidative stress.

5.1 Lifestyle Interventions

5.1.1 Exercise

Exercise is one of the promising therapeutic interventions for mitochondrial dysfunction in age-related metabolic diseases. Previous studies established that regular exercise triggers many signaling pathways involved in skeletal muscle mitochondrial biogenesis, dynamics, and metabolism.[203] The benefits of regular exercise are not restricted to younger healthy population but also paybacks to skeletal muscles in people suffering from age-related disorders. Mitochondrial dysfunction contributes to severity of many pathological conditions including skeletal muscle atrophy, diabetes, CVD and many metabolic disorders, and neurodegenerative diseases. It is evident from previous studies that increased physical activity offers many benefits through improved insulin sensitivity and mitochondrial biogenesis in skeletal muscles in T2D patients.[88,204–210] Also, a significant increase in muscle mitochondrial respiration and mitochondrial content, oxidative enzyme activity, and mitochondrial density were observed in T2D patients after exercise. Aging causes loss of muscle mass and structural changes in the neuromuscular components resulting in impaired contractile function. Exercise induces beneficial adaptations that slow down the progression of age-related muscle functional decline. Several studies established the pleiotropic effect of physical exercise on mitochondrial dynamics in aging skeletal muscle.[211]

5.1.2 Dietary Modifications

It is well established that mitochondria are major contributors to the cellular adaptations needed to prolong lifespan during restricted diet. In addition to regular exercise, calorie restriction is also an effective nutritional intervention that prolongs lifespan of a variety of organisms and helps in the prevention of age-related metabolic disorders.[212–214] However, the molecular mechanisms of calorie restrictions induced benefits in aging and related disorders are still under investigation. Further, many studies demonstrated that the calorie restriction reduces the over production of ROS and oxidative damage,[215,216] leading to enhanced mitochondrial function in humans and be an effective remedy for the treatment of obesity and insulin resistance.[217,218] Many epidemiological studies also shown that restricted diet play beneficial role in human longevity.[219,220]

5.2 Pharmacological Interventions

The recent literature revealed that oxidative stress in the cell leads to structural and functional changes in the mitochondria. These mitochondrial changes trigger cell signaling pathways and generate uncontrollable ROS which ultimately leads to organ failure and diseases. Therefore pharmaceutical drugs that can limit the overproduction of ROS in the cells may be the potential therapeutic solution to improve mitochondrial health in a wide range of diseases. Following pharmaceutical strategies have been used for the maintenance of mitochondrial health.

5.2.1 Mitochondria-Targeted Antioxidants

The pathophysiology of various metabolic disorders is associated with oxidative damage in mitochondria, the antioxidants therapies could be a potential treatment for these pathologic conditions. In the last decade, in vivo and in vitro studies carried out in experimental animals and humans established the use of various antioxidants against the oxidative stress.[10,221] In the recent years, the use of various antioxidants for the treatment of mitochondrial damage in patients with a number of pathologic conditions including neurodegenerative or metabolic diseases has received much attention. The commonly used vitamins (vitamin E and C) and other chemical compounds with antioxidant properties such as coenzyme Q, α-lipoic acid, and *N*-acetylcysteine (NAC) have been used to reduce the excess generation of ROS in various metabolic conditions.[222–224] In the recent years, antioxidant compounds incorporating ubiquinone (MitoQ) or vitamin E (MitoVit E) specifically targeted to mitochondria have been successively used against mitochondrial dysfunctions[225] (Fig. 5). MitoQ is a potent therapeutic antioxidant in which lipophilic triphenylphosphonium (TPP) cation is bound to ubiquinone antioxidant moiety of the endogenous antioxidant coenzyme Q10.[226] The lipophilic nature of TPP cation enables MitoQ to cross phospholipid bilayer that leads to its accumulation many 100-fold within mitochondria and reduce ROS in the mitochondria and protect against age-related mitochondrial insult in brain tissue.[225–227]

Recent studies established the defensive action of MitoQ in MetS by affecting the redox signaling pathways.[228,229] These findings support the understanding that increased ROS in mitochondria may contribute to the etiology of a wide variety of diseases.[230–233] Now a day, MitoQ has been extensively used as a potential therapeutic molecule for the treatment

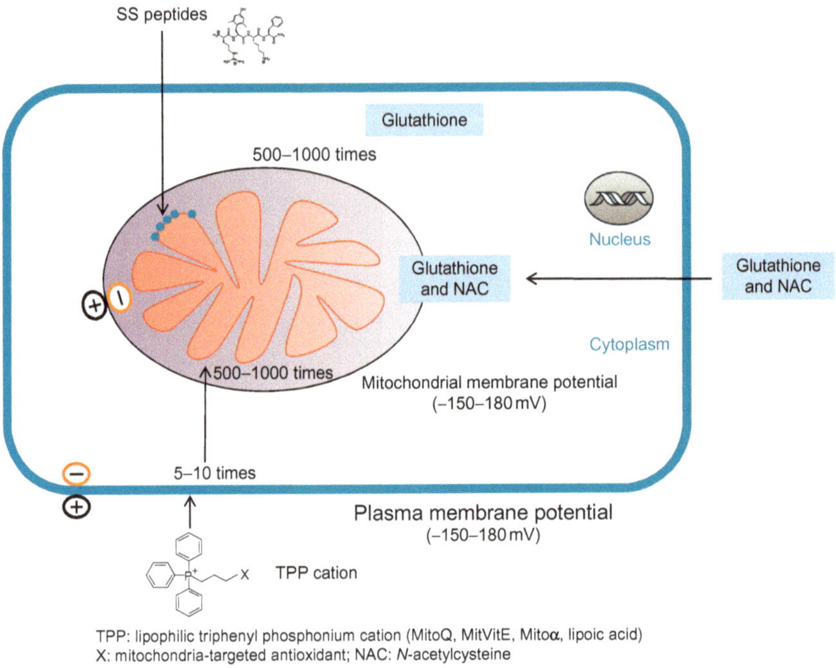

Fig. 5 Mitochondria-targeted antioxidants. A representative mitochondrial-targeted antioxidant (MitoQ) is constructed by an attachment of an antioxidant molecule—quinone to the lipophilic triphenylphosphonium cation—TPP. MitoQ accumulates several folds in the cytoplasm and further accumulates several hundred folds in the mitochondria. Mitochondrial antioxidants scavenge free radicals.

of neurodegenerative diseases.[234–236] Like MitoQ, MitoVitE, a TPP-conjugated vitamin E also protects mitochondria against oxidative damage induced by the excessive generation of ROS in many pathogenic conditions using different mechanisms.[226,227,237]

5.2.2 Sirtuins

Newer pharmacologic approaches have been proposed to improve mitochondrial function. Growing data demonstrated that NAD-dependent deacetylase family (Sirtuins), SIRT1 is involved in many cellular processes including regulation of glucose and lipid metabolism, through insulin signaling in the liver, adipose tissue, and skeletal muscles.[238–248] Resveratrol, a SIRT1 activator, found in grapes have strong antioxidant properties and improves insulin resistance. Activation of SIRT1 gene protects the cells against inflammation and oxidative stress. It activates PGC-1α that improves glucose uptake and mitochondrial biogenesis.[242,249,250] Mitochondrial fission has been implicated in various metabolic conditions; the inhibitors of mitochondrial fission may be used as therapeutic targets to treat patients with metabolic disorders.[251] Three inhibitors of mitochondrial fission have been identified as Mdivi 1, P110, and Dynasore[252–254] which play an ameliorative role against oxidative stress. Recently, a most commonly used antidiabetic drug, metformin, has been demonstrated as an antiaging medicine that inhibits ETC complex I. This bioenergetics effect of metformin is believed to inhibit hepatic gluconeogenesis.[255]

6. CONCLUDING REMARKS

Mitochondria are called the power house of the cells because they provide energy to each cell in the form of ATP by metabolizing the available nutrients. They are also responsible for many other cellular processes ranging from energy metabolism, generation of ROS, Ca^{2+} homeostasis, cell survival, and cell death. Alterations in the mitochondrial structure and functions are reported in various diseases such as cancer, MetS, including stroke, ischemia, prediabetes, diabetes, obesity, hypertension, dyslipidemia, heart disease, alcohol injury, and neurodegenerative diseases. The mitochondrial abnormalities including impaired mitochondrial dynamics, defects in mitochondrial biogenesis, mitochondrial dysfunction, and oxidative stress are largely involved in the pathophysiology of a variety of metabolic and neurodegenerative disorders. So targeting mitochondria might be a promising strategy for potential therapeutic measures to reduce and/or delay the progression of the disease. Lifestyle interventions including regular exercise and calorie restriction are the effective measures that prolong the lifespan of a variety of organisms and helps in the prevention of age-related metabolic diseases by improving the mitochondrial dynamics and mitochondrial function. It is evident that pharmaceutical drugs targeting mitochondria (sirtuins and antioxidants) are the potential therapeutic solution to improve mitochondrial health in a wide range of diseases. However, molecular links between MetS and mitochondrial structural/functional changes are not well understood. Further, genetics and genetic susceptibility to patients with MetS, in relation to aging, are poorly understood. The role of epigenetics in patients with MetS is unclear. In addition, current generalized treatments to patients with age-related MetS may not be very effective because body physiology varies from population to population. Further research is urgently needed to answer these questions. We are hopeful that understanding the mitochondrial dysfunction will empower the advancement of new therapeutics for age-related metabolic disorders.

ACKNOWLEDGMENTS

P.H.R., Ph.D., is supported by NIH grants AG042178, AG047812, and the Garrison Family Foundation. Dr. J.S.B. is financially supported by University Grants Commission, India under Raman Postdoctoral Research Fellowship in USA [F. No. 5-82/2016 (IC)].

REFERENCES

1. Trifunovic A, Larsson NG. Mitochondrial dysfunction as a cause of ageing. *J Intern Med.* 2008;263(2):167–178.
2. Leeuwenburgh C, Pamplona R, Sanz A. Mitochondria and aging. *J Aging Res.* 2011;2011:782946.
3. Kim JA, Wei Y, Sowers JR. Role of mitochondrial dysfunction in insulin resistance. *Circ Res.* 2008;102(4):401–414.
4. Halliwell B, Gutteridge JMC. *Free Radicals in Biology and Medicine.* 4th ed. Oxford: Oxford University Press; 2007.
5. Chistiakov DA, Sobenin IA, Revin VV, Orekhov AN, Bobryshev YV. Mitochondrial aging and age-related dysfunction of mitochondria. *Biomed Res Int.* 2014;2014:238463.
6. Niccoli T, Partridge L. Ageing as a risk factor for disease. *Curr Biol.* 2012;22(17): R741–R752.
7. Halliwell B. Reactive species and antioxidants. Redox biology is a fundamental theme of aerobic life. *Plant Physiol.* 2006;141(2):312–322.
8. Edgar D, Trifunovic A. The mtDNA mutator mouse: dissecting mitochondrial involvement in aging. *Aging.* 2009;1(12):1028–1032.
9. Finkel T, Holbrook NJ. Oxidants, oxidative stress and the biology of ageing. *Nature.* 2000;408(6809):239–247.
10. Bhatti J, Sidhu I, Bhatti G. Ameliorative action of melatonin on oxidative damage induced by atrazine toxicity in rat erythrocytes. *Mol Cell Biochem.* 2011;353(1–2):139–149.
11. Bhatti G, Kiran R, Sandhir R. Modulation of ethion-induced hepatotoxicity and oxidative stress by vitamin E supplementation in male Wistar rats. *Pestic Biochem Phys.* 2010;98(1):26–32.
12. Dhalla NS, Temsah RM, Netticadan T. Role of oxidative stress in cardiovascular diseases. *J Hypertens.* 2000;18(6):655–673.
13. Dalle-Donne I, Rossi R, Colombo R, Giustarini D, Milzani A. Biomarkers of oxidative damage in human disease. *Clin Chem.* 2006;52(4):601–623.
14. Jenner P. Oxidative stress in Parkinson's disease. *Ann Neurol.* 2003;53(suppl 3): S26–S36. discussion S36–S28.
15. Zhao Y, Zhao B. Oxidative stress and the pathogenesis of Alzheimer's disease. *Oxid Med Cell Longev.* 2013;2013:316523.
16. Sayre LM, Smith MA, Perry G. Chemistry and biochemistry of oxidative stress in neurodegenerative disease. *Curr Med Chem.* 2001;8(7):721–738.
17. Niedzielska E, Smaga I, Gawlik M, et al. Oxidative stress in neurodegenerative diseases. *Mol Neurobiol.* 2015;53:4094–4125.
18. Barnham KJ, Masters CL, Bush AI. Neurodegenerative diseases and oxidative stress. *Nat Rev Drug Discov.* 2004;3(3):205–214.
19. Tsutsui H, Kinugawa S, Matsushima S. Oxidative stress and heart failure. *Am J Physiol Heart Circ Physiol.* 2011;301(6):H2181–H2190.

20. Park K, Gross M, Lee DH, et al. Oxidative stress and insulin resistance: the coronary artery risk development in young adults study. *Diabetes Care.* 2009;32(7):1302–1307.
21. Thanan R, Oikawa S, Hiraku Y, et al. Oxidative stress and its significant roles in neurodegenerative diseases and cancer. *Int J Mol Sci.* 2015;16(1):193–217.
22. Roberts CK, Sindhu KK. Oxidative stress and metabolic syndrome. *Life Sci.* 2009;84(21–22):705–712.
23. Grundy SM. Metabolic syndrome pandemic. *Arterioscler Thromb Vasc Biol.* 2008;28(4):629–636.
24. Global Burden of Metabolic Risk Factors for Chronic Diseases Collaboration. Cardiovascular disease, chronic kidney disease, and diabetes mortality burden of cardiometabolic risk factors from 1980 to 2010: a comparative risk assessment. *Lancet Diabetes Endocrinol.* 2014;2(8):634–647.
25. Pan WH, Yeh WT, Weng LC. Epidemiology of metabolic syndrome in Asia. *Asia Pac J Clin Nutr.* 2008;17(suppl 1):37–42.
26. Morley JE. Diabetes and aging: epidemiologic overview. *Clin Geriatr Med.* 2008;24(3):395–405. v.
27. Schwarz PE, Reimann M, Li J, et al. The metabolic syndrome—a global challenge for prevention. *Horm Metab Res.* 2007;39(11):777–780.
28. Bhatti JS, Bhatti GK, Joshi A, et al. Identification of the risk factors for the high prevalence of type 2 diabetes and its complications in a Punjabi population: North Indian Diabetes Study: a case-control study. *Int J Diab Dev Ctries.* 2007;27(4):108–115.
29. Pollex RL, Hegele RA. Genetic determinants of the metabolic syndrome. *Nat Clin Pract Cardiovasc Med.* 2006;3(9):482–489.
30. Takahara M, Shimomura I. Metabolic syndrome and lifestyle modification. *Rev Endocr Metab Disord.* 2014;15(4):317–327.
31. Misra A, Khurana L. The metabolic syndrome in South Asians: epidemiology, determinants, and prevention. *Metab Syndr Relat Disord.* 2009;7(6):497–514.
32. Bhatti GK, Bhadada SK, Vijayvergiya R, Mastana SS, Bhatti JS. Metabolic syndrome and risk of major coronary events among the urban diabetic patients: North Indian Diabetes and Cardiovascular Disease Study-NIDCVD-2. *J Diabetes Complicat.* 2016;30(1):72–78.
33. Nojiri H, Shimizu T, Funakoshi M, et al. Oxidative stress causes heart failure with impaired mitochondrial respiration. *J Biol Chem.* 2006;281(44):33789–33801.
34. Ohta Y, Kinugawa S, Matsushima S, et al. Oxidative stress impairs insulin signal in skeletal muscle and causes insulin resistance in postinfarct heart failure. *Am J Physiol Heart Circ Physiol.* 2011;300(5):H1637–H1644.
35. Otani H. Oxidative stress as pathogenesis of cardiovascular risk associated with metabolic syndrome. *Antioxid Redox Signal.* 2011;15(7):1911–1926.
36. Ozgen IT, Tascilar ME, Bilir P, et al. Oxidative stress in obese children and its relation with insulin resistance. *J Pediatr Endocrinol Metab.* 2012;25(3–4):261–266.
37. Poitout V, Tanaka Y, Reach G, Robertson RP. Oxidative stress, insulin secretion, and insulin resistance. *Journ Annu Diabetol Hotel Dieu.* 2001;75–86.
38. Salim S, Asghar M, Chugh G, Taneja M, Xia Z, Saha K. Oxidative stress: a potential recipe for anxiety, hypertension and insulin resistance. *Brain Res.* 2010;1359:178–185.
39. Dickinson BC, Chang CJ. Chemistry and biology of reactive oxygen species in signaling or stress responses. *Nat Chem Biol.* 2011;7(8):504–511.
40. Droge W. Free radicals in the physiological control of cell function. *Physiol Rev.* 2002;82(1):47–95.
41. Finkel T, Serrano M, Blasco MA. The common biology of cancer and ageing. *Nature.* 2007;448(7155):767–774.
42. Roy M, Reddy PH, Iijima M, Sesaki H. Mitochondrial division and fusion in metabolism. *Curr Opin Cell Biol.* 2015;33:111–118.

43. Manczak M, Anekonda TS, Henson E, Park BS, Quinn J, Reddy PH. Mitochondria are a direct site of A beta accumulation in Alzheimer's disease neurons: implications for free radical generation and oxidative damage in disease progression. *Hum Mol Genet.* 2006;15(9):1437–1449.
44. Blake R, Trounce IA. Mitochondrial dysfunction and complications associated with diabetes. *Biochim Biophys Acta.* 2014;1840(4):1404–1412.
45. Lin MT, Beal MF. Mitochondrial dysfunction and oxidative stress in neurodegenerative diseases. *Nature.* 2006;443(7113):787–795.
46. Reddy PH, Reddy TP, Manczak M, Calkins MJ, Shirendeb U, Mao P. Dynamin-related protein 1 and mitochondrial fragmentation in neurodegenerative diseases. *Brain Res Rev.* 2011;67(1–2):103–118.
47. Wilson PW, D'Agostino RB, Parise H, Sullivan L, Meigs JB. Metabolic syndrome as a precursor of cardiovascular disease and type 2 diabetes mellitus. *Circulation.* 2005;112(20):3066–3072.
48. Borch-Johnsen K. The metabolic syndrome in a global perspective. The public health impact—secondary publication. *Dan Med Bull.* 2007;54(2):157–159.
49. Kaur J. A comprehensive review on metabolic syndrome. *Cardiol Res Pract.* 2014;2014:943162.
50. Alberti KG, Eckel RH, Grundy SM, et al. Harmonizing the metabolic syndrome: a joint interim statement of the International Diabetes Federation Task Force on Epidemiology and Prevention; National Heart, Lung, and Blood Institute; American Heart Association; World Heart Federation; International Atherosclerosis Society; and International Association for the Study of Obesity. *Circulation.* 2009;120(16):1640–1645.
51. O'Neill S, O'Driscoll L. Metabolic syndrome: a closer look at the growing epidemic and its associated pathologies. *Obes Rev.* 2015;16(1):1–12.
52. Alberti KG, Zimmet PZ. Definition, diagnosis and classification of diabetes mellitus and its complications. Part 1: diagnosis and classification of diabetes mellitus provisional report of a WHO consultation. *Diabet Med.* 1998;15(7):539–553.
53. Balkau B, Charles MA. Comment on the provisional report from the WHO consultation. European Group for the Study of Insulin Resistance (EGIR). *Diabet Med.* 1999;16(5):442–443.
54. National Cholesterol Education Program Expert Panel on Detection E, Treatment of High Blood Cholesterol in Adults. Third Report of the National Cholesterol Education Program (NCEP) Expert Panel on Detection, Evaluation, and Treatment of High Blood Cholesterol in Adults (Adult Treatment Panel III) final report. *Circulation.* 2002;106(25):3143–3421.
55. Einhorn D, Reaven GM, Cobin RH, et al. American College of Endocrinology position statement on the insulin resistance syndrome. *Endocr Pract.* 2003;9(3):237–252.
56. Grundy SM, Cleeman JI, Daniels SR, et al. Diagnosis and management of the metabolic syndrome: an American Heart Association/National Heart, Lung, and Blood Institute Scientific Statement. *Circulation.* 2005;112(17):2735–2752.
57. IDF consensus worldwide definition of the metabolic syndrome. Available from http://www.idf.org/webdata/docs/MetS_def_update2006.pdf. Accessed 12.09.16.
58. Beltran-Sanchez H, Harhay MO, Harhay MM, McElligott S. Prevalence and trends of metabolic syndrome in the adult U.S. population, 1999-2010. *J Am Coll Cardiol.* 2013;62(8):697–703.
59. Shah T, Jonnalagadda SS, Kicklighter JR, Diwan S, Hopkins BL. Prevalence of metabolic syndrome risk factors among young adult Asian Indians. *J Immigr Health.* 2005;7(2):117–126.
60. Deepa M, Farooq S, Datta M, Deepa R, Mohan V. Prevalence of metabolic syndrome using WHO, ATPIII and IDF definitions in Asian Indians: the Chennai Urban Rural Epidemiology Study (CURES-34). *Diabetes Metab Res Rev.* 2007;23(2):127–134.

61. Nestel P, Lyu R, Low LP, et al. Metabolic syndrome: recent prevalence in East and Southeast Asian populations. *Asia Pac J Clin Nutr*. 2007;16(2):362–367.
62. Balasubramanyam A, Rao S, Misra R, Sekhar RV, Ballantyne CM. Prevalence of metabolic syndrome and associated risk factors in Asian Indians. *J Immigr Minor Health*. 2008;10(4):313–323.
63. Misra A, Misra R, Wijesuriya M, Banerjee D. The metabolic syndrome in South Asians: continuing escalation & possible solutions. *Indian J Med Res*. 2007;125(3):345–354.
64. Gadgil MD, Anderson CA, Kandula NR, Kanaya AM. Dietary patterns are associated with metabolic risk factors in South Asians living in the United States. *J Nutr*. 2015;145(6):1211–1217.
65. Parackal S. Dietary transition in the South Asian Diaspora: implications for diabetes prevention strategies. *Curr Diabetes Rev*. 2016;12. http://dx.doi.org/10.2174/1573399812666160901094741. [Epub ahead of print].
66. Sookoian S, Pirola CJ. Metabolic syndrome: from the genetics to the pathophysiology. *Curr Hypertens Rep*. 2011;13(2):149–157.
67. Hopps E, Noto D, Caimi G, Averna MR. A novel component of the metabolic syndrome: the oxidative stress. *Nutr Metab Cardiovasc Dis*. 2010;20(1):72–77.
68. Chen XJ, Butow RA. The organization and inheritance of the mitochondrial genome. *Nat Rev Genet*. 2005;6(11):815–825.
69. Sherratt HS. Mitochondria: structure and function. *Rev Neurol*. 1991;147(6–7):417–430.
70. Chan DC. Mitochondrial fusion and fission in mammals. *Annu Rev Cell Dev Biol*. 2006;22:79–99.
71. Archer SL. Mitochondrial dynamics—mitochondrial fission and fusion in human diseases. *N Engl J Med*. 2013;369(23):2236–2251.
72. Westermann B. Mitochondrial fusion and fission in cell life and death. *Nat Rev Mol Cell Biol*. 2010;11(12):872–884.
73. Shaw JM, Nunnari J. Mitochondrial dynamics and division in budding yeast. *Trends Cell Biol*. 2002;12(4):178–184.
74. Cipolat S, Martins de Brito O, Dal Zilio B, Scorrano L. OPA1 requires mitofusin 1 to promote mitochondrial fusion. *Proc Natl Acad Sci USA*. 2004;101(45):15927–15932.
75. Hales KG, Fuller MT. Developmentally regulated mitochondrial fusion mediated by a conserved, novel, predicted GTPase. *Cell*. 1997;90(1):121–129.
76. Twig G, Hyde B, Shirihai OS. Mitochondrial fusion, fission and autophagy as a quality control axis: the bioenergetic view. *Biochim Biophys Acta*. 2008;1777(9):1092–1097.
77. Elgass K, Pakay J, Ryan MT, Palmer CS. Recent advances into the understanding of mitochondrial fission. *Biochim Biophys Acta*. 2013;1833(1):150–161.
78. Hoppins S. The regulation of mitochondrial dynamics. *Curr Opin Cell Biol*. 2014;29:46–52.
79. Hales KG. The machinery of mitochondrial fusion, division, and distribution, and emerging connections to apoptosis. *Mitochondrion*. 2004;4(4):285–308.
80. Dallner G, Sindelar PJ. Regulation of ubiquinone metabolism. *Free Radic Biol Med*. 2000;29(3–4):285–294.
81. Andreyev AY, Kushnareva YE, Starkov AA. Mitochondrial metabolism of reactive oxygen species. *Biochemistry (Mosc)*. 2005;70(2):200–214.
82. Vankoningsloo S, Piens M, Lecocq C, et al. Mitochondrial dysfunction induces triglyceride accumulation in 3T3-L1 cells: role of fatty acid beta-oxidation and glucose. *J Lipid Res*. 2005;46(6):1133–1149.
83. Choo HJ, Kim JH, Kwon OB, et al. Mitochondria are impaired in the adipocytes of type 2 diabetic mice. *Diabetologia*. 2006;49(4):784–791.
84. Ventura-Clapier R, Garnier A, Veksler V. Transcriptional control of mitochondrial biogenesis: the central role of PGC-1alpha. *Cardiovasc Res*. 2008;79(2):208–217.

85. Jornayvaz FR, Shulman GI. Regulation of mitochondrial biogenesis. *Essays Biochem.* 2010;47:69–84.
86. Hock MB, Kralli A. Transcriptional control of mitochondrial biogenesis and function. *Annu Rev Physiol.* 2009;71:177–203.
87. Virbasius JV, Scarpulla RC. Activation of the human mitochondrial transcription factor A gene by nuclear respiratory factors: a potential regulatory link between nuclear and mitochondrial gene expression in organelle biogenesis. *Proc Natl Acad Sci USA.* 1994;91(4):1309–1313.
88. Reznick RM, Shulman GI. The role of AMP-activated protein kinase in mitochondrial biogenesis. *J Physiol.* 2006;574(pt 1):33–39.
89. Qiang W, Weiqiang K, Qing Z, Pengju Z, Yi L. Aging impairs insulin-stimulated glucose uptake in rat skeletal muscle via suppressing AMPKalpha. *Exp Mol Med.* 2007;39(4):535–543.
90. Reznick RM, Zong H, Li J, et al. Aging-associated reductions in AMP-activated protein kinase activity and mitochondrial biogenesis. *Cell Metab.* 2007;5(2):151–156.
91. Ren J, Pulakat L, Whaley-Connell A, Sowers JR. Mitochondrial biogenesis in the metabolic syndrome and cardiovascular disease. *J Mol Med.* 2010;88(10):993–1001.
92. Nisoli E, Clementi E, Carruba MO, Moncada S. Defective mitochondrial biogenesis: a hallmark of the high cardiovascular risk in the metabolic syndrome? *Circ Res.* 2007;100(6):795–806.
93. Bugger H, Abel ED. Molecular mechanisms for myocardial mitochondrial dysfunction in the metabolic syndrome. *Clin Sci (Lond).* 2008;114(3):195–210.
94. Ritov VB, Menshikova EV, He J, Ferrell RE, Goodpaster BH, Kelley DE. Deficiency of subsarcolemmal mitochondria in obesity and type 2 diabetes. *Diabetes.* 2005;54(1):8–14.
95. Lowell BB, Shulman GI. Mitochondrial dysfunction and type 2 diabetes. *Science (New York, NY).* 2005;307(5708):384–387.
96. Harper ME, Bevilacqua L, Hagopian K, Weindruch R, Ramsey JJ. Ageing, oxidative stress, and mitochondrial uncoupling. *Acta Physiol Scand.* 2004;182(4):321–331.
97. Pieczenik SR, Neustadt J. Mitochondrial dysfunction and molecular pathways of disease. *Exp Mol Pathol.* 2007;83(1):84–92.
98. Reddy PH. Mitochondrial oxidative damage in aging and Alzheimer's disease: implications for mitochondrially targeted antioxidant therapeutics. *J Biomed Biotechnol.* 2006;2006(3):31372.
99. Reddy PH. Abnormal tau, mitochondrial dysfunction, impaired axonal transport of mitochondria, and synaptic deprivation in Alzheimer's disease. *Brain Res.* 2011;1415:136–148.
100. Murphy MP. How mitochondria produce reactive oxygen species. *Biochem J.* 2009;417(1):1–13.
101. Valko M, Rhodes CJ, Moncol J, Izakovic M, Mazur M. Free radicals, metals and antioxidants in oxidative stress-induced cancer. *Chem Biol Interact.* 2006;160(1):1–40.
102. Chen L, Xu WM, Zhang D. Association of abdominal obesity, insulin resistance, and oxidative stress in adipose tissue in women with polycystic ovary syndrome. *Fertil Steril.* 2014;102(4):1167–1174.
103. Das P, Biswas S, Mukherjee S, Bandyopadhyay SK. Association of oxidative stress and obesity with insulin resistance in type 2 diabetes mellitus. *Mymensingh Med J.* 2016;25(1):148–152.
104. Korkmaz GG, Altinoglu E, Civelek S, et al. The association of oxidative stress markers with conventional risk factors in the metabolic syndrome. *Metabolism.* 2013;62(6):828–835.
105. Meigs JB, Larson MG, Fox CS, Keaney Jr JF, Vasan RS, Benjamin EJ. Association of oxidative stress, insulin resistance, and diabetes risk phenotypes: the Framingham Offspring Study. *Diabetes Care.* 2007;30(10):2529–2535.

106. Calo LA, Maiolino G, Naso A, Davis PA. The association of systemic oxidative stress with insulin resistance: mechanistic insights from studies in Bartter's and Gitelman's syndromes. *Clin Endocrinol (Oxf)*. 2015;83(6):994–995.
107. Mitsuyoshi H, Itoh Y, Sumida Y, et al. Evidence of oxidative stress as a cofactor in the development of insulin resistance in patients with chronic hepatitis C. *Hepatol Res*. 2008;38(4):348–353.
108. Luque-Contreras D, Carvajal K, Toral-Rios D, Franco-Bocanegra D, Campos-Pena V. Oxidative stress and metabolic syndrome: cause or consequence of Alzheimer's disease? *Oxid Med Cell Longev*. 2014;2014:497802.
109. Keaney Jr JF, Larson MG, Vasan RS, et al. Obesity and systemic oxidative stress: clinical correlates of oxidative stress in the Framingham Study. *Arterioscler Thromb Vasc Biol*. 2003;23(3):434–439.
110. Ide T, Tsutsui H, Hayashidani S, et al. Mitochondrial DNA damage and dysfunction associated with oxidative stress in failing hearts after myocardial infarction. *Circ Res*. 2001;88(5):529–535.
111. Reddy PH, Tripathi R, Troung Q, et al. Abnormal mitochondrial dynamics and synaptic degeneration as early events in Alzheimer's disease: implications to mitochondria-targeted antioxidant therapeutics. *Biochim Biophys Acta*. 2012;1822(5): 639–649.
112. Itoh K, Nakamura K, Iijima M, Sesaki H. Mitochondrial dynamics in neurodegeneration. *Trends Cell Biol*. 2013;23(2):64–71.
113. Cho DH, Nakamura T, Lipton SA. Mitochondrial dynamics in cell death and neurodegeneration. *Cell Mol Life Sci*. 2010;67(20):3435–3447.
114. Ong SB, Hall AR, Hausenloy DJ. Mitochondrial dynamics in cardiovascular health and disease. *Antioxid Redox Signal*. 2013;19(4):400–414.
115. Jheng HF, Tsai PJ, Guo SM, et al. Mitochondrial fission contributes to mitochondrial dysfunction and insulin resistance in skeletal muscle. *Mol Cell Biol*. 2012;32(2):309–319.
116. Chen L, Knowlton AA. Mitochondrial dynamics in heart failure. *Congest Heart Fail*. 2011;17(6):257–261.
117. Shenouda SM, Widlansky ME, Chen K, et al. Altered mitochondrial dynamics contributes to endothelial dysfunction in diabetes mellitus. *Circulation*. 2011;124(4):444–453.
118. Calkins MJ, Manczak M, Mao P, Shirendeb U, Reddy PH. Impaired mitochondrial biogenesis, defective axonal transport of mitochondria, abnormal mitochondrial dynamics and synaptic degeneration in a mouse model of Alzheimer's disease. *Hum Mol Genet*. 2011;20(23):4515–4529.
119. Manczak M, Calkins MJ, Reddy PH. Impaired mitochondrial dynamics and abnormal interaction of amyloid beta with mitochondrial protein Drp1 in neurons from patients with Alzheimer's disease: implications for neuronal damage. *Hum Mol Genet*. 2011;20(13):2495–2509.
120. Reddy PH. Increased mitochondrial fission and neuronal dysfunction in Huntington's disease: implications for molecular inhibitors of excessive mitochondrial fission. *Drug Discov Today*. 2014;19(7):951–955.
121. Rocha M, Rovira-Llopis S, Banuls C, et al. Mitochondrial dysfunction and oxidative stress in insulin resistance. *Curr Pharm Des*. 2013;19(32):5730–5741.
122. Parish R, Petersen KF. Mitochondrial dysfunction and type 2 diabetes. *Curr Diab Rep*. 2005;5(3):177–183.
123. Ding WX, Yin XM. Mitophagy: mechanisms, pathophysiological roles, and analysis. *Biol Chem*. 2012;393(7):547–564.
124. Beckman KB, Ames BN. Endogenous oxidative damage of mtDNA. *Mutat Res*. 1999;424(1–2):51–58.
125. Turrens JF. Mitochondrial formation of reactive oxygen species. *J Physiol*. 2003;552(pt 2):335–344.

126. James AM, Murphy MP. How mitochondrial damage affects cell function. *J Biomed Sci.* 2002;9(6 pt 1):475–487.
127. Houstis N, Rosen ED, Lander ES. Reactive oxygen species have a causal role in multiple forms of insulin resistance. *Nature.* 2006;440(7086):944–948.
128. Ide T, Tsutsui H, Kinugawa S, et al. Mitochondrial electron transport complex I is a potential source of oxygen free radicals in the failing myocardium. *Circ Res.* 1999;85(4):357–363.
129. Madamanchi NR, Runge MS. Mitochondrial dysfunction in atherosclerosis. *Circ Res.* 2007;100(4):460–473.
130. Tsutsui H, Ide T, Hayashidani S, et al. Enhanced generation of reactive oxygen species in the limb skeletal muscles from a murine infarct model of heart failure. *Circulation.* 2001;104(2):134–136.
131. Reddy PH. Mitochondrial medicine for aging and neurodegenerative diseases. *Neuromolecular Med.* 2008;10(4):291–315.
132. Canto C, Auwerx J. PGC-1alpha, SIRT1 and AMPK, an energy sensing network that controls energy expenditure. *Curr Opin Lipidol.* 2009;20(2):98–105.
133. Mootha VK, Lindgren CM, Eriksson KF, et al. PGC-1alpha-responsive genes involved in oxidative phosphorylation are coordinately downregulated in human diabetes. *Nat Genet.* 2003;34(3):267–273.
134. Bhatti JS, Bhatti GK, Mastana SS, Ralhan S, Joshi A, Tewari R. ENPP1/PC-1 K121Q polymorphism and genetic susceptibility to type 2 diabetes in North Indians. *Mol Cell Biochem.* 2010;345(1–2):249–257.
135. Fuchsberger C, Flannick J, Teslovich TM, et al. The genetic architecture of type 2 diabetes. *Nature.* 2016;536(7614):41–47.
136. Raj R, Bhatti JS, Badada SK, Ramteke PW. Genetic basis of dyslipidemia in disease precipitation of coronary artery disease (CAD) associated type 2 diabetes mellitus (T2DM). *Diabetes Metab Res Rev.* 2014;31:663–671.
137. Ma ZA, Zhao Z, Turk J. Mitochondrial dysfunction and beta-cell failure in type 2 diabetes mellitus. *Exp Diabetes Res.* 2012;2012:703538.
138. Sivitz WI, Yorek MA. Mitochondrial dysfunction in diabetes: from molecular mechanisms to functional significance and therapeutic opportunities. *Antioxid Redox Signal.* 2010;12(4):537–577.
139. Wang CH, Wang CC, Wei YH. Mitochondrial dysfunction in insulin insensitivity: implication of mitochondrial role in type 2 diabetes. *Ann NY Acad Sci.* 2010;1201: 157–165.
140. Hojlund K, Mogensen M, Sahlin K, Beck-Nielsen H. Mitochondrial dysfunction in type 2 diabetes and obesity. *Endocrinol Metab Clin North Am.* 2008;37(3): 713–731.
141. Abdul-Ghani MA, DeFronzo RA. Mitochondrial dysfunction, insulin resistance, and type 2 diabetes mellitus. *Curr Diab Rep.* 2008;8(3):173–178.
142. Kostis JB, Sanders M. The association of heart failure with insulin resistance and the development of type 2 diabetes. *Am J Hypertens.* 2005;18(5 pt 1):731–737.
143. Reaven GM, Chen YD. Insulin resistance, its consequences, and coronary heart disease. Must we choose one culprit? *Circulation.* 1996;93(10):1780–1783.
144. Kelley DE, He J, Menshikova EV, Ritov VB. Dysfunction of mitochondria in human skeletal muscle in type 2 diabetes. *Diabetes.* 2002;51(10):2944–2950.
145. Carvalho E, Kotani K, Peroni OD, Kahn BB. Adipose-specific overexpression of GLUT4 reverses insulin resistance and diabetes in mice lacking GLUT4 selectively in muscle. *Am J Physiol Endocrinol Metab.* 2005;289(4):E551–E561.
146. Leguisamo NM, Lehnen AM, Machado UF, et al. GLUT4 content decreases along with insulin resistance and high levels of inflammatory markers in rats with metabolic syndrome. *Cardiovasc Diabetol.* 2012;11:100.

147. Pirola L, Johnston AM, Van Obberghen E. Modulation of insulin action. *Diabetologia.* 2004;47(2):170–184.
148. Hu F, Liu F. Mitochondrial stress: a bridge between mitochondrial dysfunction and metabolic diseases? *Cell Signal.* 2011;23(10):1528–1533.
149. Ashrafian H, Frenneaux MP, Opie LH. Metabolic mechanisms in heart failure. *Circulation.* 2007;116(4):434–448.
150. De Pauw A, Tejerina S, Raes M, Keijer J, Arnould T. Mitochondrial (dys)function in adipocyte (de)differentiation and systemic metabolic alterations. *Am J Pathol.* 2009;175(3):927–939.
151. Wiederkehr A, Wollheim CB. Minireview: implication of mitochondria in insulin secretion and action. *Endocrinology.* 2006;147(6):2643–2649.
152. Petersen KF, Befroy D, Dufour S, et al. Mitochondrial dysfunction in the elderly: possible role in insulin resistance. *Science.* 2003;300(5622):1140–1142.
153. Rong JX, Qiu Y, Hansen MK, et al. Adipose mitochondrial biogenesis is suppressed in db/db and high-fat diet-fed mice and improved by rosiglitazone. *Diabetes.* 2007;56(7): 1751–1760.
154. Short KR, Bigelow ML, Kahl J, et al. Decline in skeletal muscle mitochondrial function with aging in humans. *Proc Natl Acad Sci USA.* 2005;102(15):5618–5623.
155. Kowaltowski AJ, de Souza-Pinto NC, Castilho RF, Vercesi AE. Mitochondria and reactive oxygen species. *Free Radic Biol Med.* 2009;47(4):333–343.
156. Jezek P, Hlavata L. Mitochondria in homeostasis of reactive oxygen species in cell, tissues, and organism. *Int J Biochem Cell Biol.* 2005;37(12):2478–2503.
157. Irving BA, Nair KS. Aging and diabetes: mitochondrial dysfunction. *Curr Diab Rep.* 2007;7(4):249–251.
158. Mogensen M, Sahlin K, Fernstrom M, et al. Mitochondrial respiration is decreased in skeletal muscle of patients with type 2 diabetes. *Diabetes.* 2007;56(6):1592–1599.
159. Razak F, Anand SS. Impaired mitochondrial activity in the insulin-resistant offspring of patients with type 2 diabetes. Petersen KF, Dufour S, Befroy D, Garcia R, Shulman GI. N Engl J Med 2004; 350: 664–71. *Vasc Med.* 2004;9(3):223–224.
160. Morino K, Petersen KF, Dufour S, et al. Reduced mitochondrial density and increased IRS-1 serine phosphorylation in muscle of insulin-resistant offspring of type 2 diabetic parents. *J Clin Invest.* 2005;115(12):3587–3593.
161. Morino K, Petersen KF, Sono S, et al. Regulation of mitochondrial biogenesis by lipoprotein lipase in muscle of insulin-resistant offspring of parents with type 2 diabetes. *Diabetes.* 2012;61(4):877–887.
162. Patti ME, Butte AJ, Crunkhorn S, et al. Coordinated reduction of genes of oxidative metabolism in humans with insulin resistance and diabetes: potential role of PGC1 and NRF1. *Proc Natl Acad Sci USA.* 2003;100(14):8466–8471.
163. Corral-Debrinski M, Horton T, Lott MT, Shoffner JM, Beal MF, Wallace DC. Mitochondrial DNA deletions in human brain: regional variability and increase with advanced age. *Nat Genet.* 1992;2(4):324–329.
164. Krebs M, Roden M. Molecular mechanisms of lipid-induced insulin resistance in muscle, liver and vasculature. *Diabetes Obes Metab.* 2005;7(6):621–632.
165. Duncan JG, Fong JL, Medeiros DM, Finck BN, Kelly DP. Insulin-resistant heart exhibits a mitochondrial biogenic response driven by the peroxisome proliferator-activated receptor-alpha/PGC-1alpha gene regulatory pathway. *Circulation.* 2007;115(7): 909–917.
166. Wu Z, Puigserver P, Andersson U, et al. Mechanisms controlling mitochondrial biogenesis and respiration through the thermogenic coactivator PGC-1. *Cell.* 1999;98(1): 115–124.
167. Furukawa S, Fujita T, Shimabukuro M, et al. Increased oxidative stress in obesity and its impact on metabolic syndrome. *J Clin Invest.* 2004;114(12):1752–1761.

168. Wlodek D, Gonzales M. Decreased energy levels can cause and sustain obesity. *J Theor Biol*. 2003;225(1):33–44.
169. Wisloff U, Najjar SM, Ellingsen O, et al. Cardiovascular risk factors emerge after artificial selection for low aerobic capacity. *Science*. 2005;307(5708):418–420.
170. Sparks LM, Xie H, Koza RA, et al. A high-fat diet coordinately downregulates genes required for mitochondrial oxidative phosphorylation in skeletal muscle. *Diabetes*. 2005;54(7):1926–1933.
171. Gastaldi G, Giacobino JP, Ruiz J. Metabolic syndrome, a mitochondrial disease? *Rev Med Suisse*. 2008;4(160):1387–1388. 1390–1381.
172. Choksi KB, Boylston WH, Rabek JP, Widger WR, Papaconstantinou J. Oxidatively damaged proteins of heart mitochondrial electron transport complexes. *Biochim Biophys Acta*. 2004;1688(2):95–101.
173. Maassen JA, 'T Hart LM, Van Essen E, et al. Mitochondrial diabetes: molecular mechanisms and clinical presentation. *Diabetes*. 2004;53(suppl 1):S103–S109.
174. Cooper SA, Whaley-Connell A, Habibi J, et al. Renin-angiotensin-aldosterone system and oxidative stress in cardiovascular insulin resistance. *Am J Physiol Heart Circ Physiol*. 2007;293(4):H2009–H2023.
175. Smith Jr SC. Multiple risk factors for cardiovascular disease and diabetes mellitus. *Am J Med*. 2007;120(3 suppl 1):S3–S11.
176. Sowers JR. Insulin resistance and hypertension. *Am J Physiol Heart Circ Physiol*. 2004;286(5):H1597–H1602.
177. Hotamisligil GS, Spiegelman BM. Tumor necrosis factor alpha: a key component of the obesity-diabetes link. *Diabetes*. 1994;43(11):1271–1278.
178. Tzanavari T, Giannogonas P, Karalis KP. TNF-alpha and obesity. *Curr Dir Autoimmun*. 2010;11:145–156.
179. Weisberg SP, McCann D, Desai M, Rosenbaum M, Leibel RL, Ferrante Jr AW. Obesity is associated with macrophage accumulation in adipose tissue. *J Clin Invest*. 2003;112(12):1796–1808.
180. Savage DB, Petersen KF, Shulman GI. Mechanisms of insulin resistance in humans and possible links with inflammation. *Hypertension*. 2005;45(5):828–833.
181. Tsutsui H, Kinugawa S, Matsushima S. Oxidative stress and mitochondrial DNA damage in heart failure. *Circ J*. 2008;72(suppl A):A31–A37.
182. Victor VM, Apostolova N, Herance R, Hernandez-Mijares A, Rocha M. Oxidative stress and mitochondrial dysfunction in atherosclerosis: mitochondria-targeted antioxidants as potential therapy. *Curr Med Chem*. 2009;16(35):4654–4667.
183. Chen J, Mehta JL. Role of oxidative stress in coronary heart disease. *Indian Heart J*. 2004;56(2):163–173.
184. Suematsu N, Tsutsui H, Wen J, et al. Oxidative stress mediates tumor necrosis factor-alpha-induced mitochondrial DNA damage and dysfunction in cardiac myocytes. *Circulation*. 2003;107(10):1418–1423.
185. Belch JJ, Bridges AB, Scott N, Chopra M. Oxygen free radicals and congestive heart failure. *Br Heart J*. 1991;65(5):245–248.
186. Brown GC, Borutaite V. There is no evidence that mitochondria are the main source of reactive oxygen species in mammalian cells. *Mitochondrion*. 2012;12(1):1–4.
187. Bloemen PG, Van den Tweel MC, Henricks PA, et al. Stimulation of both human bronchial epithelium and neutrophils is needed for maximal interactive adhesion. *Am J Physiol*. 1996;270(1 pt 1):L80–L87.
188. Ide T, Tsutsui H, Kinugawa S, Utsumi H, Takeshita A. Amiodarone protects cardiac myocytes against oxidative injury by its free radical scavenging action. *Circulation*. 1999;100(7):690–692.
189. Ide T, Tsutsui H, Kinugawa S, et al. Direct evidence for increased hydroxyl radicals originating from superoxide in the failing myocardium. *Circ Res*. 2000;86(2):152–157.

190. Schulze-Osthoff K, Bakker AC, Vanhaesebroeck B, Beyaert R, Jacob WA, Fiers W. Cytotoxic activity of tumor necrosis factor is mediated by early damage of mitochondrial functions. Evidence for the involvement of mitochondrial radical generation. *J Biol Chem*. 1992;267(8):5317–5323.
191. Zell R, Geck P, Werdan K, Boekstegers P. TNF-alpha and IL-1 alpha inhibit both pyruvate dehydrogenase activity and mitochondrial function in cardiomyocytes: evidence for primary impairment of mitochondrial function. *Mol Cell Biochem*. 1997;177(1–2):61–67.
192. Lo EH, Dalkara T, Moskowitz MA. Mechanisms, challenges and opportunities in stroke. *Nat Rev Neurosci*. 2003;4(5):399–415.
193. Vijayan M, Reddy PH. Stroke, vascular dementia, and Alzheimer's disease: molecular links. *J Alzheimers Dis*. 2016;54(2):427–443.
194. Abramov AY, Scorziello A, Duchen MR. Three distinct mechanisms generate oxygen free radicals in neurons and contribute to cell death during anoxia and reoxygenation. *J Neurosci*. 2007;27(5):1129–1138.
195. Chan PH. Reactive oxygen radicals in signaling and damage in the ischemic brain. *J Cereb Blood Flow Metab*. 2001;21(1):2–14.
196. Doyle KP, Simon RP, Stenzel-Poore MP. Mechanisms of ischemic brain damage. *Neuropharmacology*. 2008;55(3):310–318.
197. Mehta SL, Manhas N, Raghubir R. Molecular targets in cerebral ischemia for developing novel therapeutics. *Brain Res Rev*. 2007;54(1):34–66.
198. Kim GW, Kondo T, Noshita N, Chan PH. Manganese superoxide dismutase deficiency exacerbates cerebral infarction after focal cerebral ischemia/reperfusion in mice: implications for the production and role of superoxide radicals. *Stroke*. 2002;33(3):809–815.
199. Murakami K, Kondo T, Kawase M, et al. Mitochondrial susceptibility to oxidative stress exacerbates cerebral infarction that follows permanent focal cerebral ischemia in mutant mice with manganese superoxide dismutase deficiency. *J Neurosci*. 1998;18(1):205–213.
200. Friedman JR, Nunnari J. Mitochondrial form and function. *Nature*. 2014;505(7483):335–343.
201. Reddy PH. Role of mitochondria in neurodegenerative diseases: mitochondria as a therapeutic target in Alzheimer's disease. *CNS Spectr*. 2009;14(8 suppl 7):8–13. discussion 16–18.
202. Kuzmicic J, Del Campo A, Lopez-Crisosto C, et al. Mitochondrial dynamics: a potential new therapeutic target for heart failure. *Rev Esp Cardiol*. 2011;64(10):916–923.
203. Russell AP, Foletta VC, Snow RJ, Wadley GD. Skeletal muscle mitochondria: a major player in exercise, health and disease. *Biochim Biophys Acta*. 2014;1840(4):1276–1284.
204. Rockl KS, Witczak CA, Goodyear LJ. Signaling mechanisms in skeletal muscle: acute responses and chronic adaptations to exercise. *IUBMB Life*. 2008;60(3):145–153.
205. Holloway GP. Mitochondrial function and dysfunction in exercise and insulin resistance. *Appl Physiol Nutr Metab*. 2009;34(3):440–446.
206. Joseph AM, Hood DA. Relationships between exercise, mitochondrial biogenesis and type 2 diabetes. *Med Sport Sci*. 2014;60:48–61.
207. Meex RC, Schrauwen-Hinderling VB, Moonen-Kornips E, et al. Restoration of muscle mitochondrial function and metabolic flexibility in type 2 diabetes by exercise training is paralleled by increased myocellular fat storage and improved insulin sensitivity. *Diabetes*. 2010;59(3):572–579.
208. Toledo FG, Menshikova EV, Ritov VB, et al. Effects of physical activity and weight loss on skeletal muscle mitochondria and relationship with glucose control in type 2 diabetes. *Diabetes*. 2007;56(8):2142–2147.
209. Nielsen J, Mogensen M, Vind BF, et al. Increased subsarcolemmal lipids in type 2 diabetes: effect of training on localization of lipids, mitochondria, and glycogen in sedentary human skeletal muscle. *Am J Physiol Endocrinol Metab*. 2010;298(3):E706–E713.

210. Phielix E, Meex R, Moonen-Kornips E, Hesselink MK, Schrauwen P. Exercise training increases mitochondrial content and ex vivo mitochondrial function similarly in patients with type 2 diabetes and in control individuals. *Diabetologia.* 2010;53(8): 1714–1721.
211. Barbieri E, Agostini D, Polidori E, et al. The pleiotropic effect of physical exercise on mitochondrial dynamics in aging skeletal muscle. *Oxid Med Cell Longev.* 2015;2015:917085.
212. Anderson RM, Shanmuganayagam D, Weindruch R. Caloric restriction and aging: studies in mice and monkeys. *Toxicol Pathol.* 2009;37(1):47–51.
213. Weindruch R, Walford RL, Fligiel S, Guthrie D. The retardation of aging in mice by dietary restriction: longevity, cancer, immunity and lifetime energy intake. *J Nutr.* 1986;116(4):641–654.
214. Omodei D, Fontana L. Calorie restriction and prevention of age-associated chronic disease. *FEBS Lett.* 2011;585(11):1537–1542.
215. Barja G. Endogenous oxidative stress: relationship to aging, longevity and caloric restriction. *Ageing Res Rev.* 2002;1(3):397–411.
216. Barja G, Herrero A. Oxidative damage to mitochondrial DNA is inversely related to maximum life span in the heart and brain of mammals. *FASEB J.* 2000;14(2):312–318.
217. Civitarese AE, Carling S, Heilbronn LK, et al. Calorie restriction increases muscle mitochondrial biogenesis in healthy humans. *PLoS Med.* 2007;4(3)e76.
218. Zainal TA, Oberley TD, Allison DB, Szweda LI, Weindruch R. Caloric restriction of rhesus monkeys lowers oxidative damage in skeletal muscle. *FASEB J.* 2000;14(12): 1825–1836.
219. Masoro EJ. Overview of caloric restriction and ageing. *Mech Ageing Dev.* 2005;126(9): 913–922.
220. Lee IM, Blair SN, Allison DB, et al. Epidemiologic data on the relationships of caloric intake, energy balance, and weight gain over the life span with longevity and morbidity. *J Gerontol A Biol Sci Med Sci.* 2001;56(Spec No 1):7–19.
221. Bhatti G, Bhatti J, Kiran R, Sandhir R. Biochemical and morphological perturbations in rat erythrocytes exposed to ethion: protective effect of vitamin E. *Cell Mol Biol (Noisy-le-Grand).* 2010;57(1):70–79.
222. Yorek MA. The role of oxidative stress in diabetic vascular and neural disease. *Free Radic Res.* 2003;37(5):471–480.
223. Yi X, Maeda N. alpha-Lipoic acid prevents the increase in atherosclerosis induced by diabetes in apolipoprotein E-deficient mice fed high-fat/low-cholesterol diet. *Diabetes.* 2006;55(8):2238–2244.
224. Mehta K, Van Thiel DH, Shah N, Mobarhan S. Nonalcoholic fatty liver disease: pathogenesis and the role of antioxidants. *Nutr Rev.* 2002;60(9):289–293.
225. Kelso GF, Porteous CM, Coulter CV, et al. Selective targeting of a redox-active ubiquinone to mitochondria within cells: antioxidant and antiapoptotic properties. *J Biol Chem.* 2001;276(7):4588–4596.
226. Smith RA, Porteous CM, Coulter CV, Murphy MP. Selective targeting of an antioxidant to mitochondria. *Eur J Biochem.* 1999;263(3):709–716.
227. Dhanasekaran A, Kotamraju S, Kalivendi SV, et al. Supplementation of endothelial cells with mitochondria-targeted antioxidants inhibit peroxide-induced mitochondrial iron uptake, oxidative damage, and apoptosis. *J Biol Chem.* 2004;279(36):37575–37587.
228. Mercer JR, Yu E, Figg N, et al. The mitochondria-targeted antioxidant MitoQ decreases features of the metabolic syndrome in $ATM^{+/-}/ApoE^{-/-}$ mice. *Free Radic Biol Med.* 2012;52(5):841–849.
229. Feillet-Coudray C, Fouret G, Ebabe Elle R, et al. The mitochondrial-targeted antioxidant MitoQ ameliorates metabolic syndrome features in obesogenic diet-fed rats better than apocynin or allopurinol. *Free Radic Res.* 2014;48(10):1232–1246.

230. Adlam VJ, Harrison JC, Porteous CM, et al. Targeting an antioxidant to mitochondria decreases cardiac ischemia-reperfusion injury. *FASEB J.* 2005;19(9): 1088–1095.
231. Dare AJ, Bolton EA, Pettigrew GJ, Bradley JA, Saeb-Parsy K, Murphy MP. Protection against renal ischemia-reperfusion injury in vivo by the mitochondria targeted antioxidant MitoQ. *Redox Biol.* 2015;5:163–168.
232. Galley HF. Bench-to-bedside review: targeting antioxidants to mitochondria in sepsis. *Crit Care.* 2010;14(4):230.
233. Lowes DA, Thottakam BM, Webster NR, Murphy MP, Galley HF. The mitochondria-targeted antioxidant MitoQ protects against organ damage in a lipopolysaccharide-peptidoglycan model of sepsis. *Free Radic Biol Med.* 2008;45(11): 1559–1565.
234. Yin X, Manczak M, Reddy PH. Mitochondria-targeted molecules MitoQ and SS31 reduce mutant huntingtin-induced mitochondrial toxicity and synaptic damage in Huntington's disease. *Hum Mol Genet.* 2016;25:1739–1753.
235. Mao P, Manczak M, Shirendeb UP, Reddy PH. MitoQ, a mitochondria-targeted antioxidant, delays disease progression and alleviates pathogenesis in an experimental autoimmune encephalomyelitis mouse model of multiple sclerosis. *Biochim Biophys Acta.* 2013;1832(12):2322–2331.
236. Powell RD, Swet JH, Kennedy KL, et al. MitoQ modulates oxidative stress and decreases inflammation following hemorrhage. *J Trauma Acute Care Surg.* 2015;78(3): 573–579.
237. Smith RA, Porteous CM, Gane AM, Murphy MP. Delivery of bioactive molecules to mitochondria in vivo. *Proc Natl Acad Sci USA.* 2003;100(9):5407–5412.
238. Turkmen K, Karagoz A, Kucuk A. Sirtuins as novel players in the pathogenesis of diabetes mellitus. *World J Diabetes.* 2014;5(6):894–900.
239. Huynh FK, Hershberger KA, Hirschey MD. Targeting sirtuins for the treatment of diabetes. *Diabetes Manag.* 2013;3(3):245–257.
240. Kitada M, Kume S, Kanasaki K, Takeda-Watanabe A, Koya D. Sirtuins as possible drug targets in type 2 diabetes. *Curr Drug Targets.* 2013;14(6):622–636.
241. Albiero M, Avogaro A, Fadini GP. A perspective on sirtuins in the metabolic syndrome. *Metab Syndr Relat Disord.* 2015;13(4):161–164.
242. Kumar S, Lombard DB. Mitochondrial sirtuins and their relationships with metabolic disease and cancer. *Antioxid Redox Signal.* 2015;22(12):1060–1077.
243. Chalkiadaki A, Guarente L. Sirtuins mediate mammalian metabolic responses to nutrient availability. *Nat Rev Endocrinol.* 2012;8(5):287–296.
244. Yu J, Auwerx J. The role of sirtuins in the control of metabolic homeostasis. *Ann NY Acad Sci.* 2009;1173(suppl 1):E10–E19.
245. Elliott PJ, Jirousek M. Sirtuins: novel targets for metabolic disease. *Curr Opin Investig Drugs.* 2008;9(4):371–378.
246. Guarente L. Sirtuins as potential targets for metabolic syndrome. *Nature.* 2006;444(7121):868–874.
247. Bindu S, Pillai VB, Gupta MP. Role of sirtuins in regulating pathophysiology of the heart. *Trends Endocrinol Metab.* 2016;27:563–573.
248. Tanno M, Kuno A, Horio Y, Miura T. Emerging beneficial roles of sirtuins in heart failure. *Basic Res Cardiol.* 2012;107(4):273.
249. Lagouge M, Argmann C, Gerhart-Hines Z, et al. Resveratrol improves mitochondrial function and protects against metabolic disease by activating SIRT1 and PGC-1alpha. *Cell.* 2006;127(6):1109–1122.
250. Parihar P, Solanki I, Mansuri ML, Parihar MS. Mitochondrial sirtuins: emerging roles in metabolic regulations, energy homeostasis and diseases. *Exp Gerontol.* 2015;61: 130–141.

251. Reddy PH. Inhibitors of mitochondrial fission as a therapeutic strategy for diseases with oxidative stress and mitochondrial dysfunction. *J Alzheimers Dis*. 2014;40(2):245–256.
252. Meuer K, Suppanz IE, Lingor P, et al. Cyclin-dependent kinase 5 is an upstream regulator of mitochondrial fission during neuronal apoptosis. *Cell Death Differ*. 2007;14(4):651–661.
253. Cassidy-Stone A, Chipuk JE, Ingerman E, et al. Chemical inhibition of the mitochondrial division dynamin reveals its role in Bax/Bak-dependent mitochondrial outer membrane permeabilization. *Dev Cell*. 2008;14(2):193–204.
254. Qi X, Qvit N, Su YC, Mochly-Rosen D. A novel Drp1 inhibitor diminishes aberrant mitochondrial fission and neurotoxicity. *J Cell Sci*. 2013;126(pt 3):789–802.
255. Owen MR, Doran E, Halestrap AP. Evidence that metformin exerts its anti-diabetic effects through inhibition of complex 1 of the mitochondrial respiratory chain. *Biochem J*. 2000;348(pt 3):607–614.

CHAPTER THREE

MicroRNAs as Peripheral Biomarkers in Aging and Age-Related Diseases

S. Kumar*[,1], M. Vijayan*, J.S. Bhatti*[,†], P.H. Reddy*[,‡]

*Garrison Institute on Aging, Texas Tech University Health Sciences Center, Lubbock, TX, United States
[†]Department of Biotechnology, Sri Guru Gobind Singh College, Chandigarh, India
[‡]Texas Tech University Health Sciences Center, Lubbock, TX, United States
[1]Corresponding author: e-mail address: subodh.kumar@ttuhsc.edu

Contents

Abstract

MicroRNAs (miRNAs) are found in the circulatory biofluids considering the important molecules for biomarker study in aging and age-related diseases. Blood or blood components (serum/plasma) are primary sources of circulatory miRNAs and can release these in cell-free form either bound with some protein components or encapsulated with microvesicle particles, called exosomes. miRNAs are quite stable in the peripheral circulation and can be detected by high-throughput techniques like qRT-PCR, microarray, and sequencing. Intracellular miRNAs could modulate mRNA activity through target-specific binding and play a crucial role in intercellular communications. At a pathological level, changes in cellular homeostasis lead to the modulation of molecular function of cells; as a result, miRNA expression is deregulated. Deregulated miRNAs came out

Progress in Molecular Biology and Translational Science, Volume 146
ISSN 1877-1173
http://dx.doi.org/10.1016/bs.pmbts.2016.12.013

from cells and frequently circulate in extracellular body fluids as part of various human diseases. Most common aging-associated diseases are cardiovascular disease, cancer, arthritis, dementia, cataract, osteoporosis, diabetes, hypertension, and neurodegenerative diseases such as Alzheimer's disease, Huntington's disease, Parkinson's disease, and amyotrophic lateral sclerosis. Variation in the miRNA signature in a diseased peripheral circulatory system opens up a new avenue in the field of biomarker discovery. Here, we measure the biomarker potential of circulatory miRNAs in aging and various aging-related pathologies. However, further more confirmatory researches are needed to elaborate these findings at the translation level.

ABBREVIATIONS

AD Alzheimer's disease
AFP alpha-fetoprotein
AGO Argonaut2
ALS amyotrophic lateral sclerosis
AUC area under curve
BMCs blood mononuclear cells
BMD bone mineral density
BPH benign prostatic hyperplasia
BC breast cancer
CA 19–9 carbohydrate antigen 19-9
CAD coronary artery disease
CBL casitas B-lineage lymphoma protooncogene
CD226 cluster of differentiation 226
CEA carcinoembryonic antigen
cfPWV carotid–femoral pulse wave velocity
CgA chromogranin A
COPD chronic obstructive pulmonary disease
CRC colorectal carcinoma
CRP C-reactive protein
CSF cerebrospinal fluid
CTO chronic total occlusion
CVD cardio-cerebrovascular disease
DM diabetes mellitus
ERA early rheumatoid arthritis
HD Huntington's disease
HDL high-density lipoproteins
HF heart failure
HTT Huntingtin protein
iCIMT increased CIMT
IFG impaired fasting glucose
IGF1 insulin-like growth factor 1
IGT impaired glucose tolerance
INDCs inflammatory neurological disease controls
IPAH idiopathic pulmonary hypertension

LuCa lung cancer
MCI mild cognitive impairment
miRNA microRNA
mRNA messenger RNA
MVs microvesicles
nCIMT normal carotid intima-media thickness
NDs neurodegenerative diseases
NINDCs noninflammatory neurological disease controls
NSCLC nonsmall cell lung cancer
PAG1 phosphoprotein-associated with glycosphingolipid microdomains 1
PBMCs peripheral blood mononuclear cells
PCa prostate cancer
PD Parkinson's disease
PP pulse pressure
pre-miRNA precursor miRNA
pri-miRNA primary miRNA
qRT-PCR quantitative real-time polymerase chain reaction
RA rheumatoid arthritis
RF rheumatoid factor
RISC RNA-induced silencing complex
ROC receiver operating characteristic
rRNA ribosomal RNA
sALS sporadic amyotrophic lateral sclerosis
T1DM type 1DM
T2DM type 2DM
TOB2 transducer of ERBB2, 2
VHR very high risk

1. INTRODUCTION

The first microRNAs (miRNAs) were discovered in *Caenorhabditis elegans* in 1993 by Lee and colleagues, and at present, more than 1881 precursor and 2588 mature miRNAs have been identified as updated by miRbase-21 database released in June 2014 (http://www.mirbase.org/). miRNAs are identified in various human diseases such as cancer, viral infection, diabetes, immune-related diseases, aging, and neurodegenerative disorders; their role has been established in different aspects of disease such as diagnosis, pathogenesis, and therapeutics.[1–10] miRNA synthesis processing starts in the nucleus with the formation of primary miRNA (pri-miRNA), then precursor miRNA (pre-miRNA), and finally mature miRNA generated in the cell cytoplasm.[11] Approximately 2000 human genes encode different miRNAs, annealing at

3′UTR of nearly 60% of human genes and modulating their activity at the transcription level.[1,11–13] Besides working as a gene modulator, in several circumstances, miRNAs are also secreted in the extracellular biofluids such as blood, serum, plasma, saliva, and urine.[7,14] Circulatory miRNAs have also been found to be quite stable in extracellular circulation and could be a good bioindicator for aging and age-related disease assessment.[6,8,14–17]

Aging is a multifactorial process characterized by a progressive loss of physiological integrity, leading to impaired function and increased vulnerability to death.[18] Within the aging process, cellular and molecular changes and damage lead to increased disease susceptibility and mortality.[7] A key factor for the aging process is cellular senescence. The most common cellular processes that induce senescence in normal aging are epigenetic stress, proteotoxic stress, oxidative stress, telomere damage, and DNA damage, while disease-related senescence is accompanied by smoking, telomere damage, and DNA damage.[19] Well-known aging-associated diseases are cardiovascular disease, cancer, arthritis, dementia, cataract, osteoporosis, diabetes, and hypertension and neurodegenerative diseases (NDs) such as Alzheimer's disease (AD), Huntington's disease (HD), Parkinson's disease (PD), and amyotrophic lateral sclerosis (ALS) (Fig. 1). A less invasive method to

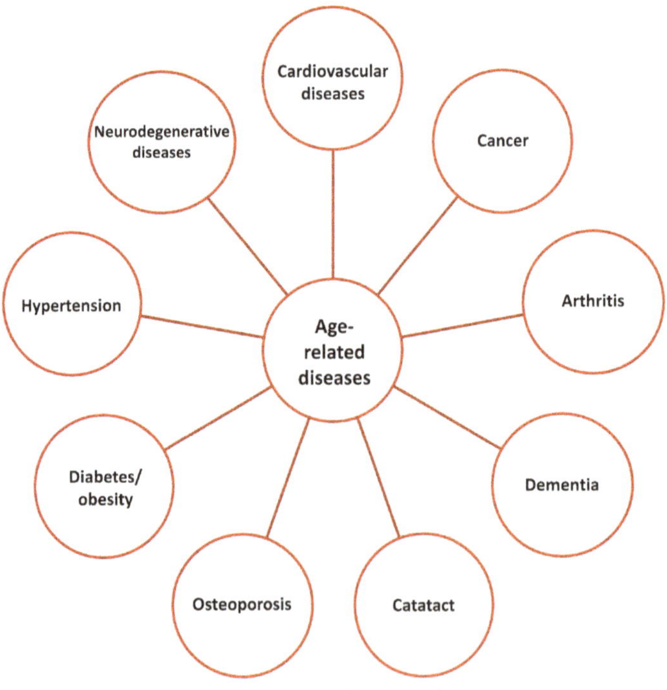

Fig. 1 Most common aging and age-related human diseases.

diagnose aging-associated pathologies and other NDs much earlier than current methods is needed, and one encouraging alternative is through the use of biomarkers. One such promising biomarker for aging and aging-related diseases is circulatory miRNAs. The purpose of this chapter is to discuss the latest developments in circulating miRNAs and their possible role in early, noninvasive identification and assessment of aging-associated diseases.

2. CIRCULATORY miRNAs

In 2010, Weber and colleagues reported that miRNAs are present in various biofluids such as blood, saliva, tears, urine, amniotic fluid, colostrum, breast milk, bronchial secretions, cerebrospinal fluid (CSF), peritoneal fluid, pleural fluid, and seminal fluid.[20] The stability and abundance of circulatory miRNAs in biofluids, such as serum and plasma, are main factors that contribute to their use as potential diagnostic and progression biomarkers in a clinical context.[21] Starting in 2007, initial observations were made that mature miRNAs are released from the cellular cytoplasm within extracellular vesicles.[22] At the time of writing, the presence of "circulatory miRNAs" has been verified in 12 different biofluids.[20]

3. miRNAs SECRETION IN CIRCULATORY BIOFLUIDS

Accumulating evidence shows that miRNAs are secreted in extracellular spaces either in the microvesicles (MVs)-encapsulated form or released in the vesicle-free form bound with some proteins or other compounds.[16,17,23] Kumar and Reddy have mentioned five different ways of miRNA transportation into extracellular circulation: (1) bound with high-density lipoproteins (HDL) particle in nonvesicle form; (2) complex form with Ago2 protein; (3) packaged within exosomes; (4) encapsulated within MVs; and (5) accumulated in apoptotic bodies.[10] Arroyo *et al.* demonstrated that the majority (90%) of circulatory miRNAs were circulated in MV-free form bound with Ago2 family proteins.[24] Ago2 protein association provides ample resistance to miRNAs from a nuclease-rich extracellular environment, because of the high stability of Ago2 protein in biofluids.[23] In some instances, particular cells showed a specific miRNA transport mechanism for certain miRNAs such as miR-122, a liver-enriched candidate exclusively detected with Ago2 complexes.[24] The HDL particle size of 8–12 nm is mainly constituted of phosphatidylcholine and apolipoprotein A-I, which forms a stable ternary structure with extracellular plasma miRNAs by divalent cation bridging.[17] HDL particles participate only in

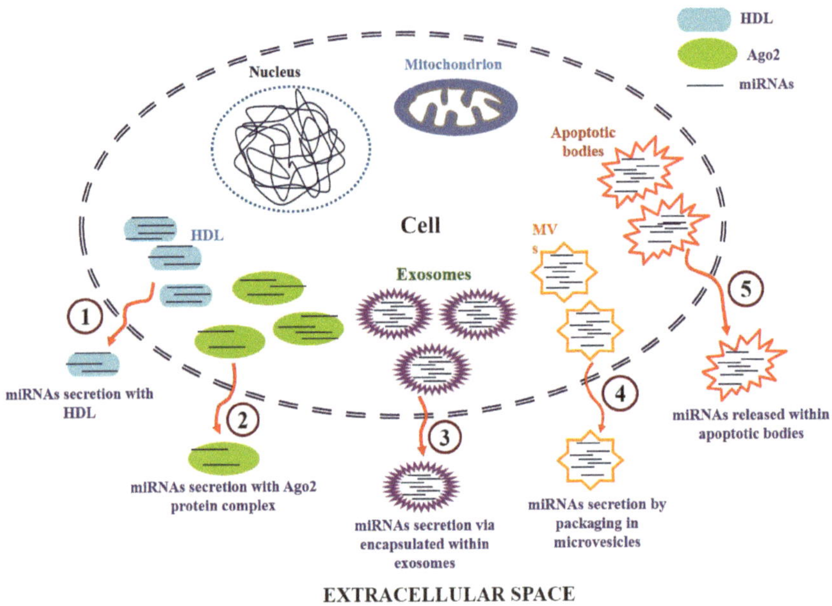

Fig. 2 Modes of circulatory miRNA secretion from cells.

a minor fraction of transportation of endogenous miRNAs in peripheral circulation. miR-223, a well-known miRNA, circulates in plasma associated with HDL in the case of atherosclerosis.[25] Another transportation mode is via miRNA encapsulation in microparticles including MVs, exosomes, and apoptotic bodies (Fig. 2). The chief functions of MVs and exosomes are intercellular communication and transportation of various bioactive components (DNA, RNA, miRNA, cytokines, and other proteins).

Exosomes are cell-derived vesicles, found in most biological fluids including cultured cells media.[26] These are round natural nanovesicles or multivasicular bodies, ~50–100 nm diameter, originated from endosomes and secreted from plasma membrane.[17] Exosomes play a precise role in search of miRNAs as diagnostic molecules due to their unique properties: (1) exosomes contain the disease-specific or deregulated miRNAs expressed during pathogenic states and nonexosomal miRNAs can be easily removed from healthy cells; (2) exosomes have the potential to cross blood barrier via transcytosis, thus easily clearing endothelial cellular layers; and (3) miRNAs within exosomes are quite a protective form of cellular RNase present in the circulatory system.[27] Sometime, it appears that miRNAs released from exosomes may be the remnants of dead cells. Cheng *et al.*'s study showed that exosomes offer protective and more enriched sources of miRNAs compared to other sources of biomarkers.[27] MVs are slightly larger vesicles (100–4000 μm diameter) generated from the cells via outward budding and blebbing of the plasma membrane. MVs are secreted by different cell types such as neurons, muscle cells, inflammatory cells, and tumor cells. However, platelets are considered to be major sources of MV production and secretion.[28] miR-150 is a well-studied miRNA that is secreted from human blood cells and monocytes via MVs.[29] Finally, apoptotic bodies are the largest MVs among all that are shed from cells during apoptosis and aid in miR-126 transportation.[30,31] miRNAs found in extracellular circulation are well packaged in macrovesicles, hence they are resistant to cellular RNase.[32] In addition to macrovesicles, Ago2 protein also formed Ago2–miRNA complex during miRNA processing into the cytoplasm, which also enhances miRNA stability and participates in miRNA binding to target sites.

4. CIRCULATORY miRNAs AS BIOMARKERS IN AGING AND AGE-ASSOCIATED DISEASES

This particular chapter introduces the concept of circulatory miRNAs as minimal invasive biomarkers for aging and age-related diseases and discusses the opportunities and challenges associated with circulatory miRNAs as biomarkers. We summarize the recently published literature on circulatory miRNAs in aging and age-related diseases, and provide an overview of the future steps necessary to develop miRNA-based biomarkers for use in clinical routine.

4.1 Aging

Aging is a complex biological process, highly regulated by multiple evolutionary conserved mechanisms.[33] Cell senescence, the key process of aging, is basically linked to complex cellular and molecular changes that occur in the cells over time. Such major biological phenomena are telomere erosion, changes in protein processing, lifestyle/epigenetics factors, and alteration in gene expression.[33] Recent studies identified miRNAs as the regulator of several pathways that are involved in aging and cellular senescence.[7,33–36] However, very few studies are available that show the significant alterations of circulatory miRNA levels during aging in the human population.[16] Hooten and colleagues quantified the miRNA expression in peripheral blood mononuclear cells (PBMCs) of young (mean age 30 years) and old (mean age 64 years) individuals. They identified three miRNAs, miR-151a-5p, miR-181a-5p, and miR-1248, that were found to be significantly downregulated in older individuals.[7] Similarly, their level also lowered in the serum samples of elderly rhesus monkeys (Table 1).[7]

Further study on human serum samples from seven elderly males (69.86 ± 1.77 years old) and females (72.43 ± 1.49 years old), and six young males (26.17 ± 0.83 years old) and females (23.17 ± 1.52 years old), identified miR-20a levels that were significantly lower in elderly male subjects when compared to young male subjects.[34] A study on serum samples from healthy subjects of different age groups, involving 21 subjects (age ±22 years), 10 subjects (age ±40 years), 10 subjects (age ±59 years), and 9 subjects (age ±70 years), identified downregulation of 5 miRNAs (miR-29b, miR-106b, miR-130b, miR-142-5p, and miR-340) and upregulation of 3 miRNAs (miR-92a, miR-222, and miR-375) in an age-dependent manner.[37] Even in C57BL/6 mice, miR-34a levels were also increased in the cochlea, auditory cortex, and plasma samples during the aging process.[36]

4.2 Cardiovascular Disease

Cardio-cerebrovascular disease (CVD) is an important age-related disease and the most common cause of mortality and morbidity worldwide. CVD is accompanied by obesity, diabetes, hyperlipidemia, and hypertension, while atherosclerosis is the most prominent indication of CVD among all complications.[38] miRNA analysis in the serum samples of atherosclerosis and preatherosclerosis (patients with hyperlipidemia, hypertension, and diabetes) patients indicated a decreased expression of miR-92a, miR-126, miR-130a, miR-222, and miR-370 levels in preatherosclerosis patients. However, in the case of atherosclerosis, miR-21, miR-122, miR-130a, and miR-211 were significantly increased, whereas miR-92a, miR-126, and miR-222 were markedly decreased compared to healthy controls (Table 1).[38]

In coronary artery disease (CAD) distinct miRNA expression profiles were observed in patients with typical unstable angina (UA) and angiographically documented CAD ($n = 13$), and in individuals with noncardiac chest pain (control, $n = 13$). Microarray analysis of plasma samples revealed a significant elevation of miR-106b/25 cluster, miR-17/92a cluster, miR-21/590-5p family, miR-126*, and miR-451 expression in UA patients compared to controls, as shown in Table 1.[39]

In heart failure (HF) patients ($n = 44$), four miRNAs, miR-103, miR-142-3p, miR-342-3p, and miR-30b, were significantly ($P = 0.002$–0.030) downregulated compared to controls ($n = 15$), chronic obstructive pulmonary

Table 1 miRNAs Expression Changes in Human and Mice Samples in Aging and Age-Related Diseases

Sample Source	Species	miRNAs Status in Disease	References
Aging			
PBMC	Human	↓miR-151a-5p, miR-181a-5p, miR-1248	Hooten *et al.*[7]
Serum	Monkey	↓miR-151a-5p, miR-181a-5p, miR-1248	
Serum	Human	↓miR-20a	Sawada *et al.*[34]
Serum	Human	↑miR-92a, miR-222, miR-375 ↓miR-29b, miR-106b, miR-130b, miR-142-5p, miR-340	Zhang *et al.*[37]
Plasma	C57BL/6 mice	↑miR-34a	Pang *et al.*[36]
Cardiovascular diseases			
Serum	Human	↑miR-21, miR-122, miR-130a, miR-211 ↓miR-92a, miR-126, miR-222	Jiang *et al.*[38]
Plasma	Human	↑miR-106b/25, miR-17/92a, miR-21/590-5p family, miR-126*, miR-451	Ren *et al.*[39]
Plasma	Human	↓miR-103, miR-142-3p, miR-342-3p, miR-30b	Ellis *et al.*[40]
Plasma	Human	↑miR-423-5p, miR-10b, miR-30d, miR-126	Hakimzadeh *et al.*[41]
Cancer			
Serum	Human	↑200c, miR-605, miR-135a* ↓miR-433, miR-106a	Alhasan *et al.*[42]
Serum	Human	↑miR-21-5p, miR-375, miR-205-5p, miR-194-5p ↓miR-382-5p, miR-376c-3p, miR-411-5p	Huo *et al.*[43]

Continued

Table 1 miRNAs Expression Changes in Human and Mice Samples in Aging and Age-Related Diseases—cont'd

Sample Source	Species	miRNAs Status in Disease	References
Serum	Human	↑miR-182, miR-183, miR-210 ↓miR-126	Zhu *et al.*[44]
Serum	Human	↑miR-429, miR-205, miR-200b, miR-203, miR-125b, miR-34b	Halvorsen *et al.*[45]
Serum	Human	↑miR-23a-3p, miR-27a-3p, miR-142-5p, miR-376c-3p	Vychytilova-Faltejskova *et al.*[46]
Serum and plasma	Human	↑miR-4270, miR-1225-5p, miR-188-5p, miR-1202, miR-4281, miR-1207-5p, miR-642b-3p, miR-1290, miR-3141	Hamam *et al.*[47]
Urine	Human	↑miR-375 ↓miR-148	Stuopelyte *et al.*[48]
Plasma	Human	↑miR-21, miR-375	Gao *et al.*[49]
Plasma	Human	↑miR-21, miR-152	Chen *et al.*[50]
Plasma	Human	↓miR-195	Su *et al.*[51]
Plasma	Human	↑miR-410-5p	Wang *et al.*[52]
Plasma	Human	↑miR-148a ↓miR-15b	Frères *et al.*[53]
Plasma	Human	↑miR-34a ↓miR-150	Aherne *et al.*[54]
Exosomes	Human	↑miR-141	Li *et al.*[55]

Exosomes	Human	↑let-7a, miR-1229, miR-1246, miR-150, miR-21, miR-223, miR-23a	Ogata-Kawata *et al.*[56]
Neutrophils	Human	↑miRs-423-3p, 148a-3p, 18a-3p, 574-3p ↓miR-26a-2-3p	Ma *et al.*[57]
Blood	human	↑miR-200c, miR-141	Antolín *et al.*[58]
Blood Tissue	Human	↑miR-193a-3p, miR-23a, miR-338-5p	Yong *et al.*[59]
Arthritis			
Serum	Human	↓miR-146a, miR-155, miR-16	Filková *et al.*[60]
Serum Blood	Human	↑miR-125b	Duroux-Richard *et al.*[61]
Plasma	Human	↑miR-4634, miR-181d, miR-4764-5p ↓miR-342-3p, miR-3926, miR-3925-3p, miR-122-3p, miR-9-5p, miR-219-2-3p	Wang *et al.*[62]
PBMCs Plasma	Human	↓miR-125b	Hruskova *et al.*[63]
Serum	Human	↑miR-16-5p, miR-23-3p, miR-125b-5p, miR-126-3p, miR-146a-5p, miR-223-3p	Castro-Villegas *et al.*[64]
Plasma	Human	↑miR-23, miR-223	

Continued

Table 1 miRNAs Expression Changes in Human and Mice Samples in Aging and Age-Related Diseases—cont'd

Sample Source	Species	miRNAs Status in Disease	References
Cataract			
Lens	Human	↑miR-34a	Chien *et al.*[65]
Epithelial cells		↑miR-15a-5p, miR-15a-3p, miR-16-1-5p ↓miR-16-1-3p	Li *et al.*[55]
Aqueous humor	Human	↑miR-4484, miR-6515-3p, miR-3663-3p, miR-4433-3p, miR-6717-5p, miR-4725-3p, miR-1202, miR-3197 ↓miR-4507, miR-3620-5p, miR-5001-5p, miR-6132, miR-4467, miR-187-5p, miR-6722-3p, miR-4749-5p, miR-1260b, miR-4634	Tanaka *et al.*[66]
Aqueous humor	Human	↑miR-451a, miR-21, miR-16	Wecker *et al.*[67]
Blood		↑miR-184, miR-4448, miR-30a, miR-29a, miR-29c, miR-19a, miR-30d, miR-205, miR-24, miR-22, miR-3074	
Osteoporosis			
Circulating monocytes	Human	↑miR-422a	Cao *et al.*[68]
Circulating monocytes	Human	↑miR-133a	Wang *et al.*[69]
Serum	Human	↑miR-152-5p, miR-335-5p, miR-320a ↓miR-30e-5p, miR-140-5p, miR-324-3p, miR-19b-3p, miR-19a-3p, miR-550a-3p, miR-186-5p, miR-532-5p, miR-93-5p, miR-378a-5p, miR-16-5p, miR-215-5p, let-7b-5p, miR-29b-3p, miR-7-5p, miR-365a-3p	Kocijan *et al.*[70]

Serum	Human	↑miR-122-5p, miR-125b-5p, miR-21-5p	Panach *et al.*[71]
Serum	Human	↑miR-21, miR-23a, miR-24, miR-93, miR-100, miR-122a, miR-124a, miR-125b, miR-148a	Seeliger *et al.*[72]
Blood	Human	↑miR-194-5p	Meng *et al.*[73]
Diabetes			
Plasma	Human	↓miR-126	Zhang *et al.*[74]
Plasma	Human	↑miR-101, miR-200a, miR-148b, miR-210, miR-155, miR-320, miR-103, miR-145, miR-21*, miR-126, miR-148a ↓miR-93, miR-146a	Assmann *et al.*[75]
Plasma	Human	↑miR-126-3p	Olivieri *et al.*[76]
Plasma	Mice	↑miR-375	Latreille *et al.*[77]
Blood	Human	↑miR-142-3p ↓miR-126a	Zhu *et al.*[78]
Blood	Human	↓miR-15a	Al-Kafaji *et al.*[79]
Serum	Human	↑miR-9, miR-29a, miR-30d, miR34a, miR-124a, miR-146a, miR-375	Kong *et al.*[80]
Serum	Human	↑miR-661, miR-571, miR-770-5p, miR-892b, miR-1303	Wang *et al.*[81]
Serum	Human	↑miR-25	Nielsen *et al.*[82]
Serum	Human	↓miR-223	Wen *et al.*[83]

Continued

Table 1 miRNAs Expression Changes in Human and Mice Samples in Aging and Age-Related Diseases—cont'd

Sample Source	Species	miRNAs Status in Disease	References
Hypertension			
Serum	Human	↑miR-1-2, miR-1957, miR-20a, miR-145, miR-27a, miR-23a, miR-23b, miR-191, miR-130 ↓miR-30c-2, miR-99a, miR-328, miR-199a, miR-330, miR-204	Sarrion *et al.*[84]
Plasma	Human	↑miR-92a	Huang *et al.*[85]
Blood	Human	↑miR-1, miR-133a, miR-26b, miR-208b, miR-499, miR-21	Parthenakis *et al.*[86]
PBMCs	Human	↓miR-9 and miR-126	Kontaraki *et al.*[87]
PBMCs	Human	↑miR-1, miR-208b, miR-499, miR-21 ↓miR-133a, miR-26b	Kontaraki *et al.*[88]
Dementia			
Plasma	Human	↑miR-128, miR-132, mir-874, miR-134, miR-323-3p, miR-382	Sheinerman *et al.*[89]
Alzheimer's disease			
PBMCs	Human	↑miR-34a, miR-181b	Schipper *et al.*[90]
Blood	Human	↑miR-26b-3p, miR-28-3p, miR-30c-5p, miR-30d-5p, miR-148b-5p, miR-151a-3p, miR-186-5p, miR-425-5p, miR-550a-5p, miR-1468, miR-4781-3p, miR-5001-3p, miR-6513-3p ↓let-7a-5p, let-7e-5p, let-7f-5p, let-7g-5p, miR-15a-5p, miR-17-3p, miR-29b-3p, miR-98-5p, miR-144-5p, miR-148a-3p, miR-502-3p, miR-660-5p, miR-1294, miR-3200-3p is	Satoh *et al.*[91]

CSF	Human	↑miR-146a, miR-100, miR-505, miR-4467, miR-766, miR-3622b-3p, miR-296 ↓miR-449, miR-1274a, miR-4674, miR-335, miR-375, miR-708, miR-219, miR-103	Denk *et al.*[2]
CSF/ECF	Human	↑miR-9, miR-125b, miR-146a, miR-155	Alexandrov *et al.*[92]
Serum/CSF	Human	↓miR-125b, miR-23a, miR-26b	Galimberti *et al.*[93]
Serum	Human	↓miR-137, miR-181c, miR-9, miR-29a, miR-29b	Geekiyanage *et al.*[94]
Serum	Human	↑miR-9 ↓miR-125b, miR-181c	Tan *et al.*[95]
Serum	Human	↑miR-3158-3p, miR-27a-3p, miR-26b-3p, miR-151b ↓miR-36, miR-98-5p, miR-885-5p, miR-485-5p, miR-483-3p, miR-342-3p, miR-30e-5p, miR-191-5p, let-7g-5p, let-7d-5p	Tan *et al.*[96]
Serum	Human	↓miR-31, miR-93, miR-143, miR-146a	Dong *et al.*[97]
Serum exosomes	Human	↑miR-361-5p, miR-30e-5p, miR-93-5p, miR-15a-5p, miR-143-3p, miR-335-5p, miR-106b-5p, miR-101-3p, miR-425-5p, miR-106a-5p, miR-18b-5p, miR-3065-5p, miR-20a-5p, miR-582-5p ↓miR-1306-5p, miR-342-3p, miR-15b-3p	Cheng *et al.*[27]
Plasma	Human	↑miR-323b-5p, miR-545-3p, miR-563, miR-600, miR-1274a, miR-1975 ↓let-7d-5p, let-7 g-5p,miR-15b-5p, miR-142-3p, miR-191-5p, miR-301a-3p, miR-545-3p,	Kumar *et al.*[98]

Continued

Table 1 miRNAs Expression Changes in Human and Mice Samples in Aging and Age-Related Diseases—cont'd

Sample Source	Species	miRNAs Status in Disease	References
Plasma exosomes	Human	↑miR-548at-5p, miR-138-5p, miR-5001-3p, miR-659-5p ↓miR-185-5p, miR-342-3p, miR-141-3p, miR-342-5p, miR-23b-3p, miR-338-3p, miR-3613-3p	Lugli *et al.*[99]
Huntington's disease			
Plasma	Human	↑miR-877-5p, miR-223-3p, miR-223-5p, miR-30d-5p, miR-128, miR-22-5p, miR-222-3p, miR-338-3p, miR-130b-3p, miR-425-5p, miR-628-3p, miR-361-5p, miR-942	Díez-Planelles *et al.*[100]
Parkinson's disease			
Serum	Human	↓miR-141, miR-214, miR-146b-5p, and miR-193a-3p	Dong *et al.*[101]
Serum	Human	↑miR-223*, miR-324-3p, mir-24 ↓miR-339-5p	Vallelunga *et al.*[102]
Amyotrophic lateral sclerosis			
Plasma	Human	↑miR-4258, miR-663b, miR-4649-5p ↓miR-26b-5p, miR-4299, let-7f-5p, miR-4419a, miR-3187-5p, miR-4496	Takahashi *et al.*[103]
Plasma Muscle Serum	Mice Human	↑miR-126	Toivonen *et al.*[104]

disease (COPD) ($n=32$), and other breathless patients ($n=59$).[40] In the case of CAD, patients with insufficient collateral network development showed a significant elevation in miR-423-5p ($P<0.05$), miR-10b ($P<0.05$), miR-30d ($P<0.05$), and miR-126 ($P<0.001$) levels in the aortic plasma compared to controls. Similarly, chronic total occlusion patients miRNA analysis also indicate significantly greater expression of miR-30d ($P<0.05$) and miR-126 ($P<0.001$) relative to healthy controls (Table 1).[41]

Current findings suggested the importance of blood-based miRNAs as biomarkers that can be monitored easily. Therefore, miRNAs potentially represent a convenient and minimally invasive tool for the diagnosis of CVD and patient stratification.

4.3 Cancer

Cancer is the leading cause of death worldwide with 8.2 million people dying from it each year (http://www.who.int/cancer/en/).[105] Uncontrolled/unwanted differentiation and proliferation of cells, localized at a particular site or invading at different places in the body, lead to cancer formation. In men, the most common types of cancers include lung, prostate, colorectal, stomach, and liver, whereas in women, common forms are breast, colorectal, lung, uterine cervix, and stomach cancer (http://www.who.int/cancer/en/). Currently, cancer detection is based on the body-specific protein biomarkers that are circulated in human blood such as alpha-fetoprotein for liver cancer, chromogranin A for neuroendocrine tumors and especially carcinoid tumors, nuclear matrix protein 22 for bladder cancer and carbohydrate antigen 125 for ovarian cancer, etc.[106] However, due to late-stage disease manifestation, patients' mortality rate is higher in a disease such as cancer. Detection of miRNAs in biological sources (such as blood, saliva, plasma, serum, and other biofluids) and their remarkable stability against RNase enzyme provide a hope for the development of a potential next-generation biomarker for cancer screening. The molecular interaction between miRNA and their target mRNA is well understood, and expression of most miRNAs is strongly deregulated in all human malignancies.[107] Recent studies have explored the role of circulatory miRNAs as biomarkers in various cancers.[6,105–107]

4.3.1 Aging and Cancer

The proportion of cancer incidence in elderly persons (65 years or older) has increased in most countries during the last few decades. According to International Agency for Research on Cancer and International Association of Cancer Registries, the proportions of all cancers among elderly men and women were 61% and 56%, respectively.[108] All cancers combined (except nonmelanoma skin cancer) were almost seven times more frequent among elderly men (2158 per 100,000 person-years), and around four times more frequent among elderly women (1192 per 100,000 person-years) than among younger persons (30–64 years old), based on the standardized rates.[108] Among elderly men, prostate cancer (451 per 100,000), lung cancer (449 per 100,000), and colon cancer (176 per 100,000) make up around half of all diagnosed cancers. Prostate cancer itself occurred around 22 times more frequently among elderly men than among younger men. The most frequent cancers among elderly women are breast cancer (248 per 100,000), colon cancer (133 per 100,000), lung cancer (118 per 100,000), and stomach cancer (75 per 100,000), with these making up 48% of all malignant cancers.[108] For most cancers, significant geographical variations in incidence rates are found among elderly individuals, reflecting socioeconomic status, differing particularly between developing and developed countries. In contrast with other major causes of death among the elderly, cancer incidence and mortality have not declined in general, indicating that primary prevention (especially cessation of tobacco smoking) remains a most valuable approach to decrease mortality; for most major cancers (prostate, colon, and breast), the causes remain almost unknown. Therefore, it is important to emphasize the increasing need for research into the prevention of cancer and the planning of treatment and care in the elderly.

Stable blood-based circulatory miRNA species have allowed for the differentiation of patients with various types of human cancers. Studies found that miR-21 has been identified as an "oncomir" in various tumors while miR-152 is a tumor suppressor.[50] Expression of both miR-21 and miR-152 were analyzed in patients with lung cancer (LuCa), colorectal carcinoma (CRC), breast cancer (BC), and prostate cancer (PCa). Quantitative real-time polymerase chain reaction (qRT-PCR) analysis of plasma samples from a total of 204 cancer patients, 159 various benign lesions, and 228 normal subjects revealed a significant elevation of miR-21 and miR-152 expression in LuCa, CRC, and BC when compared with normal controls. Upregulation of miR-21 and miR-152 levels was also observed in the plasma samples of patients with benign lesions of lung and breast, as compared to normal controls, respectively. However, no significant expression variation of these two miRNAs was observed in PCa or prostatic benign lesions as compared to controls. Receiver operating characteristic (ROC) curve analyses revealed that miR-21 and/or miR-152 can discriminate LuCa, CRC, and BC from normal controls.[50]

4.3.2 Prostate Cancer

PCa is the primary noncutaneous malignancy among men in the United States, and is the second most common cause of cancer mortality.[42] Scano-miR profiling of very high-risk (VHR) PCa patients ($n=19$) showed the deregulation of five miRNAs: miR-200c, miR-605, miR-135a*, miR-433, and miR-106a in serum samples. Among these, miR-200c had most significant elevated serum levels in all patients with VHR PCa, whereas circulating miR-605 and miR-135a* expression was reduced and miR-433 and miR-106a expression was increased in patients with VHR vs healthy volunteers (Table 1).[42] Further, biological pathway analysis showed the potential capability of these miRNAs as biomarkers for VHR aggressive PCa.

Plasma circulating miRNAs were analyzed in 149 PCa patients, 57 healthy controls, and 121 noncancer patients (with benign prostatic hyperplasia (BPH) and other urinary diseases). qRT-PCR results showed a significant elevation of circulating miR-410-5p level in the PCa patients compared to healthy controls or noncancer patients (Table 1). ROC curve analysis showed the diagnostic properties of plasma miR-410-5p with an area under curve (AUC) value of 0.8097 ($P<0.001$) in PCa patients.[52]

Further study on plasma samples from 57 PCa patients and 28 BPH patients showed upregulation of miR-21 and miR-375 by Taqman-based qRT-PCR assay. However, in the BPH group, the median relative expression levels of miR-21 and miR-375 were 0.07 and 0.55, respectively, whereas in the PCa group, the median values of miR-21 and miR-375 were 1.32 and 1.74, respectively (Table 1). ROC analysis illustrated that the AUC values for miR-21 and miR-375 were 0.799 and 0.757, respectively ($P=0.000126$). These results showed the discriminating power of circulating miR-21 and miR-375 in PCa patients from BPH controls at early stages.[49]

Analysis on 56 prostate cancerous and 16 noncancerous tissues detected differentially expressed 754 miRNAs by TaqMan Low Density Array. Highly abundant miRNAs were selected and were analyzed in urine specimens of 215 patients with BPH and 62 asymptomatic controls by qRT-PCR. Validation analysis identified most significant miR-148a and miR-375 in urine specimens (Table 1). Both miRNAs strongly improved the diagnostic power of the prostate-specific antigen test with AUC values of 0.79 and 0.84 for miR-148a and miR-375, respectively.[48] Thus, urine-circulating miRNAs may be a noninvasive tool for sensitive and specific detection of PCa.

Besides serum, plasma, and urine, exosome miRNAs are considered important sources as peripheral circulatory miRNAs. Exosomes isolated from the serum of PCa patients, patients with BPH, and healthy controls were analyzed for miR-141 expression. Exosomes contain a higher expression of miR-141 compared to whole serum samples in healthy controls. The level of serum exosomal miR-141 was found to be significantly higher in the PCa patients compared to the patients with BPH and the healthy controls (3.85-fold, $P=0.0007$ and 4.06-fold, $P=0.0005$, respectively) (Table 1). Additionally, expression of miR-141 was significantly higher in metastatic PCa compared with localized PCa ($P<0.0001$). ROC curve analysis revealed that serum exosomal miR-141 yielded an AUC value of 0.86, with 80% sensitivity and 87.1% specificity in discriminating patients with metastatic PCa from the patients with localized PCa.[55] A study suggested that the serum exosomes may serve as a more suitable material compared with the whole serum for measuring circulating miRNAs in patients.

Further, in prostate cancer patients, the concentration of miR-21-5p, miR-141-3p, miR-100-5p, and miR-375 was found to be increased in the serum samples while the expression of miR-141-3p was increased in both serum and plasma samples.[105] Beside their diagnostic value, plasma miRNAs, miR-20a-5b, miR-21-5b, and miR-145-5p, were also useful for the prediction of recurrence of PCa followed by localized treatment of cancer.[109]

4.3.3 Lung Cancer

LuCa is a leading cause of cancer-related deaths in the world, and up to 85% of lung cancer is classified as nonsmall cell lung cancer (NSCLC).[44] Circulating miRNAs have been reported to be stably expressed and detected in plasma/serum and to function as potent biomarkers in NSCLC. The diagnostic and prognostic value of miR-195 was evaluated in the plasma samples of patients with NSCLC. qRT-PCR analysis of 100 NSCLC patients and 100 healthy volunteers showed significant downregulation miR-195 in NSCLC patients compared with healthy controls ($P<0.001$). Decreased plasma miR-195 level was significantly associated with lymph node metastasis and advanced clinical stage (Table 1). Further, multivariate Cox regression analysis confirmed low plasma miR-195 expression as an independent unfavorable prognostic factor for NSCLC patients.[51] Findings indicated that plasma miR-195 might serve as a promising biomarker for the early detection and prognosis evaluation of NSCLC.

Serum-circulating miRNAs could be used as a biomarker to detect early lung cancer. The levels of miRNAs were analyzed in serum samples from 112 NSCLC patients and 104 controls (20 current smokers without lung cancer, 23 pneumonia patients, 21 gastric cancer patients, and 40 healthy controls) by Taqman probe-based qRT-PCR. Data showed significantly upregulation in serum levels of miR-182, miR-183 and miR-210, and downregulation of miR-126 level in NSCLC patients compared with healthy controls (Table 1). Further, ROC curve analysis revealed that the miR-182, miR-183, miR-210, or miR-126 level could serve as a diagnostic biomarker for NSCLC early detection, with a high sensitivity and specificity. Combination of these four miRNAs with carcinoembryonic antigen (CEA) further increased the diagnostic value, with an AUC value of 0.965 (sensitivity, 81.3%; specificity, 100.0%; and accuracy, 90.8%).[44] In addition, serum levels of miR-182, miR-183, miR-210, and miR-126 could be used to distinguish NSCLC or early-stage NSCLC from current tobacco smokers without lung cancer and pneumonia or gastric cancer patients with a high sensitivity and specificity.

Halvorsen and colleagues tested the diagnostic potential of serum miRNAs on 38 NSCLC patients, 16 patients suffering from COPD, and 16 healthy volunteers. TaqMan Low Density Arrays analysis showed deregulation of 754 unique miRNAs. Further, validation study by qRT-PCR revealed six ideal miRNAs, miR-429, miR-205, miR-200b, miR-203, miR-125b, and miR-34b, with a significantly higher abundance in NSCLC patients serum compared to controls. Further, these miRNAs revealed a significant AUC value 0.89 for stages I–IV and 0.88 for stage I/II after ROC curve analysis.[45] Thus, the accessibility and stability of these circulating miRNAs make them promising biomarkers as a supplement in future screening studies.

Peripheral neutrophils, an important component of innate and adaptive immune systems, also showed aberrant miRNA expression in NSCLC patients. miRNAs were examined in neutrophils of 15 patients with stage I NSCLC and 15 smokers without cancer, and showed deregulation of five important miRNAs: miR-423-3p, miR-148a-3p, miR-18a-3p, miR-574-3p, and miR-26a-2-3p. Further, a validation study on 82 patients with lung cancer and 73 controls showed a high level of miR-423-3p, miR-148a-3p, miR-18a-3p, and miR-574-3p, and low expression of miR-26a-2-3p in neutrophils of patients with NSCLC vs controls. However, a set of only two miRNAs, miR-26a-2-3p and miR-574-3p, were developed to produce 77.8% sensitivity and 78.1% specificity for NSCLC detection.[57]

4.3.4 Breast Cancer

BC is the most common type cancer and is the second cause of cancer-related mortality among women. Annually, the number of women who are newly diagnosed with breast cancer is currently exceeds 235,000 and there are around 40,000 deaths as a result of breast cancer in the United States.[47] Circulating miRNAs can be used to predict BC recurrence. miRNA profiling of serum samples from 48 BC patients identified significantly deregulated miRNAs by Exiqon miRCURY qRT-PCR panels. A further validation study on an independent set of sera from 20 patients with BC recurrences and 22 patients without recurrences identified four upregulated miRNAs (miR-21-5p, miR-375, miR-205-5p, and miR-194-5p), and three downregulated miRNAs (miR-382-5p, miR-376c-3p, and miR-411-5p) in recurrent patients (Table 1).[43]

A novel approach was used to isolate circulating miRNAs through an enrichment step using speed-vacuum concentration, which resulted in a fivefold increase in miRNA abundance. Microarray expression profiling on samples from 23 BC and 9 normal controls identified significantly upregulated 18 miRNAs in BC patients ($P<0.05$). Expression of nine miRNAs, miR-4270, miR-1225-5p, miR-188-5p, miR-1202, miR-4281, miR-1207-5p, miR-642b-3p, miR-1290, and miR-3141, was subsequently validated using qRT-PCR in a cohort of 46 BC and 14 controls (Table 1). Expression of these miRNAs was found to be higher in patients with stages I, II, and III, compared to stage IV, suggesting a potential utilization for early detection.[47]

A plasma miRNA profile was determined by qRT-PCR in a cohort of 378 women. Eighty-four miRNAs were found to be significantly deregulated in patients with metastatic BC compared to controls. The eight most significant miRNAs were measured in the first profiling cohort composed of 41 primary BC and 45 controls. A further validated study in diverse cohorts of 108 primary BC, 88 controls, 35 BC in remission, 31 metastatic BC, and 30 gynecologic tumors found miR-148a to be the most significantly upregulated miRNA and the most significantly downregulated one was miR-15b.[53]

Expression of miR-200c and miR-141 was examined in blood samples of 57, stages I–IV, BC patients and 20 age-matched controls by qRT-PCR. miR-200c expression was significantly downregulated ($P<0.0001$) in BC patients compared to controls and yielded an area under the ROC curve of 0.79 (90% sensitivity and 70.2% specificity). However, the miR-141 level was significantly higher in the blood of patients with stages I–III, lymph node metastasis, and HER2 negative tumors.[58] Thus, miR-200c and miR-141 were independent prognostic factors and associated with distinct outcomes of BC patients.

4.3.5 Colorectal Cancer

Over the past decade, research has shown that altered miRNA expression found in CRC and circulating miRNAs are involved in CRC detection, progression, and outcome.[110] Recently, Vychytilova-Faltejskova and colleagues did a three-phase biomarker study on 427 colon cancer patients and 276 healthy donors. Serum samples were screened for miRNA expression by using Illumina small RNA sequencing and validated by qRT-PCR. Analysis showed 54 significantly deregulated miRNAs in sera of CRC patients compared to healthy donors ($P < 0.01$). However, in a final validation study, the most significant upregulation of only four miRNAs, miR-23a-3p, miR-27a-3p, miR-142-5p, and miR-376c-3p, was observed (Table 1). Diagnostic accuracy of these miRNAs was also established with AUC value = 0.917, distinguishing patients from controls with sensitivity of 89% and specificity of 81% (AUC = 0.922).[46]

Cell-free circulatory nature of miRNAs as biomarkers in CRC was assessed by Aherne *et al.* on 48 plasma samples comprising normal, polyp, adenoma, early, and advanced cancer samples. Results showed 667 deregulated miRNAs. Three miRNAs (miR-34a, miR-150, and miR-923) were selected for further validation in a cohort of 97 subjects divided into the same five groups, and in an independent public dataset of 40 CRC samples and paired normal tissues. High levels of miR-34a and low miR-150 levels distinguished groups of patients with polyps from those with advanced cancer (AUC = 0.904), and low circulatory miR-150 levels separated patients with adenomas from those with advanced cancer (AUC = 0.875). In addition, altered expression of miR-34a and miR-150 can distinguish an independent public dataset of 40 CRC samples and paired normal tissues.[54]

Microarray analyses were performed on 88 primary CRC patients and 11 healthy controls in exosome-enriched fractions of serum samples for circulatory miRNA detection. Serum exosomal level of seven miRNAs, let-7a, miR-1229, miR-1246, miR-150, miR-21, miR-223, and miR-23a, were found to be significantly higher in primary CRC patients compared to healthy controls (Table 1). Interestingly, their levels were significantly downregulated after surgical resection of CRC tumors. Further, high sensitivities of the seven selected miRNAs showed significant AUC values by ROC curve analysis with tumor markers (CA 19-9 and CEA).[56]

CRC tissues and blood samples were analyzed for biomarker validation by stem-loop qRT-PCR. Seven miRNAs, miR-150, miR-193a-3p, miR-23a, miR-23b, miR-338-5p, miR-342-3p, and miR-483-3p, were found to be differentially expressed in both tissue and blood samples. However, significant positive correlations were observed in the tissue and blood levels only with three miRNAs (miR-193a-3p, miR-23a, and miR-338-5p). Further, ROC curve analysis of these miRNAs yielding an AUC value of 0.887 (80.0% sensitivity, 84.4% specificity, and 83.3% accuracy) confirmed them as a classifier for CRC detection.[59]

The discovery of circulatory miRNAs as biomarkers brought forward a new understanding of the basic mechanisms of carcinogenesis, and opened up exciting prospects for diagnostics and prognostics. Although still an emerging field, needing much exploration, the hope is to apply circulating miRNAs to cancer diagnosis and treatment.

4.4 Arthritis

Rheumatoid arthritis (RA) is another age-related chronic inflammatory autoimmune disease, which affects approximately 1% of the world's population.[62,63] RA is associated with persistent synovitis, leading to severe joint destruction, development of joint deformities, and increased risk of cardiovascular diseases.[63] Establishment of early-stage detection parameters and treatment response would be beneficial for patients with early rheumatoid arthritis (ERA) to prevent ongoing joint damage. Circulating miRNA expression was analyzed in the serum samples of ERA patients ($n=34$) and patients with established RA ($n=26$). Levels of three miRNAs, miR-146a, miR-155, and miR-16, were decreased in ERA patients in comparison with established RA (Table 1).[60] Analysis revealed that miR-223 may serve as a marker of disease activity, and miR-16 and miR-223 may be possible predictors for disease outcome in ERA.[60]

miRNA array analysis of plasma samples from RA patients ($n=75$, containing 44 active RA and 31 nonactive RA) unveiled differential expression of nine miRNAs as compared to controls subjects ($n=70$). miR-4634, miR-181d, and miR-4764-5p levels were increased, whereas miR-342-3p, miR-3926, miR-3925-3p, miR-122-3p, miR-9-5p, and miR-219-2-3p expression levels were decreased in RA patients. The AUC values for nine individual miRNAs ranged from 0.6254 to 0.818; however, the best miRNA candidates that showed significant differences in their expression between RA and other control groups were miR-122-3p, miR-3925-3p, miR-342-3p, and miR-4764-5p.[62]

miR-125b level was found to be significantly elevated in the total blood and serum samples of RA patients compared to osteoarthritic and healthy donors. However, miR-125b upregulation was also found in patients with other forms of chronic inflammatory arthritis. However, higher serum levels of miR-125b at disease flare were associated with good clinical response to treatment with rituximab 3 months later ($P=0.002$); hence it could be a potential predictive biomarker in response to rituximab treatment.[61] In contrast, the treatment-naïve early phase of RA patients showed significantly lowered expression of miR-125b in the PBMCs and plasma samples than healthy controls. While after 3 months of treatment, it was significantly ($P=0.042$) increased particularly in responders, ROC analysis indicated a significant ($P=0.048$) AUC value 0.652 (95% CI 0.510–0.793) in the patients after 3 months of therapy. miR-125b in PBMCs of treatment-naïve patients may present a novel biomarker for monitoring the treatment outcome during the early phase of RA.[63]

Further, miRNA expression levels in RA patients ($n=95$) before and after anti-TNFα/DMARDs combination therapy may also be potential novel biomarkers for predicting and monitoring therapy. Serum levels of six miRNAs, miR-16-5p, miR-23-3p, miR125b-5p, miR-126-3p, miR-146a-5p, and miR-223-3p, were found significantly upregulated in RA patients by anti-TNFα/DMARD combination therapy.[64] However, those miRNAs were only increased in responder patients after therapy, and paralleled with the reduction of TNFα, interleukin (IL)-6, IL-17, rheumatoid factor, and C-reactive protein. In plasma samples of RA patients, miR-23 and miR-223 may serve as both predictors and biomarkers of response to anti-TNFα/DMARDs combination therapy.[64] Churov and colleagues discussed around 10 reports and found more than 20 miRNAs that were found to be deregulated in the blood components such as serum/plasma, PBMCs, neutrophils, and PB T cells and PBCD4$^+$ cells of RA patients. The most significant circulatory miRNAs were miR-16, miR-21, miR-24, miR-26a, miR-125a-5p, miR-125b, miR-126-3p, miR-223, and miR-451.[111] These were elevated in the plasma or serum and are considered to be the most promising noninvasive biomarkers for the detection of RA.

4.5 Cataract

Cataracts, the most common cause of blindness worldwide, are significantly related to the aging process.[65] Role of miR-34a has been identified in the lens senescence in age-related cataract patients. Study on the lens epithelium samples of 110 patients with four age groups: between 55 and 64 years ($n=25$; 22.7%), between 65 and 74 years ($n=35$; 31.8%), between 75 and 84 years ($n=28$; 25.5%), and older than 85 years ($n=22$; 20%) revealed that miR-34a expression levels were significantly different between each age group and it is found to be greater in patients with older age.[65]

A further study was also conducted on the lens epithelial cells by Li and colleagues, on 60 age-related cataract patients (including 20 with cortical cataracts, 20 with nuclear cataracts, and 20 with posterior subcapsular cataracts) and 20 normal patients. Expression of miR-15a-5p, miR-15a-3p, and miR-16-1-5p was decreased in normal lens epithelial cells but was higher at significant levels in corresponding cells of patients with cortical, nuclear, or posterior subcapsular cataracts ($P<0.01$) (Table 1), whereas miR-16-1-3p expression was relatively high in normal lens epithelial cells, but significantly decreased in cells of patients from each cataract group ($P<0.01$).[55]

Next-generation sequencing (NGS) techniques of human aqueous humor samples identified 158 miRNAs in four samples; an additional 59 miRNAs were present in at least three samples. The aqueous humor miRNA profile shows some overlap with published NGS-derived inventories of circulating miRNAs in blood plasma with high prevalence of human miR-451a, miR-21, and miR-16. In contrast to blood, miR-184, miR-4448, miR-30a, miR-29a, miR-29c, miR-19a, miR-30d, miR-205, miR-24, miR-22, and miR-3074 were detected among the 20 most prevalent miRNAs in aqueous humor.[67]

Tanaka and colleagues conducted a microarray analysis of aqueous humor samples from glaucoma ($n=10$), cataract ($n=5$), and epiretinal membrane patients ($n=5$), revealing the disease-related extracellular miRNAs profiles.[66] Eight miRNAs were found to be significantly upregulated in glaucoma patients compared to controls, as follows: miR-4484, miR-6515-3p, miR-3663-3p, miR-4433-3p, miR-6717-5p, miR-4725-3p, miR-1202, and miR-3197, whereas 10 downregulated miRNAs were miR-4507, miR-3620-5p, miR-5001-5p, miR-6132, miR-4467, miR-187-5p, miR-6722-3p, miR-4749-5p, miR-1260b, and miR-4634 (Table 1). The two miRNAs miR-3620-5p and miR-6717-5p showed the maximum AUC value (0.88) to distinguish the patient and controls.[66]

4.6 Osteoporosis

Osteoporosis is a systemic skeletal disorder characterized by increased risk of bone fracture (BF) due to fragility and reduction in bone mass. BFs, particularly hip fracture, are a major concern in health care because of the associated morbidity and mortality, mainly in the elderly.[71,72] It has also been postulated that miRNAs might play important roles in age-related bone loss disorders, bone remodeling, postmenopausal osteoporosis, and osteoporotic fracture patients.[112]

Wang and colleagues identified miR-133a as a promising molecule where expression level varies between patients with low bone mineral density (BMD) ($n=10$) compared with the high BMD ($n=10$) groups during postmenopausal in Caucasian women. Microarray analysis of circulating monocytes revealed the significant ($P=0.007$) higher expression of miR-133a in patients with low BMD. Further, bioinformatic target gene analysis showed three potential osteoclast-related target genes, CXCL11, CXCR3, and SLC39A1 of miR-133a.[69]

In women with postmenopausal osteoporosis, significant miRNA signature was identified as biomarkers. miRNAs array analysis of circulating monocytes (osteoclast precursors) from 10 high BMD and 10 low BMD postmenopausal Caucasian women identified upregulation of miR-422a at the marginal significant level ($P=0.065$) in the low BMD compared with the high BMD group. A more significant upregulation of miR-422a was identified in the low BMD group by qRT-PCR analysis ($P=0.029$) (Table 1). Additionally, qRT-PCR analyses of miR-422a target genes showed the negative correlation with these five gene (CBL, CD226, IGF1, PAG1, and TOB2) expressions, suggesting miR-422a as the potential miRNA biomarker underlying postmenopausal osteoporosis.[68] Though postmenopausal osteoporosis is the most common cause of low-traumatic fractures, bone loss and low-traumatic fractures also occur in premenopausal state in women and in young males. In men, late-stage bone loss is described as "male idiopathic osteoporosis."[113] Circulating serum miRNAs showed the differential expression pattern in patients with premenopausal, postmenopausal, and male idiopathic osteoporosis. Three miRNAs were commonly upregulated, miR-152-5p, miR-335-5p, and miR-320a, while 16 were downregulated: miR-30e-5p, miR-140-5p, miR-324-3p, miR-19b-3p, miR-19a-3p, miR-550a-3p, miR-186-5p, miR-532-5p, miR-93-5p, miR-378a-5p, miR-16-5p, miR-215-5p, let-7b-5p, miR-29b-3p, miR-7-5p, and miR-365a-3p by qRT-PCR analysis among patients with prevalent low-traumatic fractures and control subjects (Table 1). Interestingly, eight miRNAs had AUC values >0.9 for the classification of fracture patients.[70]

In 2015, Meng and colleagues identified miR-194-5p as a potential biomarker for postmenopausal osteoporosis. Microarray analysis on the blood samples from osteopenia ($n=23$) and osteoporosis patients ($n=25$) identified five miRNAs (miR-130b-3p, miR-151a-3p, miR-151b, miR-194-5p, and miR-590-5p); however, only miR-194-5p expression was increased and found to be enriched in multiple osteoporosis-related pathways.[73] Further study by Panach and colleagues identified three valuable miRNAs, miR-122-5p, miR-125b-5p, and miR-21-5p, which were upregulated in serum samples of brain fracture patients with respect to controls. Remarkably, the miR-21-5p level was correlated with CTx ($r=0.76$; $P<0.00001$), a marker of bone resorption, confirming its diagnostic potential.[71]

miRNAs analysis in the serum of 30 osteoporotic and 30 nonosteoporotic patients revealed the significant upregulation of 9 miRNA candidates, namely miR-21, miR-23a, miR-24, miR-93, miR-100, miR-122a, miR-124a, miR-125b, and miR-148a in osteoporotic patients (Table 1), whereas on the bone tissue of 20 osteoporotic and 20 nonosteoporotic patients only six miRNAs (miR-21, miR-23a, miR-24, miR-25, miR-100, and miR-125b) displayed a significantly higher expression.[72] However, a total of five miRNAs display a common upregulation in both serum and bone tissue.

These studies reveal an important role for several miRNAs in osteoporotic state and in postmenopausal osteoporosis patients, and suggests that they may be used as biomarkers for diagnostic purposes and may be a target for treating bone loss and optimizing fracture healing in osteoporotic patients.

4.7 Diabetes/Obesity

Diabetes mellitus (DM) is an age-related metabolic disorder characterized by insulin secretion from pancreatic β cells that is insufficient to maintain blood glucose homeostasis. DM has a global public health issue, estimated to affect ~450 million people, and the economic cost is projected to be $490 billion/year by 2030.[114,115] It is a disease resulting from insufficient production of the insulin hormone pancreatic cells (type 1 DM, T1DM) or from ineffective insulin action (type 2 DM, T2DM).[114] For both types, T2DM is more common in humans, comprising 85%–90% of total DM cases, and it has been considered a progressive metabolic disorder. The role of miRNAs as biomarkers has been explored in both types of diabetes; however, very few studies are available in the case of T1DM.[67]

T1D is usually diagnosed when >80% of the pancreatic beta-cells are destroyed by the immune system. Plasma miRNA expression was analyzed in 25 T1D patients and 20 age- and gender-matched nondiabetic controls by using Stem-loop RT-Pre-Amp Real-time PCR. Results showed a significant two- to fivefold downregulation of miR-93* and miR-146a and 2–40fold upregulation of miR-101, miR-200a, miR-148b, miR-210, miR-155, miR-320, miR-103, miR-145, miR-21*, miR-126, and miR-148a in T1D patients (Table 1).[75]

A study by Zhang and colleagues on plasma samples identified miR-126 as a potential biomarker for early prediction of T2DM in susceptible individuals. The study included 30 subjects in each three groups: normal (fasting glucose), T2DM-susceptible, and T2DM individuals. Five miRNAs, miR-29b, miR-28-3p, miR-15a, miR-223, and miR-126, were selected for the qRT-PCR analysis. However, only miR-126 showed significantly reduced expression in susceptible individuals and T2DM patients compared to normal individuals.[74]

A comprehensive characterization of the serum miRNA profile in patients with T2DM-associated microvascular disease (T2DMC) demonstrated deregulated miRNAs expression. Serum samples obtained from 184 T2DM patients (92 with microvascular complications and 92 free of complications) and 92 age/gender-matched controls were analyzed by using a TaqMan Low Density Array. Initially, the levels of 754 miRNAs were markedly upregulated in the patients' groups; however, subsequently validated analysis by qRT-PCR identified only five ideal miRNAs (miR-661, miR-571, miR-770-5p, miR-892b, and miR-1303) that were significantly upregulated in T2DM patients ($P < 0.05$) (Table 1).[81] Thus, circulating miRNAs are an emerging class of biomarkers for T2DM.

Even in children with newly diagnosed T1D levels of sera, miRNAs were changed when compared with age-matched healthy controls and glycemic controls. Global miRNA sequencing analyses on the pooled sera samples were performed on two groups: T1D cohorts ($n = 275$ and 129, respectively), and one control group ($n = 151$). Twelve miRNAs were identified as upregulated in T1D patients (miR-152, miR-30a-5p, miR-181a, miR-24, miR-148a, miR-210, miR-27a, miR-29a, miR-26a, miR-27b, miR-25, and miR-200a) (Table 1). Several of these miRNAs were linked with important molecular pathways such as apoptosis and beta-cell networks. However, further analysis identified miR-25 as negatively associated with residual beta-cell function (est.: −0.12, $P=0.0037$), and positively associated with glycemic control (HbA1c) (est.: 0.11, $P=0.0035$) at 3 months after onset of the disease.[82] Thus, the study demonstrates that miR-25 might be a "tissue-specific" miRNA for diagnosis in new onset T1D children and may be a predictive circulating miRNA biomarker.

Another important circulatory miRNA is miR-126-3p, whose levels were found to be deregulated in T2DM patients. Plasma samples from 193 patients with T2DM aged 40–80 years, and 136 healthy subjects aged 20–90 years were used to explore the combined effect of age and glycemic state on miR-126-3p expression. Expression of miR-126-3p was significantly higher in the oldest individuals compared with the youngest controls (<45 vs >75 years) with relative expression level: 0.27 ± 0.29 vs 0.48 ± 0.39 ($P=0.047$). However, age-based comparison between controls and T2DM demonstrated significantly different miR-126-3p levels only in the oldest (0.48 ± 0.39 vs 0.22 ± 0.23, $P<0.005$). Furthermore, miR-126-3p levels were seen to be lower in patients with poor glycemic control, compared with age-matched controls. The age-related increase in plasma miR-126-3p found in controls was paralleled by a five- or sixfold increase in intra/extracellular miR-126-3p in in vitro-cultured HUVECs undergoing senescence.[76] Moreover, miR-126-3p expression was downregulated in intermediate-age HUVECs grown in a high-glucose medium until senescence.

Kong and colleagues identified seven diabetes-related serum miRNAs, miR-9, miR-29a, miR-30d, miR34a, miR-124a, miR-146a, and miR-375, having clinical significance during pathogenesis of type 2 diabetes (T2D).[80] Serum sample analysis of 56 subjects including 18 cases of newly diagnosed T2D (n-T2D) patients, 19 cases of prediabetes individuals (impaired glucose tolerance and/or impaired fasting glucose), and 19 cases of T2D-susceptible individuals with normal glucose tolerance (s-NGT) showed upregulation of these miRNAs by qRT-PCR. Furthermore, different statistical analyses showed that miR-34a was the most significant miRNA that was able to discriminate patients and controls.[80]

A study on the peripheral whole blood samples from patients with T2D ($n=24$), prediabetes individuals exhibiting impaired fasting glucose (IFG) and impaired glucose tolerance (IGT) ($n=22$), as well as healthy control subjects ($n=24$) investigated the expression miR-15a.[79] qRT-PCR analysis indicated a significant downregulation of miR-15a in patients with T2D and IFG/IGT individuals, compared with healthy control subjects ($P<0.05$).

Multivariate logistic regression analysis showed a significant association of lower miR-15a expression with T2D and prediabetes ($P < 0.05$). Furthermore, ROC curve analysis revealed that blood miR-15a was able to distinguish patients with T2D and IFG/IGT individuals from healthy controls (AUC; 0.864).[79] Thus, the miR-15a level in peripheral whole blood may serve as a potential biomarker for T2D and prediabetes.

A study on the mice model (375KO) identified miR-375 as an important regulator of β-cell mass and function. Mice overexpressing miR-375 exhibit normal β-cell mass and function.[77] Analysis of plasma samples from 375KO indicated an elevation of the miR-375 level after acute and profound β-cell destruction. Furthermore, these findings are supported by higher expression of miR-375 levels in the circulation of T1D subjects, but not mature onset diabetes of the young and T2D patients.[77] Altogether, the study suggests an essential role for miR-375 in the maintenance of β-cell mass, and total plasma miR-375 levels make this miRNA an unlikely biomarker for β-cell function, but suggest a utility for the detection of acute β-cell death for autoimmune diabetes.

An interesting meta-analysis on 38 miRNA expression profiling studies selected some potent miRNAs as a biomarker for T2DM.[78] The top upregulated miRNA in T2DM patients was miR-142-3p and the top downregulated miRNA was miR-126a. The dysregulation of miR-199a-3p and miR-223 was highly pancreas-specific and liver-specific. miR-30e was downregulated in patients with T2DM as well, while miR-92a was downregulated in animal models of diabetes. Meta-analysis confirmed that miR-29a, miR-34a, miR-375, miR-103, miR-107, miR-132, miR-142-3p, and miR-144 are potential circulating biomarkers of type 2 diabetes. In addition, miR-199a-3p and miR-223 are potential tissue biomarkers of T2DM.[78]

Obesity is also an age-related health complication and a serious risk factor for many metabolic disorders, especially diabetes. Over the past decade, the prevalence of obesity has increased dramatically across the world, especially in developed countries.[83] Technically, obesity results from a chronic imbalance between energy intake and energy expenditure. Recent studies have proposed that miRNA expression is deregulated in obese patients, and miRNAs are the potent regulator of many diseases related to obesity.[83]

A study on 13 patients with type 2 diabetes, 20 obese patients, 16 obese patients with type 2 diabetes, and 20 healthy controls detected three serum miRNAs, miR-138, miR-15b, and miR-376a, that were found to have potential as predictive biomarkers in obesity. miR-138 and miR-376a are potential predictive tools for distinguishing obese patients from normal healthy controls, diabetic patients, and obese diabetic patients. In addition, the combination of miR-503 and miR-138 can be used to distinguish diabetic from obese diabetic patients.[116]

Wen and colleagues identified miR-223 as a potent regulator of obesity. A study on 121 subjects, including 41 normal weight, 40 overweight, and 40 obesity subjects quantified the miR-223 expression in the serum samples by real-time PCR. The miR-223 expression was lower in both overweight and obesity subjects compared with normal-weight control (1.06 vs 7.54, $P<0.001$; 4.56 vs 7.54, $P<0.001$, respectively). However, after 3 months, lifestyle intervention circulating miR-223 level was increased significantly in both overweight and obese groups.[83]

Taken together, we have demonstrated a group of diabetes-related circulatory miRNAs with biomarker properties; however, several technical and scientific obstacles need to be overcome for miRNAs to become a part of the diagnostic arsenal to identify individuals with diabetes mellitus and its devastating complications.

4.8 Hypertension

Hypertension is a leading cause of cardiovascular disease, including CAD, HF, chronic kidney disease, peripheral vascular disease, and stroke.[117] Idiopathic pulmonary hypertension (IPAH) is a rare disease characterized by a progressive increase in pulmonary vascular resistance leading to HF.

Serum microarray expression profiling of circulating miRNAs in 12 well-characterized IPAH patients and 10 healthy volunteer showed significant changes in 61 miRNAs. Nine miRNAs (miR-1-2, miR-1957, miR-20a, miR-145, miR-27a, miR-23a, miR-23b, miR-191, and miR-130) were upregulated, whereas six miRNAs (miR-30c-2, miR-99a, miR-328, miR-199a, miR-330, and miR-204) were downregulated (Table 1). However, the important one was miR-23a because it was correlated with the patients' pulmonary function as well as controlling the expression of 17% of the significantly changed mRNAs including PGC1α, which was recently associated with the progression of IPAH. Furthermore, the silencing of miR-23a leads to an increase of PGC1α expression.[84]

Parthenakis and colleagues evaluated the overexpression of six miRNAs, miR-1, miR-133a, miR-26b, miR-208b, miR-499, and miR-21, in peripheral blood of patients with well-controlled essential hypertension in relation to arterial stiffness.[86] However, after 1 year of effective antihypertensive therapy, only the miR-21 level showed a significant decrease in patients, and it was correlated with changes in both carotid femoral pulse wave velocity (cfPWV) and carotid radial pulse wave velocity (crPWV) independent of blood pressure levels ($r = -0.56$ and $r = -0.46$, respectively; $P < 0.001$ for both).[86] Furthermore, low levels of miR-21 are strongly associated with an improvement in arterial stiffness in patients with well-controlled essential hypertension, independent of their blood pressure levels.

Kontaraki and colleagues evaluated the expression of miR-9 and miR-126 in 60 patients with untreated essential hypertension and in 29 healthy individuals. qRT-PCR analysis of PBMCs RNA showed significantly lower miR-9 and miR-126 ($P < 0.001$) expression levels in hypertensive patients compared with healthy controls (Table 1). Interestingly, miR-9 levels showed a significant positive correlation with the left ventricular mass index. Furthermore, both miR-9 and miR-126 expression levels showed significant positive correlations with the 24-h mean pulse pressure (PP) in hypertensive patients.[87]

A further study on 102 patients with essential hypertension and 30 healthy individuals showed the deregulation of six miRNAs' expression in PBMCs by qRT-PCR. Hypertensive patients showed significantly lower level of miR-133a and miR-26b, and higher expression of miR-1, miR-208b, miR-499, and miR-21 compared with healthy controls. Essentially, significant negative correlations in miR-1 and miR-133a were observed with the left ventricular mass index, while miR-208b, miR-26b, miR-499, and miR-21 expression showed a positive correlation with this index.[88]

Plasma miR-92a expression was analyzed in 240 participants, including 60 healthy volunteers with normal carotid intima-media thickness (nCIMT), 60 healthy volunteers with increased CIMT (iCIMT), 60 hypertensive patients with nCIMT, and 60 hypertensive patients with iCIMT by qRT-PCR.[85] miR-92a expression was significantly lowered (24.59 ± 1.30 vs 27.76 ± 2.13 vs 29.29 ± 1.89 vs 33.76 ± 2.08; $P < 0.001$) in healthy controls with nCIMT, followed by healthy controls with iCIMT, then hypertensive patients with nCIMT and the highest expression in hypertensive patients with iCIMT (Table 1). miR-92a levels also showed a significant positive correlation with 24-h mean systolic BP, 24-h mean diastolic BP, 24-h mean PP, 24-h daytime PP, 24-h nighttime PP, CIMT, and cfPWV.[85] This evidence suggests that possibilities of circulating miR-92a represent a potential noninvasive atherosclerosis marker in essential hypertensive patients. Thus, results point to the utility of circulating miRNA expression as a biomarker of disease progression.

4.9 Neurodegenerative Diseases

The most common NDs include AD, mild cognitive impairment (MCI), PD, ALS, and HD. Besides AD, very few studies have demonstrated the potential role of miRNAs as a noninvasive biomarker in other NDs. miRNA profiling of whole blood samples of PD patients showed the downregulation of miR-1, miR-22p, and miR-29a in the patients compared to controls, and differential expression of miRNAs signatures such as miR-16-2-3p, miR-26a-2-3p, and miR-30a were used to differentiate between treated and untreated patients (Table 1).[118] Furthermore, miRNA analysis of plasma samples indicated upregulation of miR-181c, miR-331-5p, miR-193a-5p, miR-196b, miR-454, miR-125a-3p, and miR-137 in PD patients.[119] In 2014, a study conducted by Batta-Orfila and colleagues indicated significant suppression of miR-19b, miR-29a, and miR-29c in the serum samples of PD patients.[120] ALS is also a fatal neurodegenerative disease that progressively weakens neuronal cells that leads to degeneration of upper and lower motor neurons.[104] A study on the SOD1-G93A mice, an ALS mouse model, showed upregulation of miR-206 in skeletal muscle and plasma through microarrays analysis. Even human ALS patients' serum samples also revealed upregulation of miR-206.[104] A recent study by Takahashi and colleagues covered the plasma samples of two cohort of ALS patients: (1) ALS patients ($n=16$) and healthy controls ($n=10$); and (2) 48 ALS patients ($n=48$), healthy controls ($n=47$), and disease controls ($n=30$), the discovery through microarray analysis and validation by qRT-PCR found the upregulation of miR-4649-5, and downregulation of miR-4299 in patients compared to controls (Table 1).[103]

4.9.1 Dementia

MCI is a syndrome characteristic of early stages of many NDs. Recently, we have identified two sets of circulating brain-enriched miRNAs: the miR-132 family (miR-128, miR-132, and miR-874) normalized per miR-491-5p and the miR-134 family (miR-134, miR-323-3p, and miR-382) normalized per miR-370, capable of differentiating MCI from age-matched control with high accuracy (Table 1). Here, we report a biomarker validation study of the identified miRNA pairs using larger independent sets of age- and gender-matched plasma samples. Biomarker pairs detected MCI with sensitivity, specificity, and overall accuracy similar to those obtained in the first study. The miR-132 family biomarkers differentiated MCI from AMC with 84%–94% sensitivity and 96%–98% specificity, and the miR-134 family biomarkers demonstrated 74%–88% sensitivity and 80%–92% specificity. When miRNAs of the same family were combined, miR-132 and miR-134 family biomarkers demonstrated 96% and 87% overall accuracy, respectively.[89]

4.9.2 Alzheimer's Disease

AD is the most important age-related neurological disorder and occurs in elderly individuals. AD pathogenesis is associated with gradual loss of neurons, synapses and synaptic function, abnormalities in mitochondrial function, and inflammatory responses.[121] In order to search for the miRNAs as a promising biomarker to monitor the AD pathogenesis particularly in its presymptomatic state, very few reports are available on human biofluid samples such as serum, plasma, CSF, and exosomes derived from serum and plasma as well (Table 1). Most of the studies were conducted on human subjects having MCI and AD dementia, and an almost equal number of healthy controls were also included.

4.9.2.1 Circulatory miRNAs in Whole Blood

Human blood is the most vital and widely used specimen for human disease assessments, and its testing is also minimally invasive. Reanalysis of a publically available small RNA-Seq data set identified differential expression of 27 miRNAs in 48 AD patients and 22 normal subjects.[91] Whole-blood specimens were analyzed for miRNAs expression by single-end sequencing on Hiseq 2000 (Illumina). Thirteen miRNAs (miR-26b-3p, miR-28-3p, miR-30c-5p, miR-30d-5p, miR-148b-5p, miR-151a-3p, miR-186-5p, miR-425-5p, miR-550a-5p, miR-1468, miR-4781-3p, miR-5001-3p, and miR-6513-3p) were upregulated and 14 miRNAs (let-7a-5p, let-7e-5p, let-7f-5p, let-7g-5p, miR-15a-5p, miR-17-3p, miR-29b-3p, miR-98–5p, miR-144-5p, miR-148a-3p, miR-502-3p, miR-660-5p, miR-1294, and miR-3200-3p) were found to be downregulated in AD patients compared to controls (Table 1). Further, ROC curve analysis revealed a significant discrimination potential of these 27 miRNAs for AD and controls.[91] However, the potential role of these miRNAs needs to be established in a large population group.

4.9.2.2 Blood Mononuclear Cells as a Source of miRNAs

Early studies on blood mononuclear cells (BMCs) as a source of circulatory miRNAs were conducted by Schipper and colleagues on 16 AD cases and 16 negative controls. Expression profiling of RNA samples showed significant upregulation of miR-34a and miR-181b through microarray and qRT-PCR analysis.[90] However, the study did not reveal these miRNAs as a biomarker for AD, because the BMCs could not be good sources for cell-free miRNAs. However, this study generated information about the augmented level of miRNA expression in BMC and identification of putative gene targets of these miRNAs and their probable role in AD pathogenesis.

4.9.2.3 Serum as Sources of Circulatory miRNAs

Circulatory miRNAs as a biomarker for AD were mostly studied on the patients' sera samples (Table 1), as serum is considered to be the most suitable and gentle circulatory biofluid and an appropriate source for cell-free miRNAs. A study on seven AD patients and seven healthy controls sera showed downregulation of five miRNA candidates, miR-137, miR-181c, miR-9, miR-19a, miR-29b, by qRT-PCR when compared to negative controls.[94] Opposite to upregulation of miR-181b in BMCs, the sera level of miR-181c was downregulated in AD patients. In the same direction, Galimberti and colleagues also investigated serum samples from a cohort consisting of 7 AD and 6 noninflammatory neurological disease control (NINDC) subjects. The eighty-four most abundantly expressed miRNAs were analyzed by a miRNA PCR array. The results showed a significant downregulation of miR-125b, miR-223, miR-23a, and miR-26b in AD compared to negative controls (Table 1).[93] This was further validated in a large cohort of 15 AD, 12 NINDCs, 8 inflammatory neurological disease controls (INDCs), and 10 frontotemporal dementia, demonstrating significant downregulation only in miR-125b, miR-23a, and miR-26b in AD patients. Additionally, expression analysis of these miRNAs in CSF of AD and NINDCs showed a low level of only miR-125b and miR-26b in AD patients. Interestingly, miR-26b also showed a negative correlation with tau and Ptau protein level in AD patients. Even ROC curve analysis showed a significant AUC value (0.77) for miR-26b; however, the miR-125b AUC value was more accurate (0.82). This observation further substantiates the diagnostic potential of miR-26b and miR-125b to distinguish AD from NINDCs.[93]

A different study on the serum samples also indicated downregulation of miR-125b in AD patients compared to healthy controls with more significant AUC values (0.85, $P \leq 0.0001$).[95] Downregulation of miR-181c and upregulation of miR-9 were also observed in AD patients by qRT-PCR analysis. Significant AUC values of miR-181c and miR-9 (0.74 and 0.62, respectively) also revealed their importance as biomarkers for AD.[95] However, the main drawback of such studies was the small population cohort and these were very preliminary data; hence, these observations need to be replicated in a larger population.

Tan and colleagues conducted an important genome-wide expression profiling of serum miRNAs on AD patients.[96] Their study included primary screening of cohort 1: AD ($n=50$) and negative control ($n=50$), through high-throughput sequencing of miRNAs.[96] Unique miRNAs were identified and examined against the miRbase database Release 19, which detected only 90 miRNAs that were found to be significantly modulated in probable AD patients. Out of only 96, miR-36 is identified as novel miRNA as it was not listed on miRbase-19. The authors chose another 14 ideal miRNAs that were expressed differentially (±twofold) in AD patients and controls. Among them four miRNAs (miR-3158-3p, miR-27a-3p, miR-26b-3p, and miR-151b) were upregulated, whereas 10 miRNAs (miR-36, miR-98-5p, miR-885-5p, miR-485-5p, miR-483-3p, miR-342-3p, miR-30e-5p, miR-191-5p, let-7g-5p, and let-7d-5p) were downregulated in AD patients compared to controls.[96] Further validation by qRT-PCR on a large cohort (2) AD ($n=158$) and negative control ($n=155$) revealed downregulation of only six miRNAs (miR-483-3p, miR-342-3p, miR-98-5p, miR-191-5p, miR-885-5p, and let-7d-5p) in the probable AD patients (Table 1). Additionally, ROC curve analysis of these miRNAs indicated the highest diagnostic accuracy of miR-342-3p with an AUC value of 0.84, and a cut-off value of 0.93.[96] Such studies are recommended to expand on a large cohort in a different ethnic population for better investigation of the disease.

Dong and colleagues examined the AD patients, individuals with MCI, and vascular dementia (VD) along with nondemented controls for the expression profiling of serum miRNAs by Solexa sequencing and qRT-PCR analysis.[97] Four miRNAs (miR-31, miR-93, miR-143, and miR-146a) were downregulated in the AD patients compared to controls in both the discovery and the validation set (Table 1). AUC values (0.72, 0.69, 0.70, and 0.70, respectively) through ROC curve analysis also showed their significant discriminating power to AD patients from controls. However, expression of these miRNAs was not consistent in MCI and VD individuals where miR-93 and miR-146a were significantly elevated in MCI cases, whereas miR-143 expression was decreased and miR-31 showed no change compared to controls. In VD cases, miR-143 expression was decreased, and miR-31, miR-93, and miR-146a levels were significantly higher compared to controls.[97] Hence, differential expression of these miRNAs can discriminate AD cases, but their ambiguous expression pattern in MCI and VD cases could not explain the AD progression.

4.9.2.4 Serum Exosomal miRNAs

Analysis of serum exosomes for circulatory miRNA detection is supposed to be more feasible and more fertile than whole serum. Exosomes, the cargo, may offer an enriched population of miRNAs that were found to be free from endogenous RNA contaminants, e.g., ribosomal RNA. Hence, exosomes may be considered a prominent house of disease-specific miRNA signatures.[27] Exosomes were prepared from serum samples of AD patients, MCI individuals, and healthy controls using specific kits, and were processed for sequencing analysis of miRNAs differentially expressed in three groups. An initial screening of first cohort indicated a significant upregulation of 14 miRNAs (miR-361-5p, miR-30e-5p, miR-93-5p, miR-15a-5p, miR-143-3p, miR-335-5p, miR-106b-5p, miR-101-3p, miR-425-5p, miR-106a-5p, miR-18b-5p, miR-3065-5p, miR-20a-5p, and miR-582-5p) and downregulation of three miRNAs (miR-1306-5p, miR-342-3p, and miR-15b-3p) (Table 1). Validation analysis through qRT-PCR of the second cohort also confirmed the above observations, though the study lacked a ROC curve analysis of deregulated miRNAs for diagnostic accuracy. Analysis of exosomal miRNA profiling is also a good approach in order to look for disease-specific miRNA signatures for AD.

4.9.2.5 Plasma as Sources of Circulatory miRNAs

Blood-based plasma samples are another important sources of circulatory miRNAs. Plasma is the largest component of human blood, making up about 55% of its overall content. Important constituents of blood plasma are immunoglobulins (antibodies), clotting factors, proteins albumin and fibrinogen, enzymes, and water. The main function of plasma is the transportation of cellular nutrients, hormones, and proteins to the different parts of the body. Kumar and colleagues did a plasma miRNAs analysis in AD, MCI, and healthy control patients. Primary testing by nCounter miRNA assay (Nanostring Technology, Seattle, WA, USA) of cohort 1 revealed upregulation of six miRNAs, miR-323b-5p, miR-545-3p, miR-563, miR-600, miR-1274a, and miR-1975, and downregulation of seven miRNAs, let-7d-5p, let-7g-5p, miR-15b-5p, miR-142-3p, miR-191-5p, miR-301a-3p, and miR-545-3p, was also observed in AD and MCI cases compared to healthy controls (Table 1). Validation studies on cohort

2 confirm the downregulation of these miRNAs expression in AD patients by qRT-PCR. However, none of the miRNA candidate showed upregulation by qRT-PCR analysis.[98] ROC curve analysis showed that the best miRNA signatures were miR-142-3p and miR-301a-3p, having 100% specificity. However, miR-142-3p has better sensitivity (0.65) than miR-301a-3p (0.25). The combination of two different miRNAs (miR-545-3p and miR-15b) also displayed very significant diagnostic accuracy with AUC value=0.96, sensitivity=0.9, and specificity=0.94. However, based on AUC value, sensitivity, and specificity, the best-characterized individual miRNAs were miR-191-5p, miR-15b-5p, and let-7d-5p.[98] Furthermore, to diagnose the AD at a preclinical early stage, the next step would be needed to analyze the longitudinal plasma samples from a large cohort having the MCI.

4.9.2.6 Plasma Exosomal miRNAs

Like serum, plasma exosomes also transport the disease-associated miRNAs. A recent study by Lugli and colleagues identified differential expression of plasma exosomal miRNAs in AD patients and controls upon Illumina deep sequencing.[99] A total of 20 miRNAs were found to be deregulated, where four (miR-548at-5p, miR-138-5p, miR-5001-3p, and miR-659-5p) were upregulated and seven (miR-185-5p, miR-342-3p, miR-141-3p, miR-342-5p, miR-23b-3p, miR-338-3p, and miR-3613-3p) were significantly downregulated in AD patients compared to controls (Table 1). Among those, miR-242-3p was more interesting because its brain-enriched nature and its expression was reduced at a more significant level, as confirmed by a *t*-test.[99]

4.9.2.7 CSF and Extracellular Fluid Circulatory miRNAs

CSF is a clear biofluid, secreted by the choroid plexus and circulates into the brain ventricles across the blood–brain barrier. CSF plays important role in intercerebral transportation.[27] CSF is collected from the brain by a sophisticated lumber puncture procedure. Studies showed that CSF is also a source of circulatory miRNAs for assessment of neurological and neurodegenerative disorders. Microarray analysis of both CSF and extracellular fluid (ECF) samples indicated significant overexpression of miR-9, miR-125b, miR-146a, and miR-155 in AD cases compared to healthy controls (Table 1).[92] A recent study on CSF samples from a large cohort of AD ($n=22$) and healthy controls ($n=28$) identified 1178 miRNAs by Open Array qRT-PCR.[2] Analysis showed upregulation of seven miRNAs (miR-146a, miR-100, miR-505, miR-4467, miR-766, miR-3622b-3p, miR-296) and downregulation of eight miRNAs (miR-449, miR-1274a, miR-4674, miR-335, miR-375, miR-708, miR-219, and miR-103) in AD patients compared to controls (Table 1). The diagnostic accuracy of these miRNAs was measured by ROC curve analysis, but only miR-146a, miR-375, miR-103, and miR-100 showed significant AUC values (0.97, 0.99, 0.87, and 0.72, respectively).[2] Expression analysis of miRNAs in CSF and ECF also provides the informative biomarkers that can be used to compare and detect AD from heterogeneous controls.

As discussed earlier, multiple serum/plasma/exosomal miRNAs have been identified, and these miRNA may be useful in determining circulatory biomarkers for AD. Furthermore, screening of large population with different stages of disease progression is needed with a universally standardized procedure.

4.9.3 Huntington's Disease

Huntington's disease (HD) is an inherited neurodegenerative disorder which is caused by an unstable CAG triplet expansion in the HD gene, encoding for a polyglutamine tract in the huntingtin protein (HTT).[100] Circulatory miRNA profile was analyzed in the plasma samples from 15 symptomatic patients, with 40–45 CAG repeats in the HTT gene, and 7 healthy matched controls. A total of 752 human mature miRNAs had sequences against human miRNome panels. Further analysis showed alteration of 168 plasma miRNAs in symptomatic patients. However, statistical analysis indicated significant upregulation of 13 miRNAs (miR-877-5p, miR-223-3p, miR-223-5p, miR-30d-5p, miR-128, miR-22-5p, miR-222-3p, miR-338-3p, miR-130b-3p, miR-425-5p, miR-628-3p, miR-361-5p, and miR-942) in HD patients as compared with controls (Table 1).[100]

4.9.4 Parkinson's Disease

PD is the second most common neurodegenerative disorder in the United States, affecting approximately 1 million Americans and 5 million people worldwide.[122] Very few studies are available that identify several dysregulated circulating miRNAs in PD patients. Recently, Dong and colleagues identified novel circulating miRNAs by screening of 169 PD patients and 180 healthy controls by Solexa sequencing technology and qRT-PCR. Analysis showed a significant decreased in four serum miRNAs (miR-141, miR-214, miR-146b-5p, and miR-193a-3p) in PD patients compared with controls (Table 1). This 4-miRNA panel could be used to differentiate

HY stage 1 and 2 PD patients from controls and thus may be novel biomarkers for the early detection of PD.[101]

The miRNA expression also varies in PD and similar atypical conditions such as multiple system atrophy (MSA), and circulating miRNAs could be used to distinguish PD patients from MSA and healthy individuals. Serum samples were processed by TaqMan Low Density Array technology, and 754 miRNAs were analyzed. The nine most significant circulatory miRNAs were identified that expressed differentially in 25 PD and 25 MSA patients as compared to 25 controls. However, a validation study found four more specific miRNAs: three were upregulated miR-223*, miR-324-3p, and miR-24, whereas miR-339-5p was downregulated in both diseases (Table 1). Specifically, miR-30c and miR-148b were downregulated in PD and miR-148b was upregulated in MSA. However, comparison of MSA and PD showed three upregulated miRNAs (miR-24, miR-34b, and miR-148b) in MSA serum.[102]

4.9.5 Amyotrophic Lateral Sclerosis

Amyotrophic Lateral Sclerosis (ALS) is a lethal motor neuron disease that progressively debilitates neuronal cells that control voluntary muscle activity.[104] miRNAs from the plasma of sporadic amyotrophic lateral sclerosis (sALS) patients and healthy controls were analyzed using two cohorts: a discovery cohort analyzed with microarray (16 sALS patients and 10 healthy controls) and a validation cohort confirmed with qPCR (48 sALS patients, 47 healthy controls, and 30 disease controls). Three miRNAs were upregulated: miR-4258, miR-663b, and miR-4649-5p, whereas six were downregulated significantly: miR-26b-5p, miR-4299, let-7f-5p, miR-4419a, miR-3187-5p, and miR-4496 in the discovery cohort (Table 1). Interestingly, upregulation of miR-4649-5p and downregulation of miR-4299 was not influenced by clinical characteristics, hence they have the potential to be ALS diagnosis biomarkers.[103]

To find biomarkers for ALS, miRNA alterations were studied in skeletal muscle and plasma of mutated human superoxide dismutase 1 (SOD1-G93A) mice, and subsequently miRNAs levels were tested in the serum from human ALS patients. Muscles tissues from symptomatic SOD1-G93A mice (age 90 days) and their control littermates were first studied using miRNA microarrays, and then evaluated with quantitative PCR from five age groups from neonatal to the terminal disease stage (10–120 days). The only miR-206 was found to be consistently altered in relative to various age/gender/muscle groups and during the course of the disease pathology.[104]

5. CONCLUDING REMARKS

To date, accumulating evidence has shown that changes in serum/plasma/CSF/ECF/urine and other biofluids' miRNA levels are correlated with certain biological conditions such as aging and aging-related diseases including CVD, cancer, arthritis, dementia, cataract, osteoporosis, diabetes, hypertension, and NDs. Specific cellular and molecular changes in miRNA transcription levels or at miRNA secretory levels have been linked to the development and progression of human diseases. Experimental observations indicate their novel informative biomarkers nature and/or therapeutic targets with higher sensitivity and specificity for such diseases. Nevertheless, potential biomarker applications will require a more refined understanding of the mechanisms regarding how circulatory miRNAs are changing with disease development and progression. Additionally, the analysis of circulatory miRNAs as a biomarker has several preanalytical as well as analytical challenges during application. Therefore, some strengths and weaknesses still exist in the path of miRNAs as a futuristic biomarker (Fig. 3). To overcome these challenges, more population-based studies with constant analytical standardization is further recommended to decide the clinical utility of miRNAs in the management of aging diseases.

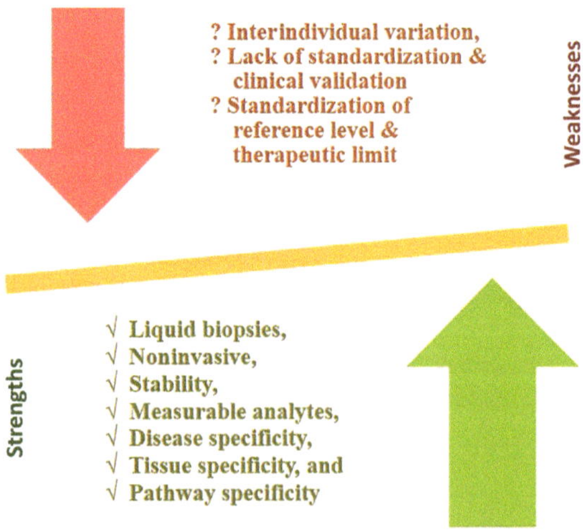

Fig. 3 Summary of strengths and weaknesses of circulatory miRNAs as biomarkers in human diseases.

ACKNOWLEDGMENTS

P.H.R. is supported by NIH Grants AG042178, AG047812, and the Garrison Family Foundation.

REFERENCES

1. Bartel DP. MicroRNAs: genomics, biogenesis, mechanism, and function. *Cell.* 2004;116(2):281–297.
2. Denk J, Boelmans K, Siegismund C, Lassner D, Arlt S, Jahn H. MicroRNA profiling of CSF reveals potential biomarkers to detect Alzheimer‘s disease. *PLoS One*. 2015;10(5): e0126423.
3. Kumar S, Chawla YK, Ghosh S, Chakraborti A. Severity of hepatitis C virus (genotype-3) infection positively correlates with circulating microRNA-122 in patients sera. *Dis Markers*. 2014;2014:435476.
4. Li Y, Kowdley KV. MicroRNAs in common human diseases. *Genomics Proteomics Bioinformatics*. 2012;10(5):246–253.
5. Zhang L, Xu Y, Jin X, et al. A circulating miRNA signature as a diagnostic biomarker for non-invasive early detection of breast cancer. *Breast Cancer Res Treat.* 2015;154(2):423–434.
6. Schwarzenbach H, Nishida N, Calin GA, Pantel K. Clinical relevance of circulating cell-free microRNAs in cancer. *Nat Rev Clin Oncol.* 2014;11(3):145–156.
7. Noren Hooten N, Fitzpatrick M, Wood 3rd WH, et al. Age-related changes in microRNA levels in serum. *Aging (Albany, NY).* 2013;5(10):725–740.
8. Vijayan M, Reddy PH. Peripheral biomarkers of stroke: focus on circulatory microRNAs. *Biochim Biophys Acta.* 2016;1862(10):1984–1993.
9. Reddy PH, Tonk S, Kumar S, et al. A critical evaluation of neuroprotective and neurodegenerative microRNAs in Alzheimer's disease. *Biochem Biophys Res Commun.* 2016 [Epub ahead of print].
10. Kumar S, Reddy PH. Are circulating microRNAs peripheral biomarkers for Alzheimer's disease? *Biochim Biophys Acta.* 2016;1862(9):1617–1627.
11. Femminella GD, Ferrara N, Rengo G. The emerging role of microRNAs in Alzheimer's disease. *Front Physiol.* 2015;6:40.
12. Harries LW. MicroRNAs as mediators of the ageing process. *Genes (Basel).* 2014;5(3):656–670.
13. Kawahara Y. Human diseases caused by germline and somatic abnormalities in microRNA and microRNA-related genes. *Congenit Anom (Kyoto).* 2014;54(1):12–21.
14. Machida T, Tomofuji T, Ekuni D, et al. MicroRNAs in salivary exosome as potential biomarkers of aging. *Int J Mol Sci.* 2015;16(9):21294–21309.
15. Grasso M, Piscopo P, Confaloni A, Denti MA. Circulating miRNAs as biomarkers for neurodegenerative disorders. *Molecules.* 2014;19(5):6891–6910.
16. Jung HJ, Suh Y. Circulating miRNAs in ageing and ageing-related diseases. *J Genet Genomics.* 2014;41(9):465–472.
17. Creemers EE, Tijsen AJ, Pinto YM. Circulating microRNAs: novel biomarkers and extracellular communicators in cardiovascular disease? *Circ Res.* 2012;110(3):483–495.
18. Bao Q, Pan J, Qi H, et al. Aging and age-related diseases—from endocrine therapy to target therapy. *Mol Cell Endocrinol.* 2014;394(1–2):115–118.
19. Childs BG, Durik M, Baker DJ, van Deursen JM. Cellular senescence in aging and age-related disease: from mechanisms to therapy. *Nat Med.* 2015;21(12):1424–1435.
20. Weber JA, Baxter DH, Zhang S, et al. The microRNA spectrum in 12 body fluids. *Clin Chem.* 2010;56(11):1733–1741.

21. Ardila-Molano J, Vizcaíno M, Serrano ML. Circulating microRNAs as potential cancer biomarkers. *Rev Colomb Cancerol.* 2015;19(4):229–238.
22. Valadi H, Ekstrom K, Bossios A, Sjostrand M, Lee JJ, Lotvall JO. Exosome-mediated transfer of mRNAs and microRNAs is a novel mechanism of genetic exchange between cells. *Nat Cell Biol.* 2007;9(6):654–659.
23. Turchinovich A, Samatov TR, Tonevitsky AG, Burwinkel B. Circulating miRNAs: cell-cell communication function? *Front Genet.* 2013;4:119.
24. Arroyo JD, Chevillet JR, Kroh EM, et al. Argonaute2 complexes carry a population of circulating microRNAs independent of vesicles in human plasma. *Proc Natl Acad Sci USA.* 2011;108(12):5003–5008.
25. Vickers KC, Palmisano BT, Shoucri BM, Shamburek RD, Remaley AT. MicroRNAs are transported in plasma and delivered to recipient cells by high-density lipoproteins. *Nat Cell Biol.* 2011;13(4):423–433.
26. Keller S, Sanderson MP, Stoeck A, Altevogt P. Exosomes: from biogenesis and secretion to biological function. *Immunol Lett.* 2006;107(2):102–108.
27. Cheng L, Doecke JD, Sharples RA, et al. Prognostic serum miRNA biomarkers associated with Alzheimer's disease shows concordance with neuropsychological and neuroimaging assessment. *Mol Psychiatry.* 2015;20(10):1188–1196.
28. Boon RA, Vickers KC. Intercellular transport of microRNAs. *Arterioscler Thromb Vasc Biol.* 2013;33(2):186–192.
29. Zhang Y, Liu D, Chen X, et al. Secreted monocytic miR-150 enhances targeted endothelial cell migration. *Mol Cell.* 2010;39(1):133–144.
30. Zernecke A, Bidzhekov K, Noels H, et al. Delivery of microRNA-126 by apoptotic bodies induces CXCL12-dependent vascular protection. *Sci Signal.* 2009;2(100):ra81.
31. Kosaka N, Iguchi H, Yoshioka Y, Takeshita F, Matsuki Y, Ochiya T. Secretory mechanisms and intercellular transfer of microRNAs in living cells. *J Biol Chem.* 2010;285(23):17442–17452.
32. Smith-Vikos T, Slack FJ. MicroRNAs circulate around Alzheimer's disease. *Genome Biol.* 2013;14(7):125.
33. Dimmeler S, Nicotera P. MicroRNAs in age-related diseases. *EMBO Mol Med.* 2013;5(2):180–190.
34. Sawada S, Akimoto T, Takahashi M, et al. Effect of aging and sex on circulating microRNAs in humans. *Adv Aging Res.* 2014;3(2):152–159.
35. Ameling S, Kacprowski T, Chilukoti RK, et al. Associations of circulating plasma microRNAs with age, body mass index and sex in a population-based study. *BMC Med Genomics.* 2015;8:61.
36. Pang J, Xiong H, Yang H, et al. Circulating miR-34a levels correlate with age-related hearing loss in mice and humans. *Exp Gerontol.* 2016;76:58–67.
37. Zhang H, Yang H, Zhang C, et al. Investigation of microRNA expression in human serum during the aging process. *J Gerontol A Biol Sci Med Sci.* 2015;70(1):102–109.
38. Jiang Y, Wang HY, Li Y, Guo SH, Zhang L, Cai JH. Peripheral blood miRNAs as a biomarker for chronic cardiovascular diseases. *Sci Rep.* 2014;4:5026.
39. Ren J, Zhang J, Xu N, et al. Signature of circulating microRNAs as potential biomarkers in vulnerable coronary artery disease. *PLoS One.* 2013;8(12):e80738.
40. Ellis KL, Cameron VA, Troughton RW, Frampton CM, Ellmers LJ, Richards AM. Circulating microRNAs as candidate markers to distinguish heart failure in breathless patients. *Eur J Heart Fail.* 2013;15(10):1138–1147.
41. Hakimzadeh N, Nossent AY, van der Laan AM, et al. Circulating microRNAs characterizing patients with insufficient coronary collateral artery function. *PLoS One.* 2015;10(9):e0137035.
42. Alhasan AH, Scott AW, Wu JJ, et al. Circulating microRNA signature for the diagnosis of very high-risk prostate cancer. *Proc Natl Acad Sci USA.* 2016;113(38):10655–10660.

43. Huo D, Clayton WM, Yoshimatsu TF, Chen J, Olopade OI. Identification of a circulating microRNA signature to distinguish recurrence in breast cancer patients. *Oncotarget*. 2016;7(34):55231–55248.
44. Zhu W, Zhou K, Zha Y, et al. Diagnostic value of serum miR-182, miR-183, miR-210, and miR-126 levels in patients with early-stage non-small cell lung cancer. *PLoS One*. 2016;11(4):e0153046.
45. Halvorsen AR, Bjaanaes M, LeBlanc M, et al. A unique set of 6 circulating microRNAs for early detection of non-small cell lung cancer. *Oncotarget*. 2016;7(24): 37250–37259.
46. Vychytilova-Faltejskova P, Radova L, Sachlova M, et al. Serum-based microRNA signatures in early diagnosis and prognosis prediction of colon cancer. *Carcinogenesis*. 2016;37:941–950.
47. Hamam R, Ali AM, Alsaleh KA, et al. MicroRNA expression profiling on individual breast cancer patients identifies novel panel of circulating microRNA for early detection. *Sci Rep*. 2016;6:25997.
48. Stuopelyte K, Daniunaite K, Bakavicius A, Lazutka JR, Jankevicius F, Jarmalaite S. The utility of urine-circulating miRNAs for detection of prostate cancer. *Br J Cancer*. 2016;115(6):707–715.
49. Gao Y, Guo Y, Wang Z, et al. Analysis of circulating miRNAs 21 and 375 as potential biomarkers for early diagnosis of prostate cancer. *Neoplasma*. 2016;63(4):623–628.
50. Chen H, Liu H, Zou H, et al. Evaluation of plasma miR-21 and miR-152 as diagnostic biomarkers for common types of human cancers. *J Cancer*. 2016;7(5):490–499.
51. Su K, Zhang T, Wang Y, Hao G. Diagnostic and prognostic value of plasma microRNA-195 in patients with non-small cell lung cancer. *World J Surg Oncol*. 2016;14(1):224.
52. Wang J, Ye H, Zhang D, et al. MicroRNA-410-5p as a potential serum biomarker for the diagnosis of prostate cancer. *Cancer Cell Int*. 2016;16:12.
53. Freres P, Wenric S, Boukerroucha M, et al. Circulating microRNA-based screening tool for breast cancer. *Oncotarget*. 2016;7(5):5416–5428.
54. Aherne ST, Madden SF, Hughes DJ, et al. Circulating miRNAs miR-34a and miR-150 associated with colorectal cancer progression. *BMC Cancer*. 2015;15:329.
55. Li Y, Liu S, Zhang F, Jiang P, Wu X, Liang Y. Expression of the microRNAs hsa-miR-15a and hsa-miR-16-1 in lens epithelial cells of patients with age-related cataract. *Int J Clin Exp Med*. 2015;8(2):2405–2410.
56. Ogata-Kawata H, Izumiya M, Kurioka D, et al. Circulating exosomal microRNAs as biomarkers of colon cancer. *PLoS One*. 2014;9(4):e92921.
57. Ma J, Li N, Lin Y, Gupta C, Jiang F. Circulating neutrophil microRNAs as biomarkers for the detection of lung cancer. *Biomark Cancer*. 2016;8:1–7.
58. Antolin S, Calvo L, Blanco-Calvo M, et al. Circulating miR-200c and miR-141 and outcomes in patients with breast cancer. *BMC Cancer*. 2015;15:297.
59. Yong FL, Law CW, Wang CW. Potentiality of a triple microRNA classifier: miR-193a-3p, miR-23a and miR-338-5p for early detection of colorectal cancer. *BMC Cancer*. 2013;13:280.
60. Filkova M, Aradi B, Senolt L, et al. Association of circulating miR-223 and miR-16 with disease activity in patients with early rheumatoid arthritis. *Ann Rheum Dis*. 2014;73(10):1898–1904.
61. Duroux-Richard I, Pers YM, Fabre S, et al. Circulating miRNA-125b is a potential biomarker predicting response to rituximab in rheumatoid arthritis. *Mediators Inflamm*. 2014;2014:342524.
62. Wang W, Zhang Y, Zhu B, et al. Plasma microRNA expression profiles in Chinese patients with rheumatoid arthritis. *Oncotarget*. 2015;6(40):42557–42568.

63. Hruskova V, Jandova R, Vernerova L, et al. MicroRNA-125b: association with disease activity and the treatment response of patients with early rheumatoid arthritis. *Arthritis Res Ther*. 2016;18(1):124.
64. Castro-Villegas C, Perez-Sanchez C, Escudero A, et al. Circulating miRNAs as potential biomarkers of therapy effectiveness in rheumatoid arthritis patients treated with anti-TNFalpha. *Arthritis Res Ther*. 2015;17:49.
65. Chien KH, Chen SJ, Liu JH, et al. Correlation between microRNA-34a levels and lens opacity severity in age-related cataracts. *Eye (Lond)*. 2013;27(7):883–888.
66. Tanaka Y, Tsuda S, Kunikata H, et al. Profiles of extracellular miRNAs in the aqueous humor of glaucoma patients assessed with a microarray system. *Sci Rep*. 2014;4:5089.
67. Wecker T, Hoffmeier K, Plotner A, et al. MicroRNA profiling in aqueous humor of individual human eyes by next-generation sequencing. *Invest Ophthalmol Vis Sci*. 2016;57(4):1706–1713.
68. Cao Z, Moore BT, Wang Y, et al. MiR-422a as a potential cellular microRNA biomarker for postmenopausal osteoporosis. *PLoS One*. 2014;9(5):e97098.
69. Wang Y, Li L, Moore BT, et al. MiR-133a in human circulating monocytes: a potential biomarker associated with postmenopausal osteoporosis. *PLoS One*. 2012;7(4):e34641.
70. Kocijan R, Muschitz C, Geiger E, et al. Circulating microRNA signatures in patients with idiopathic and postmenopausal osteoporosis and fragility fractures. *J Clin Endocrinol Metab*. 2016;101(11):4125–4134.
71. Panach L, Mifsut D, Tarin JJ, Cano A, Garcia-Perez MA. Serum circulating microRNAs as biomarkers of osteoporotic fracture. *Calcif Tissue Int*. 2015;97(5):495–505.
72. Seeliger C, Karpinski K, Haug AT, et al. Five freely circulating miRNAs and bone tissue miRNAs are associated with osteoporotic fractures. *J Bone Miner Res*. 2014;29(8):1718–1728.
73. Meng J, Zhang D, Pan N, et al. Identification of miR-194-5p as a potential biomarker for postmenopausal osteoporosis. *PeerJ*. 2015;3:e971.
74. Zhang T, Lv C, Li L, et al. Plasma miR-126 is a potential biomarker for early prediction of type 2 diabetes mellitus in susceptible individuals. *Biomed Res Int*. 2013;2013:761617.
75. Assmann T, Coutinho M, Tschiedel B, Canani L, Crispim D. Circulating microRNAs as biomarkers for type 1 diabetes mellitus. *Diabetol Metab Syndr*. 2015;7(suppl 1):A206.
76. Olivieri F, Bonafe M, Spazzafumo L, et al. Age- and glycemia-related miR-126-3p levels in plasma and endothelial cells. *Aging (Albany, NY)*. 2014;6(9):771–787.
77. Latreille M, Herrmanns K, Renwick N, et al. miR-375 gene dosage in pancreatic beta-cells: implications for regulation of beta-cell mass and biomarker development. *J Mol Med*. 2015;93(10):1159–1169.
78. Zhu H, Leung SW. Identification of microRNA biomarkers in type 2 diabetes: a meta-analysis of controlled profiling studies. *Diabetologia*. 2015;58(5):900–911.
79. Al-Kafaji G, Al-Mahroos G, Alsayed NA, Hasan ZA, Nawaz S, Bakhiet M. Peripheral blood microRNA-15a is a potential biomarker for type 2 diabetes mellitus and pre-diabetes. *Mol Med Rep*. 2015;12(5):7485–7490.
80. Kong L, Zhu J, Han W, et al. Significance of serum microRNAs in pre-diabetes and newly diagnosed type 2 diabetes: a clinical study. *Acta Diabetol*. 2011;48(1):61–69.
81. Wang C, Wan S, Yang T, et al. Increased serum microRNAs are closely associated with the presence of microvascular complications in type 2 diabetes mellitus. *Sci Rep*. 2016;6:20032.
82. Nielsen LB, Wang C, Sorensen K, et al. Circulating levels of microRNA from children with newly diagnosed type 1 diabetes and healthy controls: evidence that miR-25 associates to residual beta-cell function and glycaemic control during disease progression. *Exp Diabetes Res*. 2012;2012:896362.
83. Wen D, Qiao P, Wang L. Circulating microRNA-223 as a potential biomarker for obesity. *Obes Res Clin Pract*. 2015;9(4):398–404.

84. Sarrion I, Milian L, Juan G, et al. Role of circulating miRNAs as biomarkers in idiopathic pulmonary arterial hypertension: possible relevance of miR-23a. *Oxid Med Cell Longev*. 2015;2015:792846.
85. Huang Y, Tang S, Ji-Yan C, et al. Circulating miR-92a expression level in patients with essential hypertension: a potential marker of atherosclerosis. *J Hum Hypertens*. 2016. [Epub ahead of print].
86. Parthenakis F, Marketou M, Kontaraki J, et al. Low levels of microRNA-21 are a marker of reduced arterial stiffness in well-controlled hypertension. *J Clin Hypertens (Greenwich)*. 2016. [Epub ahead of print].
87. Kontaraki JE, Marketou ME, Zacharis EA, Parthenakis FI, Vardas PE. MicroRNA-9 and microRNA-126 expression levels in patients with essential hypertension: potential markers of target-organ damage. *J Am Soc Hypertens*. 2014;8(6):368–375.
88. Kontaraki JE, Marketou ME, Parthenakis FI, et al. Hypertrophic and antihypertrophic microRNA levels in peripheral blood mononuclear cells and their relationship to left ventricular hypertrophy in patients with essential hypertension. *J Am Soc Hypertens*. 2015;9(10):802–810.
89. Sheinerman KS, Tsivinsky VG, Abdullah L, Crawford F, Umansky SR. Plasma microRNA biomarkers for detection of mild cognitive impairment: biomarker validation study. *Aging (Albany, NY)*. 2013;5(12):925–938.
90. Schipper HM, Maes OC, Chertkow HM, Wang E. MicroRNA expression in Alzheimer blood mononuclear cells. *Gene Regul Syst Bio*. 2007;1:263–274.
91. Satoh J, Kino Y, Niida S. MicroRNA-seq data analysis pipeline to identify blood biomarkers for Alzheimer's disease from public data. *Biomark Insights*. 2015;10:21–31.
92. Alexandrov PN, Dua P, Hill JM, Bhattacharjee S, Zhao Y, Lukiw WJ. MicroRNA (miRNA) speciation in Alzheimer's disease (AD) cerebrospinal fluid (CSF) and extracellular fluid (ECF). *Int J Biochem Mol Biol*. 2012;3(4):365–373.
93. Galimberti D, Villa C, Fenoglio C, et al. Circulating miRNAs as potential biomarkers in Alzheimer's disease. *J Alzheimers Dis*. 2014;42(4):1261–1267.
94. Geekiyanage H, Jicha GA, Nelson PT, Chan C. Blood serum miRNA: non-invasive biomarkers for Alzheimer's disease. *Exp Neurol*. 2012;235(2):491–496.
95. Tan L, Yu JT, Liu QY, et al. Circulating miR-125b as a biomarker of Alzheimer's disease. *J Neurol Sci*. 2014;336(1–2):52–56.
96. Tan L, Yu JT, Tan MS, et al. Genome-wide serum microRNA expression profiling identifies serum biomarkers for Alzheimer's disease. *J Alzheimers Dis*. 2014;40(4): 1017–1027.
97. Dong H, Li J, Huang L, et al. Serum microRNA profiles serve as novel biomarkers for the diagnosis of Alzheimer's disease. *Dis Markers*. 2015;2015:625659.
98. Kumar P, Dezso Z, MacKenzie C, et al. Circulating miRNA biomarkers for Alzheimer's disease. *PLoS One*. 2013;8(7):e69807.
99. Lugli G, Cohen AM, Bennett DA, et al. Plasma exosomal miRNAs in persons with and without Alzheimer disease: altered expression and prospects for biomarkers. *PLoS One*. 2015;10(10):e0139233.
100. Diez-Planelles C, Sanchez-Lozano P, Crespo MC, et al. Circulating microRNAs in Huntington's disease: emerging mediators in metabolic impairment. *Pharmacol Res*. 2016;108:102–110.
101. Dong H, Wang C, Lu S, et al. A panel of four decreased serum microRNAs as a novel biomarker for early Parkinson's disease. *Biomarkers*. 2016;21(2):129–137.
102. Vallelunga A, Ragusa M, Di Mauro S, et al. Identification of circulating microRNAs for the differential diagnosis of Parkinson's disease and multiple system atrophy. *Front Cell Neurosci*. 2014;8:156.
103. Takahashi I, Hama Y, Matsushima M, et al. Identification of plasma microRNAs as a biomarker of sporadic amyotrophic lateral sclerosis. *Mol Brain*. 2015;8(1):67.

104. Toivonen JM, Manzano R, Olivan S, Zaragoza P, Garcia-Redondo A, Osta R. MicroRNA-206: a potential circulating biomarker candidate for amyotrophic lateral sclerosis. *PLoS One*. 2014;9(2):e89065.
105. He Y, Lin J, Kong D, et al. Current state of circulating microRNAs as cancer biomarkers. *Clin Chem*. 2015;61(9):1138–1155.
106. Ma R, Jiang T, Kang X. Circulating microRNAs in cancer: origin, function and application. *J Exp Clin Cancer Res*. 2012;31:38.
107. Khoury S, Tran N. Circulating microRNAs: potential biomarkers for common malignancies. *Biomark Med*. 2015;9(2):131–151.
108. Hansen J. Common cancers in the elderly. *Drugs Aging*. 1998;13(6):467–478.
109. Shen J, Hruby GW, McKiernan JM, et al. Dysregulation of circulating microRNAs and prediction of aggressive prostate cancer. *Prostate*. 2012;72(13):1469–1477.
110. Hollis M, Nair K, Vyas A, Chaturvedi LS, Gambhir S, Vyas D. MicroRNAs potential utility in colon cancer: early detection, prognosis, and chemosensitivity. *World J Gastroenterol*. 2015;21(27):8284–8292.
111. Churov AV, Oleinik EK, Knip M. MicroRNAs in rheumatoid arthritis: altered expression and diagnostic potential. *Autoimmun Rev*. 2015;14(11):1029–1037.
112. Sun M, Zhou X, Chen L, et al. The regulatory roles of microRNAs in bone remodeling and perspectives as biomarkers in osteoporosis. *Biomed Res Int*. 2016;2016:1652417.
113. Gennari L, Bilezikian JP. Idiopathic osteoporosis in men. *Curr Osteoporos Rep*. 2013;11(4):286–298.
114. Chien HY, Lee TP, Chen CY, et al. Circulating microRNA as a diagnostic marker in populations with type 2 diabetes mellitus and diabetic complications. *J Chin Med Assoc*. 2015;78(4):204–211.
115. Guay C, Regazzi R. Circulating microRNAs as novel biomarkers for diabetes mellitus. *Nat Rev Endocrinol*. 2013;9(9):513–521.
116. Pescador N, Perez-Barba M, Ibarra JM, Corbaton A, Martinez-Larrad MT, Serrano-Rios M. Serum circulating microRNA profiling for identification of potential type 2 diabetes and obesity biomarkers. *PLoS One*. 2013;8(10):e77251.
117. Romaine SP, Charchar FJ, Samani NJ, Tomaszewski M. Circulating microRNAs and hypertension—from new insights into blood pressure regulation to biomarkers of cardiovascular risk. *Curr Opin Pharmacol*. 2016;27:1–7.
118. Margis R, Margis R, Rieder CR. Identification of blood microRNAs associated to Parkinsonis disease. *J Biotechnol*. 2011;152(3):96–101.
119. Cardo LF, Coto E, de Mena L, et al. Profile of microRNAs in the plasma of Parkinson's disease patients and healthy controls. *J Neurol*. 2013;260(5):1420–1422.
120. Botta-Orfila T, Morato X, Compta Y, et al. Identification of blood serum micro-RNAs associated with idiopathic and LRRK2 Parkinson's disease. *J Neurosci Res*. 2014;92(8):1071–1077.
121. Reddy PH. Abnormal tau, mitochondrial dysfunction, impaired axonal transport of mitochondria, and synaptic deprivation in Alzheimer's disease. *Brain Res*. 2011;1415:136–148.
122. Khoo SK, Petillo D, Kang UJ, et al. Plasma-based circulating microRNA biomarkers for Parkinson's disease. *J Parkinsons Dis*. 2012;2(4):321–331.

CHAPTER FOUR

Molecular Links and Biomarkers of Stroke, Vascular Dementia, and Alzheimer's Disease

M. Vijayan*[,1], S. Kumar*, J.S. Bhatti*[,†], P.H. Reddy*[,‡]
*Garrison Institute on Aging, Texas Tech University Health Sciences Center, Lubbock, TX, United States
[†]Department of Biotechnology, Sri Guru Gobind Singh College, Chandigarh, India
[‡]Texas Tech University Health Sciences Center, Lubbock, TX, United States
[1]Corresponding author: e-mail address: murali.vijayan@ttuhsc.edu

Contents

Abstract

Stroke is a very common neurological disease, and it occurs when the blood supply to part of the brain is interrupted and the subsequent shortage of oxygen and nutrients causes damage to the brain tissue. Stroke is the second leading cause of death and the third leading cause of disability-adjusted life years. The occurrence of stroke increases with age, but anyone at any age can suffer a stroke. Stroke can be broadly classified in two major clinical types: ischemic stroke (IS) and hemorrhagic stroke. Research also revealed that stroke, vascular dementia (VaD), and Alzheimer's disease (AD) increase with a number of modifiable factors, and most strokes can be prevented and/or controlled through pharmacological or surgical interventions and lifestyle changes. The pathophysiology of stroke, VaD, and AD is complex, and recent molecular and postmortem brain studies have revealed that multiple cellular changes have been implicated, including inflammatory responses, micro-RNA alterations, and marked changes in brain proteins. These molecular and cellular

Progress in Molecular Biology and Translational Science, Volume 146
ISSN 1877-1173
http://dx.doi.org/10.1016/bs.pmbts.2016.12.014

changes provide new information for developing therapeutic strategies for stroke and related vascular disorders treatment. IS is the major risk factor for VaD and AD. This chapter summarizes the (1) links among stroke–VaD–AD; (2) updates the latest developments of research in identifying protein biomarkers in peripheral and central nervous system tissues; and (3) critically evaluates miRNA profile and function in human blood samples, animal, and postmortem brains.

ABBREVIATIONS

aAbs auto antibodies
AD Alzheimer's disease
Apo C1 apolipoprotein C1
Apo C3 apolipoprotein C3
Aβ amyloid beta
BBB blood–brain barrier
CDK5 cell division protein kinase 5
CNS central nervous system
CRP C-reactive protein
CSF cerebrospinal fluid
CT computed tomography
HS hemorrhagic stroke
ICH intracerebral hemorrhage
IL-6 interleukin-6
IS ischemic stroke
Lp-PLA2 lipoprotein-associated phospholipase A2
MBP myelin basic protein
miRNA microRNA
MRI magnetic resonance imaging
MRS modified ranking scale
NMDA-R-Ab *N*-methyl-D-aspartate receptor antibody
S100B S100 calcium-binding protein B
SAH subarachnoid hemorrhage
T2DM type 2 diabetes mellitus
TIA transient ischemic attack
TNF-α tumor necrosis factor-α
VaD vascular dementia
WHO World Health Organization

1. INTRODUCTION

1.1 Stroke

Stroke is a common neurological disease that occurs when the blood supply to the brain is interrupted, resulting in a shortage of oxygen and nutrients to brain tissue. It is the second leading cause of death and the third leading cause of disability-adjusted life years worldwide. World Health Organization (WHO) defined stroke as "rapidly developing clinical signs of focal or global disturbance of cerebral function, with symptoms lasting 24 h or longer, or leading to death with no apparent cause other than vascular origin".[1] The risk of having a stroke increases after the age of 55, but it can occur at any age. Stroke can be further classified into two types, i.e., ischemic stroke (IS) and hemorrhagic stroke (HS).[2] The effects of ISs and HSs depend on the part of the brain that is injured and the severity of the injury. Patients having the same type of stroke can have differing clinical symptoms. Similarly, patients with the same clinical symptoms can have different etiopathologies. Due to its multifactorial nature, stroke may be classified as a syndrome, not as a single disease. Modern neuroimaging, typically with computed tomography (CT) or magnetic resonance imaging (MRI), is now used to accurately diagnose a stroke. IS can be described as a lack of blood supply and oxygen availability to an area of the brain due to partial or complete obstruction of an artery leading to or within the brain, accounts for 87% of all strokes worldwide.[3] According to the Trial of Org 10172 in Acute Stroke Treatment (TOAST) diagnostic criteria for the stroke, IS can be classified into five clinical subtypes: large-vessel disease, small-vessel disease, cardioembolic stroke, stroke of another determined etiology, and stroke of undetermined etiology.[4–6] HS is defined as an acute neurologic injury occurring as a result of bleeding into the head, and it accounts for 13% of all strokes worldwide.[7] HS is either a brain aneurysm that bursts or a weakened blood vessel that leaks. Based on the origin and site of the bleeding, HS can involve an intracerebral hemorrhage (ICH) or a subarachnoid hemorrhage (SAH). HS is more frequently lethal than IS.[8] ICH is described as bleeding that occurs from a broken blood vessel within the brain. Other kinds of stroke can also translate to an ICH and are especially common for embolic strokes that are related to a heart valve infection. SAH is defined as bleeding from a damaged blood vessel which causes blood to accumulate at the surface of the brain.[9–11] According to the WHO, stroke affects 15 million people worldwide. Of these, about 5 million patients suffer from permanent disability and about 5.5 million patients succumb to their disabilities.[12] Globally, the prevalence rate of stroke is about 400–800/1,000,000 persons.[13]

1.2 Dementia

Dementia is a clinical disorder triggered by neurodegeneration. There are more than 100 disease conditions that can lead to dementia, the most common of which are stroke and Alzheimer's disease (AD). In 2015, it was estimated that 47.5 million people have dementia worldwide, and the numbers are projected to rise to 75.6 million by 2030 and to 131.5 million by 2050.[14] Dementia is a progressive or long-lasting condition that results in cognitive changes, such as progressively worsening cognitive skills, memory loss, changes in learning capacity, mood changes, impaired judgment, and problems in understanding language and performing routine activities, such as paying bills or preparing meals.[15] However, memory loss alone does not indicate dementia.[16,17] Common progressive diseases or conditions that are characterized by dementia are AD, vascular dementia (VaD), Lewy body dementia, and frontotemporal dementia. Other diseases or conditions associated with dementia are Huntington's disease, traumatic brain injury, Creutzfeldt–Jakob disease, and Parkinson's disease.

1.2.1 Vascular Dementia

Next to AD, VaD is the most seriously dementing illness, accounting for about 10% of all dementia cases. VaD is characterized by a decline in thinking skills caused by conditions that block or reduce blood flow to the brain, depriving brain cells of vital oxygen and nutrients. About 50% of individuals with dementia have pathologic indications of VaD. In most cases, the infarcts exist with AD pathology.[18] VaD includes a set of varied dementing disorders owing to cerebrovascular inadequacy. Four types of VaD are stroke-related dementia, single- and multiinfarct dementia, subcortical dementia, and mixed dementia. VaD may result from brain damage caused by numerous strokes or minored blood clots in the heart or neck arteries that block a branch of a blood vessel in the brain.[19] The annual incidence of VaD may range from 20 to 40 per 100,000 in individuals aged 60–69 years, to 200–700 per 100,000 in individuals over age 80 years.[20]

1.2.2 Alzheimer's Disease

AD is a multifactorial, late-onset neurodegenerative disease characterized by memory loss, multiple cognitive impairments, and progressive degeneration of behavioral and functional capacities.[21–25] Some of these cases involve exclusively AD pathology; many have an indication of pathologic variations linked to other dementias. AD is associated with the loss of synapses, synaptic dysfunction, mitochondrial structural, and functional abnormalities, inflammatory responses, in addition to extracellular neuritic plaques and intracellular neurofibrillary tangles. There are two different types of AD: early-onset familial AD and late-onset sporadic AD. Type 2 diabetes, traumatic brain injury, and IS are other contributing factors for the development of AD.[26] AD accounts for more than 80% of dementia cases worldwide in people older than 65.[27] In the United States, one in nine Americans over 65 years of age has AD, which is ranked as the sixth most common cause of death in the United States. By 2050, there could be as many as 7 million people age 85 years and older with AD.[14]

2. RISK FACTORS FOR IS, VaD, AND AD

Stroke is associated with risk factors, some of which may be nonmodifiable (e.g., age, sex, ethnicity, and heredity) and some which are modifiable (e.g., hypertension, smoking, diabetes, atrial fibrillation, and occupational/environmental exposure)[28] (Fig. 1).

Age is the significant risk factor for the stroke with sudden increases in incidence rates with increasing age. The risk of having a first-time stroke increases exponentially from about 30 per 100,000 individuals at 30–39 years of age, to about 2000–3000 per 100,000 at ages above 85.[29] IS, VaD, and AD occur predominantly in elderly adults. Stroke incidental rate is higher in men than in women but only at younger and middle-age groups, but this not true

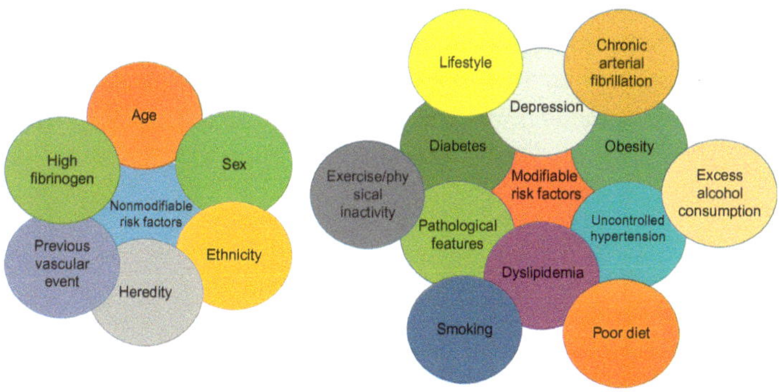

Fig. 1 List of modifiable and nonmodifiable risk factors associated with stroke, vascular dementia, and Alzheimer's disease. Stroke is pathologically heterogeneous and the risk factor profiles leading to different types of stroke-related disorders.

in the oldest age groups, incidence rates in women are about equivalent or even greater than in men.[8,30] It is unclear indistinct whether women have a higher risk than men for developing dementia or AD at a given age. Numerous European studies have proposed that women have a higher incidence rate of dementia or AD than men. However, studies in the United States have not revealed a difference or the difference has diversified with age.[31] The race has been reported being an independent predictor of stroke severity and the subtype of stroke.[32] Stroke incidence differs across racial groups, with black individuals at higher incidence rates of strokes compared to Caucasians.[33,34] There are more non-Hispanic whites existing with AD and other dementias than people of any added racial or ethnic group in the United States, older African-Americans, and Hispanics are more likely than older whites to have AD and other dementias.[35] The proportion of VaD was diverse from that in Europe and other Asian countries. There was a higher prevalence of VaD in the urban than the rural areas.[36]

About 50% of victims of IS have high blood pressure, making high blood pressure is the highest risk factor for the stroke.[37,38] Diabetes mellitus is the main risk factor for the stroke. Type 2 diabetes mellitus (T2DM) has been associated with vascular diseases, ultimately leading to cognitive dysfunction and VaD,[39] but recent studies have established that T2DM is also associated with AD, possibly due to T2DM accelerating AD-associated pathologies through insulin resistance. When people with diabetes have a stroke, the effect of the stroke on them is far worse than on individuals without diabetes.

The risk of stroke increases with the number of modifiable risk factors that an individual has.[8] Both "modifiable" and "nonmodifiable" risk factors have been connected to stroke, AD, and VaD. Modifiable risk factors can be controlled through pharmacological or surgical interventions and lifestyle changes, as primary or secondary stroke prevention strategies. These factors can be controlled and prevent and/or delay IS, VaD, and AD in elderly individuals.

3. MOLECULAR LINKS AND PATHWAYS

IS, VaD, and AD place a huge burden on universal health care. These three are among the leading causes of developed disability worldwide,[40,41] and the lifetime risk of AD or stroke is as high as one in two for women and one in three for men.[42] An estimated 24.3 million people were thought to have dementia in 2001, and this is expected to rise to 81.1 million by 2040.[41,43]

After a few minutes or even a few seconds following IS, the IS cascade begins and can continue for hours until the disease ceases. A brain ischemia leads to a cascade of pathophysiological processes, which contribute to ischemic cell damage. Stimulation of the inflammatory process, free radical production, excitotoxicity, disruption of sodium and calcium influx, enzymatic changes, endothelin release, delayed coagulation, activation of platelets and leukocytes, and endothelial dysfunction were the pathophysiological reactions that separately and/or together contribute to the brain injury resulting the onset of stroke. Dementia syndromes established after stroke were typically considered to be vascular in origin and poststroke dementia might be the significance of the effects of stroke and degenerative changes.[44] Increasing evidence suggests epidemiological and pathological links between IS, VaD, and AD. Many studies have shown IS to be a risk factor for developing VaD and AD,[45–47] suggesting that shared pathological processes may be involved in both conditions. Other studies have indicated a synergistic relationship between IS and AD, with the combination of both leading to an increased risk of cognitive decline and dementia. Studies have also shown that cerebrovascular events lead to more-rapid cognitive decline in patients with AD.[48] Postmortem studies have shown that individuals with cerebral infarcts as well as neuropathological AD had a markedly increased risk of dementia in life compared with those with AD pathology without infarcts.[49–51]

Research linking stroke and dementia has focused on shared vascular risk factors, ameliorated by lifestyle activities or medication. Aging is the most important risk factor for the stroke and dementia. Dementia occurs in up to one-third of elderly patients with stroke, a subset of whom have AD rather than a pure VaD. A mixed etiology of dementia and cerebrovascular disease is thought to become more common with increasing age, although no clinical criteria for the diagnosis of dementia with cerebrovascular diseases are currently available.[52] Stroke doubles the risk for dementia (poststroke dementia), and approximately 30% of stroke patients go on to develop cognitive dysfunction within 3 years.[53,54] The association between stroke and dementia was also observed in patients younger than 50 years, up to 50% of whom exhibited cognitive deficits after a decade.[55]

VaD and AD are important because these results from a variety of causes, including cerebrovascular dysfunction, but the evidence of their association with other neurodegenerative disorders is limited.[56] VaD, AD, and stroke have common risk factors including hypertension, insulin resistance, diabetes, obesity, hyperhomocysteinemia, and hyperlipidemia. Cerebrovascular disease has been suggested contributing to AD neuropathological changes including selective brain atrophy and accumulation of abnormal proteins such as amyloid beta (Aβ).[57] Recent clinical–pathological studies have focused on cognitive impairment and increased the risk of dementia in patients with cerebrovascular disease.[58] In addition, VaD is the severest form of vascular cognitive impairment,[59] and it results from subclinical vascular brain injury and stroke. Levels of the toxic form of Aβ that accumulates in the brains of AD patients rise in the presence of an activator of CDK5 called p25, which increases after an IS. Mouse model that overexpresses p25 to enhance CDK5 activity, observed that the animals show an increase in BACE, or beta secretase, and high levels of Aβ peptide in brain. The animals do not have any overt cognitive deficits and do not develop tangles, but pathway might be important in understanding AD and how stroke might put people at increased risk. This pathway has been implicated in both conditions.[60]

The molecular links between the stroke and VaD, and stroke and AD are currently not clearly understood.[61,62] There are many questions raised in research linking stroke and dementia which are largely unanswered. Hence, it is important to understand the early events of stroke, and stroke leading to VaD and AD. Very little can be done after disease onset starts. Therefore, therapy needs to be initiated as soon as possible, with its immediate goal to normalize a perfusion and to mediate any biochemical dysfunction to recover the obscurity as early as possible.[63]

4. MOLECULAR BIOMARKERS

A biomarker—such as a protein, nucleic acid, or metabolite—is a quantification of a definite biological state, typically one relevant to the risk, occurrence, severity, prognosis, or projected therapeutic response of disease. Biomarkers may be useful in identifying different diseases, such as stroke, VaD, AD, cancer, and diabetes, and disease severity. Identification of biomarkers can inform researchers in their attempts to develop early detectable peripheral biomarkers. Identification of biomarkers can contribute to a better understanding of the etiologies and mechanisms underlying particular diseases, such as stroke, VaD, and AD. To identify peripheral biomarkers of stroke, multiple approaches have been developed, including circulatory microRNAs (miRNAs), blood-based protein markers.

4.1 Protein Biomarkers in Stroke, VaD, and AD

A single biomarker might not be adequate to identify underlying complexities known to underlie cellular changes linked to disease and to discriminate diseased from healthy individuals. A biomarker panel that reflects diverse pathophysiological characteristics of a disease or syndrome might be needed to capture the complexities of a particular disease. For example, a biomarker panel for stroke could provide information about inflammation, atherosclerosis, cerebral ischemia, blood–brain barrier (BBB) disruption, thrombus formation, oxidative stress, and endothelial injury. Biomarker panels have been sought to improve the diagnosis of stroke, VaD, AD, and its cause. A biomarker needs to identify a particular feature of disease state as accurately and specificity as possible and to be presented as clearly as possible for use by clinicians.[64] Some methods have been combined to identify protein biomarkers, such as Western blot, enzyme-linked immune sorbent assay, immunohistochemical staining, and mass spectrometry.

Use of these procedures facilitates the identification of a wide variety of proteins as possible biomarkers for stroke, VaD, and AD. These proteins have a crucial role in central nervous system (CNS) tissue injury, inflammatory, and coagulation/thrombosis biomarkers for stroke, VaD, and AD. Biomarkers identifying the coagulation cascade have been linked to stroke, VaD, AD, and severe thrombus in cerebral circulation; such biomarkers include the fibrogens vWF, apolipoprotein C1 (Apo C1), apolipoprotein C3 (Apo C3), D-dimer, lipoprotein-associated phospholipase A2 (Lp-PLA2), and ApoA2. Among these fibrinogens, Lp-PLA2 and Apo C1/3 have been described the most in recent years. Fibrinogen, a blood-borne glycoprotein, is a coagulation factor responsible for blood clotting. Fibrinogen has been significantly associated with stroke, coronary heart disease, and other diseases that cause vascular and nonvascular mortality[65] (Fig. 2).

The CardioVascular Disease Risk Factors Two-township Study focused on identifying biomarkers for cardiovascular diseases and their risk factors. In one such study, a dose–response relationship was associated with the risk of IS and tertiles of fibrinogen. A 72% increase (hazard ratio, 1.72; 1.02–2.90) in the risk of IS was found in individuals with fibrinogen at least 8.79 μmol/L compared with those individuals with fibrinogen less than 7.03 μmol/L.[66] In one study of first IS or transient ischemic attack (TIA) patients and healthy controls (aged 18–75 years old), fibrinogen γ/total fibrinogen ratio was higher in acute phase of stroke patients than in controls and lower in the convalescent phase (3 months after stroke).[67] A number of studies were also documented that found elevated levels of fibrinogen in

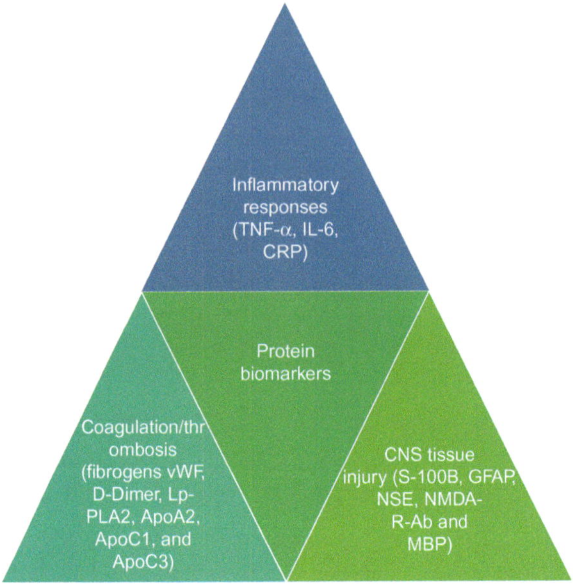

Fig. 2 List of protein biomarkers associated with stroke, vascular dementia, and Alzheimer's disease. A biomarker panel that reflects diverse pathophysiological characteristics of a disease or syndrome might be needed to capture the complexities of a particular disease.

patients with risks of recurrent IS events, but the elevated levels were not as high as in previous studies of fibrinogen.[67,68]

Many studies reported an association between plasma levels of inflammation markers and the risk of dementia. Study based on the prospective population-based Rotterdam Study stated that individuals with higher levels of fibrinogen had an increased risk of dementia. Further, high fibrinogen levels were associated with an increased risk of both AD and VaD and suggested that the increased risk of dementia associated with fibrinogen was because of the hemostatic rather than the inflammatory properties of fibrinogen.[69] Another study was aimed to investigate the relationship between plasma fibrinogen level and risk for cognitive decline and dementia in patients with mild cognitive impairment (MCI). Patients with hyperfibrinogenemia had an increased risk for dementia and VaD compared with patients with normal level of plasma fibrinogen. Further, it concluded that plasma fibrinogen level might be associated with cognitive decline, and hyperfibrinogenemia might increase risk for dementia in patients with MCI.[70] Fibrinogen and β-amyloid association alters thrombosis and fibrinolysis, a possible contributing factor to AD. Further, depletion of fibrinogen lessened cerebral amyloid angiopathy pathology and reduced cognitive impairment in AD mice.[71]

Lp-PLA2 and Apo C1/3 were reported to be associated with increased risk for stroke, VaD, and AD. Lp-PLA2, 45-kDa protein of 441 amino acids, catalyzes the degradation of platelet-activating factor to biologically inactive products. Lp-PLA2 is an enzyme expressed primarily by leukocytes that is active in the metabolism of low density lipoprotein to proinflammatory mediators. It is also a vascular specific inflammatory biomarker extremely expressed in the necrotic core of atherosclerotic plaque and is linked to plaque inflammation and variability. Level of Lp-PLA2 might be a risk factor to identify middle-aged individuals at increased risk for their first IS event and the might be complementary beyond traditional risk factors in identifying middle-aged individuals at increased risk for IS. Further, future studies should regulate whether discriminating inhibition of Lp-PLA2 reduces IS and whether statins and/or fibrates were more effective for stroke prevention in patients with elevated levels of Lp-PLA2.[72] Another prospective population-based cohort study found that Lp-PLA2 is an independent predictor of IS in the general population.[73] Similarly, Lp-PLA2 (adjusted hazard ratio, 2.08; 95% confidence interval, 1.04–4.18) might be a stronger predictor of recurrent stroke risk.[74] Apo C1 and Apo C3 were the potential markers in blood plasma to discriminate between IS and HS.[75] In a recent study, the researchers used a reaction monitoring assay to a panel of nine apolipoproteins. They found that the plasma levels of specific apolipoproteins, including Apo C1 and Apo C3, distinguished (with high sensitivity and specificity) IS, HS, and normal patient sample groups from each other.[76]

The associations between Lp-PLA2 mass and activity with risk of dementia and its subtypes with 3320 participants of the Cardiovascular Health Study revealed that higher Lp-PLA2 mass and activity were related to increased risk of dementia and further the data supported Lp-PLA2 as a risk factor for dementia independent of CVD and its risk factors.[77] Using a case–control design, Doody and colleagues examined Lp-PLA2, homocysteine independent effects and interactions with cardiovascular disease equivalent, on AD risk. Higher Lp-PLA2 and Hcy are independently associated with AD. The association of Lp-PLA2 with AD may be mediated through vascular damage.[78]

Numerous biomarkers have been associated with CNS tissue injury, including NSE, *N*-methyl-D-aspartate receptor antibody (NMDA-R-Ab), S-100B, GFAP, and myelin basic protein (MBP). These might also be valuable in forecasting clinical consequences in patients with IS, VaD, and AD. BBB limits discharge of CNS biomarkers into systemic flow. As an outcome, biomarker levels might not associate with infarct volume or stroke severity given that the breakdown of the BBB is flexible between IS and the anatomic site of stroke and has dissimilar clinical impressions. NMDA is a glutamate-gated ion channel protein family. NMDA receptors are both ligand-gated and voltage-dependent and involve long-term potentiation, an activity-dependent increase in the efficiency of synaptic transmission that supposedly triggers definite classes of memory and learning.[79]

The diagnostic accuracy of serum auto antibodies (aAbs) to NR2A/2B, a subtype of NMDA receptors, in evaluating TIA and IS and its ability to discriminate IS from ICH in 105 TIA/stroke patients and 255 age- and sex-matched healthy controls.[80] NR2A/2B aAbs were independent and sensitive serologic markers capable of detecting TIA with a high posttest probability and, in conjunction with neurologic observation and neuroimaging, ruling out ICH. Further, they demonstrated that some NMDA receptors were able to differentiate acute IS from ICH patients.[80] In 2015, Stanca and coworkers sought to determine protein markers, using 49 subjects with IS, 23 subjects who had ICH, and 52 controls. Their data revealed that NMDA has significantly higher levels during an entire IS episode at all time points, and a quantification of NMDA in IS patients might sufficiently distinguish IS patients from ICH patients. When these researchers used NMDA in combination with GFAP, also a marker, they could differentiate between ischemic and hemorrhagic, at 12 h after stroke with a sensitivity and specificity of 94% and 91%, respectively.[81]

Serum NMDAR antibodies of IgM, IgA, or IgG subtypes were detected in 16.1% of dementia patients and in 2.8% of cognitively healthy controls. Further, serum IgA/IgM NMDAR antibodies occur in a significant number of patients with dementia.[82] Busse and colleagues examined the prevalence of NR1a NMDA-R autoantibodies in the serum and cerebrospinal fluid (CSF) of 24 patients with AD, 20 patients with subcortical ischemic vascular dementia (SIVD), and 274 volunteers without neuropsychiatric disorder. Analysis of the patient samples showed that four patients with AD and three patients with SIVD had positive NMDA-R IgM, IgG, and/or IgA autoantibody titers in serum. Further, concluded that the seroprevalence of NMDA-R-directed autoantibodies was age related.[83]

Another possible biomarker to identify the onset of IS is MBP, a hydrophilic protein found in myelin sheaths. Higher serum levels of MBP were found in a range of acute neurological disorders. A preliminary prospective cohort study to determine whether a panel of biochemical markers could distinguish acute IS cases, found elevated levels of MBP in only 39% patients, and peak level of MBP did not significantly correlate with discharge of modified ranking scale (MRS).[84] In 2006, Jauch and coworkers used an NIH stroke scale to determine stroke markers. They found that a higher 24-h peak concentration of MBP was associated with higher National Institutes of Health Stroke Scale baseline scores ($r=0.186$, $P<0.0001$) and also that MBP became elevated within the first 24 h after stroke, although they did not peak until some days after stroke.[85]

Myelin loss as one of the features of white matter abnormalities across three common dementing disorders such as VaD, AD, and dementia with Lewy bodies. This study was attested by the use of protein biomarker, suggested that myelin loss may evolve in parallel with shrunken oligodendrocytes in VaD but their increased density in AD, highlighting partially different mechanisms were associated with myelin degeneration, which could originate from hypoxic–ischemic damage to oligodendrocytes in VaD, whereas secondary to axonal degeneration in AD.[86]

S110B is an astroglial protein that has been studied as a serum marker for cerebral injury and disruption of the BBB. The quantity of S100 calcium-binding protein B (S100B) varies under normal conditions, but during an ischemic injury, S100B increases.[87] S100B has also been used as an independent predictor and diagnostic marker for stroke, VaD, and AD. Lynch and research group enrolled 65 IS patients and 157 controls and analyzed 26 blood-borne biochemical markers that were hypothesized to play a crucial role in the IS cascade. Out of these 26 markers, S100B correlated highly with stroke and with other inflammation and thrombosis biomarkers.[88] Retrospective study with 275 patients with IS (mean age 69 ± 13 years; 46% female) who had received thrombolytic therapy within 6 h of symptom onset revealed, elevated S100B serum levels before thrombolytic therapy constituted an independent risk factor for hemorrhagic transformation in patients with acute stroke.[89]

Levada and Trailin evaluated serum level of S100B in subcortical VaD ($n=11$) (SVD) and subcortical vascular mild cognitive impairment ($n=19$) (SVMCI). They found that the serum S100B level significantly increased and concluded that the serum level of S100B could be used as marker of progression SVMCI into SVD and therapy effectiveness.[90] Another study stated that the serum levels of S100B might be a marker for brain functional condition.[91]

There is extensive literature supporting the role of inflammatory responses playing a central role in IS pathogenesis. Key factors in the inflammatory responses are the transcriptional regulators, and adhesion and signaling molecules. These biomarkers are used in stroke diagnosis to differentiate clinical subtypes of stroke. TNF-α, an acute-phase protein, is involved in systemic inflammation and regulation of immune cells. Cytokines are involved in pathogenesis of IS. Furthermore, they found that TNF-α was activated in experimental ischemia. They also observed increased levels of TNF-α in patients who had been diagnosed with experimental brain ischemia.[92] A study to determine the role of TNF-α in identifying protein marker using IS, HS, and healthy controls found, plasma TNF-α levels ($P < 0.001$, $r = 0.503$, CI: 18.197–1672.950) to be significantly elevated in stroke patients, in IS and HS subtypes, indicating that TNF-α is a promising protein marker for IS and HS patients.[93]

In a recent study, TNF-α and (interleukin-6) IL6 were independently predictive of all-cause death. They found that IL-6 could become a proinflammatory cytokine and antiinflammatory cytokine. High levels of IL-6 were observed in stroke patients and served the vital role of a messenger molecule among leucocytes, the vascular endothelium, and parenchyma resident cells.[94] Another study measured serum IL-6 levels in patients with acute IS, found a significant positive correlation among IL-6, NIHSS, and MRS ($P < 0.001$, $r = 0.6$), and a significant correlation between IL-6 and infarction size, as determined by an MRI scan of the brain. They concluded that IL-6 is associated with the severity of IS as well as clinical outcomes.[95] Smith and coworkers sought to determine inflammatory response protein markers in stroke patients. They found that the concentration of IL-6 in the blood plasma significantly correlated ($P < 0.001$) with the infarct volume of CT brain infarct volume ($r = 0.75$) and MRS at 3 months poststroke ($r = 0.72$).[96] A series of studies from different populations have also revealed that IL-6 was associated with stroke.[97–99]

There is a growing evidence which supports the hypothesis of defective immune regulation and autoimmunity or inflammatory processes as viable mechanisms of the pathogenesis of AD. Cojocaru and colleagues aimed to evaluate the IL-6 level in serum of patients with AD. The results suggested that increased production of IL-6 cytokine was found in AD patients, and further it concluded that high peripheral IL-6 secretion levels might be responsible for acute-phase proteins observed in the serum of AD patients.[100] The other study documented high IL-6 plasma levels are associated with functional impairment in older patients with VaD.[101] Further, it was proved in CSF of VaD patients.[102]

Another protein that has been studied as a possible biomarker for IS is the protein C-reactive protein (CRP). CRP has already been identified as a strong biomarker for inflammation in various diseases, such as stroke, cancer, diabetes, and coronary artery disease. It is produced mainly by the liver and is regulated by inflammatory cytokines. CRP is also associated with measures of clinical stroke severity, as a major predictor of both death and functional outcomes after stroke. Prospective controlled clinical study of 200 IS patients and 50 control subjects found an association between raised levels of CRP and atherothrombotic and cardio embolic strokes, suggesting that CRP might be characteristic of both the response at the acute phase of stroke and endothelial inflammatory processes.[103] The other study concluded that the level of CRP level is a good prognostic indicator of IS patients at the time of discharge and exhibits increased utility as a biomarker to identify. Further, increased levels of CRP might predict future outcome of stroke in terms of mortality and morbidity.[104] These findings were in agreement with many other studies that sought to determine protein markers.[105–109]

In 2010, O'Bryant and coworkers evaluated CRP levels in 192 patients diagnosed with probable AD as compared to 174 nondemented controls. Mean CRP levels were found to be significantly decreased in AD vs controls. Further they have concluded that elevated CRP continues to predict increased dementia severity suggestive of a possible proinflammatory endophenotype in AD.[110] Yarchoan and colleagues measured plasma CRP in AD, MCI, and control subjects and administered annual Mini-Mental State Examinations (MMSE) during a 3-year follow-up period to investigate CRP's relationship with diagnosis and progression of cognitive decline. The results supported previous reports of reduced levels of plasma CRP in AD and indicate its potential utility as a biomarker for the diagnosis of AD.[111]

Overall, findings from the earlier studies clearly suggest that inflammatory responses can be used as biomarkers of stroke, VaD, and AD. However, further research is needed to identify precise protein markers linked to inflammatory responses, in the early stages of stroke, VaD, and AD.

4.2 miRNAs as Peripheral Biomarkers in Stroke, VaD, and AD

miRNAs are composed of a group of endogenous and small nonprotein coding genes present in virtually all animals, plants, and some viruses. miRNAs are important regulators of several biological processes, such as cell

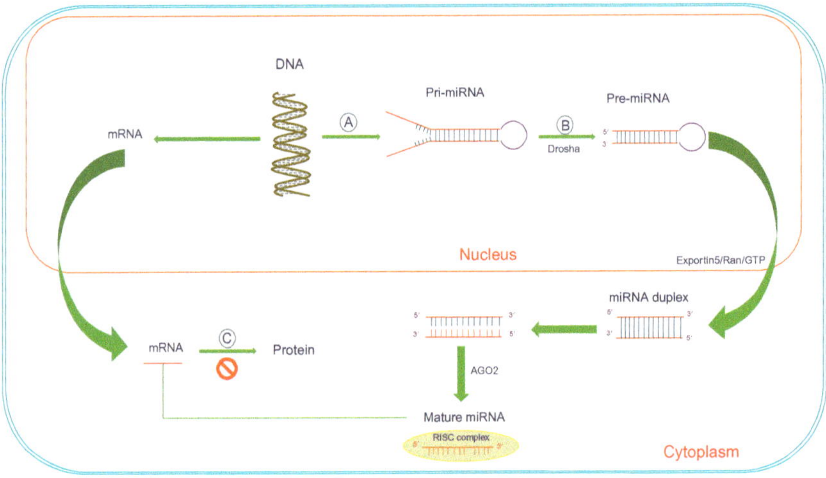

Fig. 3 MiRNA processing and function: The primary miRNA transcript (pri-miRNA) is transcribed from DNA and excised by Drosha to produce the pre-miRNA. (A) Transcription, (B) microprocessing, and (C) translation.

growth, apoptosis, cell proliferation, embryonic development, and tissue differentiation[112] (Fig. 3). These genes encode long RNAs with a hairpin structure of 17–25 nucleotides and act as an antisense regulator of other RNAs.[113] It is copied from genes that lie inside recognized exons and introns or other intergenic regions of the genome.[114] Sequence variations in miRNAs are known to alter miRNA regulations and have been associated with human disorders. miRNAs have also been found to improve the gene-regulatory processes in cerebrovascular diseases.[115] At present, the diagnosis of IS, VaD, and AD depends on the clinical examination and neuroimaging techniques. There are no reliable circulating biomarkers for acute IS, VaD, and AD risk prediction and diagnosis. IS, VaD, and AD clinical diagnosis with biomarkers should be fast, cost-effective, specific, and sensitive. The serum miRNAs have been reported to be reproducible and steady among persons.[116] miRNAs are a novel class of small, noncoding, single-stranded RNA that negatively regulates gene expression via translational inhibition or mRNA degradation followed by protein synthesis repression. miRNAs are present in serum and have the potential to serve as disease biomarkers. Increasing number of miRNAs have been proven to be critical for the pathogenesis of neurological diseases.[117] As such, it is important to explore the clinical value of miRNAs in serum as biomarkers for IS, VaD, and AD and influence on the pathogenesis of IS, VaD, and AD.[118–120] Given the structure and localization of miRNAs, it has been suggested that miRNAs might be useful in determining peripheral biomarkers and treating human diseases.[121] Many studies have shown that miRNAs altered after CNS injury moderate processes that stimulate neuronal death with inflammation, apoptosis, and oxidative stress. Furthermore, miRNAs can act as sensitive biomarkers of secondary brain damage.[122] Table 1 summarizes human studies investigating the role of miRNAs

Table 1 Overview of Circulating miRNAs in Stroke, Vascular Dementia, and Alzheimer's Disease

Diseases Type	miRNAs	Relationship	References
Stroke	miR-25, -181a, -513a-5p, -550, -602, -665, -891a, -923, -933, -939, -1184, -1246, -1261, -1275, -1285, -1290, -let-7e	↑	123
	miR-15b, -126, -142-3p, -186, -519e, -768-5p, -1259, -let-7f	↓	
	hsa-miR-1258, -125a-5p, -1260, -1273, -149, -220b, -23a, -26b, -29b-1, -302e, -488, -490-3p, -506, -659, -890, -920, -934	↑	124
	miR-25, -34b, -483-5p, miR-498	↓	
	miR-363, miR-487b	↑	125
	miR-122, -148a, -let-7i, miR-19a, -320d	↓	
	hsa-miR-106b-5P, hsa-miR-4306	↑	126
	hsa-miR-320e, hsa-miR-320d	↓	
	miR-30a, miR-126	↓	127
	miR-125b-2*, miR-27a*, miR-422a, miR-488, miR-627	↑	128
Vascular dementia and Alzheimer's disease	miR-10b*, miR-29a-3p, miR-130b-3p	↓	129
	miR-29a/b-1 cluster, miR-212/132, miR23a/b, miR-26b	↑	130–132
	miR-30a-5p, miR-206, miR-125, miR-1229-3p, miR-124, miR-29	↓	133–138

(upregulated and downregulated) in the progression of stroke, VaD, and AD patients.

4.2.1 miRNA Profile and Function in Human Samples

Recently Ragusa and colleagues analyzed the expression of miRNAs in plasma samples of patients with VaD in order to identify potential miRNA biomarker profiles to separate VaD from other types of dementia. miR-10b*, miR29a-3p, and miR-130b-3p were discovered and validated as significantly downregulated differentially expressed circulatory miRNAs in VaD patients compared to unaffected controls. These miRNAs also were found to be significantly downregulated in a matched cohort of AD patients. A negative correlation was detected between miR-29a and miR-130b expression and cognitive impairment in VaD and AD. Further, they have concluded that these miRNAs cover the way to translational applications to molecular VaD diagnosis.[129] The circulating miRNA profile expression in patients with early-onset poststroke depression study revealed that four miRNAs were upregulated (hsa-miR-22-3p, miR-4476, miR-486-5p, and miR-92a-3p), and 21 miRNAs were downregulated (hsa-miR-187-5p, 5571-5p, 4310, 3202, 133a, 548ai, 4769-5p, 4716-3p, 4738-3p, 1247-3p, 183-3p, 3615, 629-3p, 887, 3184-3p, 665, 4714-5p, 636, 1234-5p, 3667-5p, and 1185-1-3p). Using microarray analysis of hsa-miR-133, -92a-3p, and -187-5p, they validated their microarray results.[139] A research group from Denmark analyzed the expression of miRNAs in CSF and blood of patients with AD and other neurodegenerative disorders in order to identify potential miRNA biomarker candidates able to separate AD from other types of dementia. Fifty-two miRNAs were detected in CSF. Among these, two miRNAs (let-7i-5p and miR-15a-5p) were found significantly upregulated, and one miRNA (miR-29c-3p) was found significantly downregulated in patients with AD. Further, they have concluded that the deregulated miRNAs in CSF of AD patients may be associated with relevant target genes related to AD pathology, including amyloid precursor protein (APP) and β-site APP cleaving enzyme 1 (BACE1), which suggests that miRNAs are interesting candidates for AD biomarkers in the future.[140]

Several miRNAs have been studied in stroke using experimental models and small groups of patients. miRNAs related to atherosclerosis or hypoxia would be altered in the blood of acute IS patients and their initial expression levels will be able to reflect atherosclerosis activity and to predict future vascular event. A total of 120 patients were included in the study, with 83 acute stroke patients and 37 controls. They have measured five miRNAs (miR-17, miR-21, miR-106a, miR-126, and miR-200b), which had been reported to be related to atherosclerosis, in which miR-17 level was elevated in acute IS and associated with future stroke recurrence.[141] Another study screened differentially expressed serum miRNAs from IS and normal persons by miRNA microarray analysis, and validated the expression of candidate miRNAs using quantitative reverse transcriptase polymerase chain reaction assays. They have revealed that 115 miRNAs were differentially expressed in IS, among which miR-32-3p, miR-106-5p, and miR-532-5p were first found to be associated with IS and found to be a potential diagnostic biomarkers for IS.[118] Dong and coworkers examined the candidate miRNAs in the serum samples of patients with MCI and VaD. The results showed that four miRNAs (miR-31, miR-93, miR-143, and miR-146a) were markedly decreased in AD patients. MiR-31, miR-93, and miR-146a could be used to discriminate AD from VaD.[142]

Circulating miRNAs in blood plasma from subjects with acute stroke and control subjects can serve as possible biomarkers for acute stroke in humans.[126] Using miRNA microarrays and real-time PCR analyses, they found that hsa-miR-106b-5P and hsa-miR-4306 were present in high abundance in patients of acute stroke, whereas hsa-miR-320e and hsa-miR-320d were present in low abundance in control subjects. The following four miRNAs were upregulated in acute stroke patients compared to the control subjects: hsa-miR-106b-5P (3.63-fold in MRI(–) patients and 23.90-fold in MRI(+) patients), hsa-miR-4306 (3.19-fold in MRI(–) patients and 5.30-fold in MRI(+) patients), hsa-miR-320e (0.33-fold in MRI(–) patients and 0.13-fold in MRI(+) patients), and hsa-miR-320d (0.23-fold in MRI(–) patients and 0.07-fold in MRI(+) patients). Based on the upregulation of these miRNAs, Wang *et al.* suggested that circulatory miRNAs in blood plasma might be promising biomarkers for the early detection of acute stroke in humans. [126] In 2015, Denk and colleagues applied Open Array technology to profile the expression of 1178 unique miRNAs in CSF samples of AD patients ($n=22$) and controls ($n=28$). Discrimination analysis revealed that miR-100, miR-103, and miR-375 were able to detect AD in CSF by positively classifying controls, respectively. Further, they could identify a set of AD-associated genes that were targeted by these miRNAs. Highly predicted targets included genes involved in the regulation of tau and amyloid pathways in AD like MAPT, BACE1, and mTOR.[143] Genome-wide serum miRNA expression analysis was used to investigate the value of serum miRNAs as biomarkers for the diagnosis of AD. MiR-98-5p, miR-885-

5p, miR-483-3p, miR-342-3p, miR-191-5p, and miR-let-7d-5p displayed significantly different expression levels in AD patients.[144] In investigations of circulatory mRNAs in peripheral blood samples, using a customized TaqMan Low Density Array, Sepramaniam research group examined a panel of 32 miRNAs that were hypothesized to distinguish stroke etiologies during the acute phase of stroke. Five miRNAs were consistently altered in blood specimens from acute stroke patients, irrespective of age at the time of stroke, severity of the stroke, and confounding metabolic complications: miR-125b-2*, miR-125b-27a*, miR-125b-422a, miR-125b-488, and miR-125b-627. These five miRNAs were found to be possible biomarkers for diagnosis of stroke.[128]

In 2013, the Tan research group characterized miRNA profiles in patients with low/no risk IS, who did not have preexisting risk factors for stroke. They correlated the expressions of miRNAs to cerebrovascular lesions caused by cerebral ischemia. They found that 21 miRNAs exhibited similar expression levels in all IS patients (hsa-miR-1258, -125a-5p, -1260, -1273, -149, -220b, -23a*, -25*, -26b*, -29b-1*, -302e, -34b, -483-5p, -488, -490-3p, -498, -506, -659, -890, -920, and -934). Among the 21, 17 were upregulated and 4 were downregulated (miR25, 34b, 483-5p, and miR-498). They also found that miR25 was downregulated in all IS patients, and even the expression level of miR25 was found to be upregulated in their previous study,[123] suggesting that miR25 might be specific for stroke pathogenesis in low-risk stroke patients and might present a different molecular mechanism for its stroke pathogenesis.[124] Sala Frigerio and coworkers evaluated miRNAs as potential biomarkers for AD by analyzing the expression level of miRNAs in CSF of patients with AD dementia and nonaffected controls. The study further highlighted hsa-miR-27a-3p as a candidate biomarker for AD.[145]

In 2012, Gan research group also studied the role of circulatory miRNA-145 expression in IS patients ($n=32$) and 14 healthy control subjects ($n=14$) who had no identifiable risk factors for stroke and no history of cardiovascular and cerebrovascular diseases. Using TaqMan Real-Time PCR, they found that circulatory miRNA-145 expression was upregulated in the IS patients but not in the controls. This finding argues for miRNA-145 being a possible biomarker for IS and useful to elucidate mechanotransduction of IS in stroke patients.[146] MiRNA-9, miRNA-125b, miRNA-146a, and miRNA-155 were CSF abundant, NF-κB-sensitive proinflammatory miRNAs, and their enrichment in circulating CSF suggested that they might be involved in the modulation or proliferation of miRNA-triggered pathogenic signaling throughout the brain and CNS.[147]

Zeng and colleagues conducted the study to evaluate whether miRNA-210 could be a blood biomarker for acute cerebral ischemia because miRNA-210 is a master and pleiotropic hypoxia-miRNA, and it plays multiple roles in brain ischemia.[148] Using quantitative PCR, they measured miRNA-210 in blood samples from stroke patients ($n=112$) and healthy controls ($n=60$). They found that, compared to healthy controls, miRNA-210 was significantly decreased in stroke patients, especially at 7 and 14 days after stroke onset (0.56 vs 1.36; $P=0.001$, respectively, and 0.50 vs 1.36; $P=0.001$, respectively). They also found that the level of miR-210 in blood drawn from stroke patients was significantly higher than in blood samples from patients who never had a stroke. These findings suggest that miR-210 in blood samples from acute IS patients might be useful in diagnosing and prognosing stroke, and it might also be useful in predicting the response of stroke patients.

The first study to identify the expression of miRNAs in normal healthy persons ($n=5$) and in IS patients ($n=19$) conducted by Tan and colleagues.[123] They found that among the 836 miRNAs present on the array chip, 157 miRNAs were differentially regulated in the stroke subjects. Of the highly expressed miRNAs, 17 were upregulated (miR-25, -181a, -513a-5p, -550, -602, -665, -891a, -923, -933, -939, -1184, -1246, -1261, -1275, -1285, -1290, and -let-7e), and of the poorly expressed miRNAs, 8 were downregulated (miR-15b, -126, -142-3p, -186, -519e, -768-5p, -1259, and -let-7f). The researchers found that analysis of miRNA profiling revealed the following key events that occur during stroke recovery: regulation of hypoxia, angiogenesis, and erythropoiesis/hematopoiesis. They concluded that these miRNAs could be used to differentiate large artery, small artery, and cardio embolic strokes from each other.[123] The study results of miRNA expression in Alzheimer blood mononuclear cells revealed that miR-34a and 181b were significantly upregulated in AD patients. Further, induction of miRNA expression in blood mononuclear cells might contribute to the aberrant systemic decline in mRNA levels in sporadic AD.[149]

4.2.2 MiRNA Profile and Function in Animal and Postmortem Brain

Postmortem human brain tissue was being used for quantifying cellular and molecular markers of neural courses with the area of improved understanding the variations in the brain caused by neurological diseases.[150] Studies of postmortem brain tissue have previously established their effectiveness in drug finding. The majority of the animal models used that were phylogenetically separated from humans loads of years before.[151] New research techniques such as gene expression profiling and proteomics using postmortem brain tissues are providing exciting new avenues for research on human subjects.

Many studies found differentially expressed miRNAs with ischemic brain damage, which were identified using miRNA profiling techniques in mice and rat MCA occlusion (MCAO) models.[45] MiRNAs are a newly discovered group of noncoding small RNA molecules that negatively regulate target gene expression and are involved in the regulation of cell proliferation and cell apoptosis.[152]

Peng and colleagues reported for the first time that the downregulated miR-181b in N2A cells after OGD in vitro and mouse brain following the MCAO model of stroke in vivo induces neuroprotection against ischemic injury through upregulating HSPA5 and UCHL1 protein levels. In addition, they found that miR-181b could be a potential therapeutic target for the treatment of IS. Further, they have concluded that the experiments using neuron-specific miR-181b transgenic and knockout animal models are needed for validation of using miR-181b in treating ischemic brain damage.[153]

The mechanisms of neuroprotection induced by fastigial nucleus stimulation (FNS) were not entirely understood. MiR-29c was decreased after FNS, and it attenuates ischemic neuronal apoptosis by negatively regulating apoptotic proteins Birc2 and Bak1. Therefore, miR-29c might be involved in apoptosis processes of neuroprotection induced by FNS in stroke.[154] MiR-223 is also expressed in the nervous system and controls the expression and function of GluR2 and NR2B subunits of the glutamate receptor. MiR-223 is a neuroprotective miRNA using in vivo and in vitro models of ischemic reperfusion brain injury and excitotoxic neuronal death. Further, concluded that a therapeutic role for miR-223 in stroke and other excitotoxic neuronal disorders.[155]

Liu and coworkers investigated the role of miR-424 in transient cerebral ischemia in mice with a focus on oxidative stress-induced neuronal injury. MiR-424 levels in the periinfarct cortex increased at 1 and 4 h then decreased at 24 h after reperfusion. Further, they have concluded that miR-424 protects against transient cerebral ischemia/reperfusion injury by inhibiting oxidative stress.[156]

A group of researchers from China showed that miR-134 expression levels increased in mouse brain from 12 h to 7 days reoxygenation/reperfusion after 1 h MCAO treatment. MiR-134 overexpression endorsed neuronal cell death and apoptosis by decreasing HSPA12B protein levels. Conclusively, miR-134 might impact neuronal cell survival against ischemic injury in mouse brain with IS by negatively modulating HSPA12B protein expression in a posttranscriptional manner.[45]

Neuroprotective effect of miRNAs in stroke and to set up a valid therapeutical approach able to contrast the role of specific miRNAs that regulate NCX expression under experimental conditions imitating stroke. NCX1 physiological expression was dramatically reduced when cortical neurons were treated with mir-103-1. AntimiR-103-1 prevented NCX1 protein downregulation induced by the increase in miR-103-1 after brain ischemia, thereby reducing brain damage and neurological deficits. Further, they concluded that blocking mir-103-1 by miRNA inhibitors was a reasonable strategy to stop neurodetrimental regulation of NCX occurring during ischemic conditions.[157]

Several studies reported that miRNAs can regulate APP and BACE1 expressions. Ai and coworkers in 2013 evaluated the effect of miRNA on memory impairment in rats induced by chronic brain hypoperfusion (CBH). qRT-PCR analysis showed that miR-195 was downregulated in both the hippocampus and cortex of rats following CBH, and in the plasma of dementia patients. APP and BACE1 proteins were downregulated by miR-195 overexpression, upregulated by miR-195 inhibition, and unchanged by binding-site mutation or miR-masks, indicating that APP and BACE1 were two potential targets for miR-195 and might be a potentially valuable antidementia approach.[158]

The p35/CDK5 active complex plays a fundamental role in brain development and functioning, but its deregulated activity has also been implicated in various neurodegenerative disorders, including AD. Downregulation of the miR-15/107 family might have a role in the pathogenesis of AD by increasing the levels of CDK5R1/p35 and consequently enhancing CDK5 activity.[159] Upregulation of miR-26b in neurons causes pleiotropic phenotypes that are also observed in AD. Elevated levels of miR-26b might contribute to the AD neuronal pathology.[130] Neuropeptide Y modulates brain-derived neurotrophic factor and its regulating miRNA miR-30a-5p in opposite direction with a mechanism that possibly contributes to the neuroprotective effect of NPY in rat cortical neurons exposed to Aβ.[133] Increased expression of miR-34a gene and miR 34a-mediated concurrent repression of its target genes in neural networks might result in dysfunction of synaptic plasticity, energy metabolism, and resting state network activity.[160]

Overall, findings from the earlier studies are useful and provide new information about circulating miRNAs, indicating that miRNAs are potential peripheral biomarkers for stroke, VaD, and AD.

5. CONCLUDING REMARKS

In the last few years, noteworthy development has been made in understanding the pathophysiology that triggers stroke and its related disorders. Cerebral abnormalities in the stroke, particularly IS, may lead to biochemical dysfunction in the brain, ultimately leading to VaD and AD. In this chapter, we have described in detail about the molecule links and molecular biomarkers for stroke, VaD, and AD. Multiple approaches have been developed to identify biomarkers, including circulatory miRNAs, blood-based protein markers, coagulation, and thrombosis biomarkers. Among these, circulatory miRNAs are reported to be promising peripheral biomarkers in stroke and stroke-linked VaD and AD. Although much research has been done on IS and its molecular and cellular links with VaD (1) we still do not know whether stroke-associated circulatory miRNAs can be used for VaD and AD, (2) we still do not have complete understanding of the genetic basis of IS leading to VaD and AD, and (3) we still do not know for sure but this is the clearest mechanism linked with stroke–VaD–AD. Further research is needed to answer these important questions.

ACKNOWLEDGMENTS

P.H.R., Ph.D. is supported by NIH Grants—AG042178 and AG47812, and the Garrison Family Foundation.

REFERENCES

1. WHO MONICA Project. The World Health Organization MONICA Project (monitoring trends and determinants in cardiovascular disease): a major international collaboration. WHO MONICA Project Principal Investigators. *J Clin Epidemiol.* 1988;41(2):105–114.
2. Vijayan M, Chinniah R, Ravi PM, et al. ACE-II genotype and I allele predicts ischemic stroke among males in south India. *Meta Gene.* 2014;2:661–669.
3. Murali V, Rathika C, Ramgopal S, et al. Susceptible and protective associations of HLA DRB1*/DQB1* alleles and haplotypes with ischaemic stroke. *Int J Immunogenet.* 2016;43(3):159–165.
4. Vijayan M, Chinniah R, Ravi PM, et al. MTHFR (C677T) CT genotype and CT-apoE3/3 genotypic combination predisposes the risk of ischemic stroke. *Gene.* 2016;591(2):465–470.
5. Rosamond W, Flegal K, Furie K, et al. Heart disease and stroke statistics—2008 update: a report from the American Heart Association statistics committee and stroke statistics subcommittee. *Circulation.* 2008;117(4):e25–e146.

6. Adams Jr HP, Bendixen BH, Kappelle LJ, et al. Classification of subtype of acute ischemic stroke. Definitions for use in a multicenter clinical trial. TOAST. Trial of Org 10172 in acute stroke treatment. *Stroke*. 1993;24(1):35–41.
7. Feigin VL, Lawes CM, Bennett DA, Barker-Collo SL, Parag V. Worldwide stroke incidence and early case fatality reported in 56 population-based studies: a systematic review. *Lancet Neurol*. 2009;8(4):355–369.
8. Lloyd-Jones D, Adams R, Carnethon M, et al. Heart disease and stroke statistics—2009 update: a report from the American Heart Association statistics committee and stroke statistics subcommittee. *Circulation*. 2009;119(3):e21–e181.
9. Smith SD, Eskey CJ. Hemorrhagic stroke. *Radiol Clin North Am*. 2011;49(1):27–45.
10. Lozano R, Naghavi M, Foreman K, et al. Global and regional mortality from 235 causes of death for 20 age groups in 1990 and 2010: a systematic analysis for the Global Burden of Disease Study 2010. *Lancet*. 2012;380(9859):2095–2128.
11. Murray CJ, Vos T, Lozano R, et al. Disability-adjusted life years (DALYs) for 291 diseases and injuries in 21 regions, 1990–2010: a systematic analysis for the Global Burden of Disease Study 2010. *Lancet*. 2012;380(9859):2197–2223.
12. WHO. *The Atlas of Heart Disease and Stroke*. 2011. Retrieved 19.10.2011 from, http://www.who.int/cardiovascular_diseases/resources/atlas/en/.
13. Banarjee TK, Roy MK, Bhoi KK. Is stroke increasing in India—preventive measures that need to be implemented. *J Indian Med Assoc*. 2005;103:162–166.
14. Alzheimer's Disease International. *World Alzheimer Report (2015)*. London: Alzheimer's Disease International; 2015.
15. American Psychiatric Association. *Diagnostic and Statistical Manual of Mental Disorders*. 4th ed. text revision. Washington, DC: American Psychiatric Association; 2000.
16. Bozoki A, Giordani B, Heidebrink JL, Berent S, Foster NL. Mild cognitive impairments predict dementia in nondemented elderly patients with memory loss. *Arch Neurol*. 2001;58(3):411–416.
17. Steinberg M, Sheppard JM, Tschanz JT, et al. The incidence of mental and behavioral disturbances in dementia: the cache county study. *J Neuropsychiatry Clin Neurosci*. 2003;15(3):340–345.
18. Fernando MS, Ince PG, Function MRCC. Ageing Neuropathology Study G. Vascular pathologies and cognition in a population-based cohort of elderly people. *J Neurol Sci*. 2004;226(1–2):13–17.
19. Gorelick PB, Scuteri A, Black SE, et al. Vascular contributions to cognitive impairment and dementia: a statement for healthcare professionals from the American heart association/American stroke association. *Stroke*. 2011;42(9):2672–2713.
20. Brayne C, Richardson K, Matthews FE, et al. Neuropathological correlates of dementia in over-80-year-old brain donors from the population-based Cambridge city over-75 s cohort (CC75C) study. *J Alzheimers Dis*. 2009;18(3):645–658.
21. Selkoe DJ. Alzheimer's disease: genes, proteins, and therapy. *Physiol Rev*. 2001;81(2):741–766.
22. Mattson MP. Pathways towards and away from Alzheimer's disease. *Nature*. 2004;430(7000):631–639.
23. LaFerla FM, Green KN, Oddo S. Intracellular amyloid-beta in Alzheimer's disease. *Nat Rev Neurosci*. 2007;8(7):499–509.
24. Reddy PH, Manczak M, Mao P, Calkins MJ, Reddy AP, Shirendeb U. Amyloid-beta and mitochondria in aging and Alzheimer's disease: implications for synaptic damage and cognitive decline. *J Alzheimers Dis*. 2010;20(suppl 2):S499–S512.
25. Anand R, Gill KD, Mahdi AA. Therapeutics of Alzheimer's disease: past, present and future. *Neuropharmacology*. 2014;76(pt A):27–50.
26. Duthey B. *Background Paper 6.11. Alzheimer Disease and Other Dementias;* 2013. Available from: http://www.who.int/medicines/areas/priority_medicines/BP6_11Alzheimer.pdf. Accessed May 15, 2014.

27. Reitz C, Mayeux R. Alzheimer disease: epidemiology, diagnostic criteria, risk factors and biomarkers. *Biochem Pharmacol.* 2014;88(4):640–651.
28. Goldstein LB, Adams R, Becker K, et al. Primary prevention of ischemic stroke: a statement for healthcare professionals from the Stroke Council of the American Heart Association. *Stroke.* 2001;32(1):280–299.
29. Fagius J, Aquilonius SM. Neurologi. 4 suppl. [Neurology 4 issue] Stockholm: Liber AB; 2006:258–76.
30. Mozaffarian D, Benjamin EJ, Go AS, et al. Heart disease and stroke statistics—2015 update: a report from the American Heart Association. *Circulation.* 2015;131(4): e29–e322.
31. Mielke MM, Vemuri P, Rocca WA. Clinical epidemiology of Alzheimer's disease: assessing sex and gender differences. *Clin Epidemiol.* 2014;6:37–48.
32. Kirshner HS. Differentiating ischemic stroke subtypes: risk factors and secondary prevention. *J Neurol Sci.* 2009;279(1–2):1–8.
33. Bao DQ, Mori TA, Burke V, Puddey IB, Beilin LJ. Effects of dietary fish and weight reduction on ambulatory blood pressure in overweight hypertensives. *Hypertension.* 1998;32(4):710–717.
34. Hackam DG, Spence JD. Combining multiple approaches for the secondary prevention of vascular events after stroke: a quantitative modeling study. *Stroke.* 2007;38(6):1881–1885.
35. Dilworth-Anderson P, Hendrie HC, Manly JJ, et al. Diagnosis and assessment of Alzheimer's disease in diverse populations. *Alzheimers Dement.* 2008;4(4):305–309.
36. Xu WL, Qiu CX, Wahlin A, Winblad B, Fratiglioni L. Diabetes mellitus and risk of dementia in the Kungsholmen project: a 6-year follow-up study. *Neurology.* 2004;63(7):1181–1186.
37. Dickinson CJ. Strokes and their relationship to hypertension. *Curr Opin Nephrol Hypertens.* 2003;12(1):91–96.
38. National Stroke Association 2012: Available from http://www.stroke.org/site/PageServer?pagename=aamer, Accessed June 28, 2012
39. Kodl CT, Seaquist ER. Cognitive dysfunction and diabetes mellitus. *Endocr Rev.* 2008;29(4):494–511.
40. Mukherjee D, Chirag PG. Epidemiology and the global burden of stroke. *World Neurosurg.* 2011;76:S85–S90.
41. Ballard C, Gauthier S, Corbett A, Brayne C, Aarsland D, Jones E. Alzheimer's disease. *Lancet.* 2011;377(9770):1019–1031.
42. Seshadri S, Wolf PA. Lifetime risk of stroke and dementia: current concepts, and estimates from the Framingham Study. *Lancet Neurol.* 2007;6(12):1106–1114.
43. Ferri CP, Prince M, Brayne C, et al. Global prevalence of dementia: a Delphi consensus study. *Lancet.* 2005;366(9503):2112–2117.
44. Henon H, Pasquier F, Durieu I, et al. Preexisting dementia in stroke patients. Baseline frequency, associated factors, and outcome. *Stroke.* 1997;28(12):2429–2436.
45. Chi NF, Chien LN, Ku HL, Hu CJ, Chiou HY. Alzheimer disease and risk of stroke: a population-based cohort study. *Neurology.* 2013;80(8):705–711.
46. Tolppanen AM, Lavikainen P, Solomon A, Kivipelto M, Soininen H, Hartikainen S. Incidence of stroke in people with Alzheimer disease: a national register-based approach. *Neurology.* 2013;80(4):353–358.
47. Gamaldo A, Moghekar A, Kilada S, Resnick SM, Zonderman AB, O'Brien R. Effect of a clinical stroke on the risk of dementia in a prospective cohort. *Neurology.* 2006;67(8):1363–1369.
48. Regan C, Katona C, Walker Z, Hooper J, Donovan J, Livingston G. Relationship of vascular risk to the progression of Alzheimer disease. *Neurology.* 2006;67(8):1357–1362.

49. Snowdon DA, Nun S. Healthy aging and dementia: findings from the Nun Study. *Ann Intern Med*. 2003;139(5 pt 2):450–454.
50. Snowdon DA, Greiner LH, Mortimer JA, Riley KP, Greiner PA, Markesbery WR. Brain infarction and the clinical expression of Alzheimer disease. The Nun Study. *JAMA*. 1997;277(10):813–817.
51. Esiri MM, Nagy Z, Smith MZ, Barnetson L, Smith AD. Cerebrovascular disease and threshold for dementia in the early stages of Alzheimer's disease. *Lancet*. 1999;354(9182):919–920.
52. Selnes OA, Vinters HV. Vascular cognitive impairment. *Nat Clin Pract Neurol*. 2006;2(10):538–547.
53. Pendlebury ST, Rothwell PM. Prevalence, incidence, and factors associated with pre-stroke and post-stroke dementia: a systematic review and meta-analysis. *Lancet Neurol*. 2009;8(11):1006–1018.
54. Allan LM, Rowan EN, Firbank MJ, et al. Long term incidence of dementia, predictors of mortality and pathological diagnosis in older stroke survivors. *Brain*. 2011;134(pt 12): 3716–3727.
55. Schaapsmeerders P, Maaijwee NA, van Dijk EJ, et al. Long-term cognitive impairment after first-ever ischemic stroke in young adults. *Stroke*. 2013;44(6): 1621–1628.
56. Toledo JB, Arnold SE, Raible K, et al. Contribution of cerebrovascular disease in autopsy confirmed neurodegenerative disease cases in the National Alzheimer's Coordinating Centre. *Brain*. 2013;136(pt 9):2697–2706.
57. Kalaria RN, Akinyemi R, Ihara M. Does vascular pathology contribute to Alzheimer changes? *J Neurol Sci*. 2012;322(1–2):141–147.
58. Schneider JA, Arvanitakis Z, Bang W, Bennett DA. Mixed brain pathologies account for most dementia cases in community-dwelling older persons. *Neurology*. 2007;69(24):2197–2204.
59. Rizzi L, Rosset I, Roriz-Cruz M. Global epidemiology of dementia: Alzheimer's and vascular types. *Biomed Res Int*. 2014;2014:908915.
60. Wen Y, Yu WH, Maloney B, et al. Transcriptional regulation of beta-secretase by p25/cdk5 leads to enhanced amyloidogenic processing. *Neuron*. 2008;57(5):680–690.
61. Honig LS, Tang MX, Albert S, et al. Stroke and the risk of Alzheimer disease. *Arch Neurol*. 2003;60(12):1707–1712.
62. Vijayan M, Reddy PH. Stroke, vascular dementia, and Alzheimer's disease: molecular links. *J Alzheimers Dis*. 2016;54(2):427–443.
63. Doyle KP, Simon RP, Stenzel-Poore MP. Mechanisms of ischemic brain damage. *Neuropharmacology*. 2008;55(3):310–318.
64. Alvarez-Llamas G, de la Cuesta F, Barderas ME, Darde V, Padial LR, Vivanco F. Recent advances in atherosclerosis-based proteomics: new biomarkers and a future perspective. *Expert Rev Proteomics*. 2008;5(5):679–691.
65. Collaboration FS. Plasma fibrinogen level and the risk of major cardiovascular diseases and nonvascular mortality, an individual participant meta-analysis. *JAMA*. 2005;294:1799–1809.
66. Chuang SY, Bai CH, Chen WH, Lien LM, Pan WH. Fibrinogen independently predicts the development of ischemic stroke in a Taiwanese population: CVDFACTS study. *Stroke*. 2009;40(5):1578–1584.
67. Rothwell PM, Howard SC, Power DA, et al. Fibrinogen concentration and risk of ischemic stroke and acute coronary events in 5113 patients with transient ischemic attack and minor ischemic stroke. *Stroke*. 2004;35(10):2300–2305.
68. Di Napoli M, Papa F, Bocola V. Prognostic influence of increased C-reactive protein and fibrinogen levels in ischemic stroke. *Stroke*. 2001;32(1):133–138.

69. van Oijen M, Witteman JC, Hofman A, Koudstaal PJ, Breteler MM. Fibrinogen is associated with an increased risk of Alzheimer disease and vascular dementia. *Stroke*. 2005;36(12):2637–2641.
70. Xu G, Zhang H, Zhang S, Fan X, Liu X. Plasma fibrinogen is associated with cognitive decline and risk for dementia in patients with mild cognitive impairment. *Int J Clin Pract*. 2008;62(7):1070–1075.
71. Cortes-Canteli M, Paul J, Norris EH, et al. Fibrinogen and beta-amyloid association alters thrombosis and fibrinolysis: a possible contributing factor to Alzheimer's disease. *Neuron*. 2010;66(5):695–709.
72. Ballantyne CM, Hoogeveen RC, Bang H, et al. Lipoprotein-associated phospholipase A2, high-sensitivity C-reactive protein, and risk for incident ischemic stroke in middle-aged men and women in the Atherosclerosis Risk in Communities (ARIC) study. *Arch Intern Med*. 2005;165(21):2479–2484.
73. Oei HH, van der Meer IM, Hofman A, et al. Lipoprotein-associated phospholipase A2 activity is associated with risk of coronary heart disease and ischemic stroke: the Rotterdam Study. *Circulation*. 2005;111(5):570–575.
74. Elkind MS, Tai W, Coates K, Paik MC, Sacco RL. High-sensitivity C-reactive protein, lipoprotein-associated phospholipase A2, and outcome after ischemic stroke. *Arch Intern Med*. 2006;166(19):2073–2080.
75. Allard L, Lescuyer P, Burgess J, et al. ApoC-I and ApoC-III as potential plasmatic markers to distinguish between ischemic and hemorrhagic stroke. *Proteomics*. 2004;4(8):2242–2251.
76. Lopez MF, Sarracino DA, Prakash A, et al. Discrimination of ischemic and hemorrhagic strokes using a multiplexed, mass spectrometry-based assay for serum apolipoproteins coupled to multi-marker ROC algorithm. *Proteomics Clin Appl*. 2012;6(3–4):190–200.
77. Fitzpatrick AL, Irizarry MC, Cushman M, Jenny NS, Chi GC, Koro C. Lipoprotein-associated phospholipase A2 and risk of dementia in the Cardiovascular Health Study. *Atherosclerosis*. 2014;235(2):384–391.
78. Doody RS, Demirovic J, Ballantyne CM, et al. Lipoprotein-associated phospholipase A2, homocysteine, and Alzheimer's disease. *Alzheimers Dement (Amst)*. 2015;1(4):464–471.
79. Luscher C, Malenka RC. NMDA receptor-dependent long-term potentiation and long-term depression (LTP/LTD). *Cold Spring Harb Perspect Biol*. 2012;4(6). pii: a005710. http://dx.doi.org/10.1101/cshperspect.a005710, PMID: 22510460, PMCID: PMC3367554.
80. Dambinova SA, Khounteev GA, Izykenova GA, Zavolokov IG, Ilyukhina AY, Skoromets AA. Blood test detecting autoantibodies to N-methyl-D-aspartate neuroreceptors for evaluation of patients with transient ischemic attack and stroke. *Clin Chem*. 2003;49(10):1752–1762.
81. Stanca DM, Marginean IC, Soritau O, et al. GFAP and antibodies against NMDA receptor subunit NR2 as biomarkers for acute cerebrovascular diseases. *J Cell Mol Med*. 2015;19(9):2253–2261.
82. Doss S, Wandinger KP, Hyman BT, et al. High prevalence of NMDA receptor IgA/IgM antibodies in different dementia types. *Ann Clin Transl Neurol*. 2014;1(10):822–832.
83. Busse S, Busse M, Brix B, et al. Seroprevalence of N-methyl-D-aspartate glutamate receptor (NMDA-R) autoantibodies in aging subjects without neuropsychiatric disorders and in dementia patients. *Eur Arch Psychiatry Clin Neurosci*. 2014;264(6):545–550.
84. Hill MD, Jackowski G, Bayer N, Lawrence M, Jaeschke R. Biochemical markers in acute ischemic stroke. *CMAJ*. 2000;162(8):1139–1140.
85. Jauch EC, Lindsell C, Broderick J, et al. Association of serial biochemical markers with acute ischemic stroke: the National Institute of Neurological Disorders and Stroke

recombinant tissue plasminogen activator Stroke Study. *Stroke*. 2006;37(10): 2508–2513.
86. Ihara M, Polvikoski TM, Hall R, et al. Quantification of myelin loss in frontal lobe white matter in vascular dementia, Alzheimer's disease, and dementia with Lewy bodies. *Acta Neuropathol*. 2010;119(5):579–589.
87. van Munster BC, Korse CM, de Rooij SE, Bonfrer JM, Zwinderman AH, Korevaar JC. Markers of cerebral damage during delirium in elderly patients with hip fracture. *BMC Neurol*. 2009;9:21.
88. Lynch JR, Blessing R, White WD, Grocott HP, Newman MF, Laskowitz DT. Novel diagnostic test for acute stroke. *Stroke*. 2004;35(1):57–63.
89. Foerch C, Wunderlich MT, Dvorak F, et al. Elevated serum S100B levels indicate a higher risk of hemorrhagic transformation after thrombolytic therapy in acute stroke. *Stroke*. 2007;38(9):2491–2495.
90. Levada OA, Trailin AV. Serum level of S100B as a marker of progression of vascular mild cognitive impairment into subcortical vascular dementia and therapy effectiveness. *Lik Sprava*. 2012;(3–4):53–59, Article in Ukrainian. PMID: 23356138.
91. Chaves ML, Camozzato AL, Ferreira ED, et al. Serum levels of S100B and NSE proteins in Alzheimer's disease patients. *J Neuroinflammation*. 2010;7:6.
92. Tuttolomondo A, Di Raimondo D, di Sciacca R, Pinto A, Licata G. Inflammatory cytokines in acute ischemic stroke. *Curr Pharm Des*. 2008;14(33):3574–3589.
93. Bokhari FA, Shakoori TA, Butt A, Ghafoor F. TNF-alpha: a risk factor for ischemic stroke. *J Ayub Med Coll Abbottabad*. 2014;26(2):111–114.
94. Greisenegger S, Segal HC, Burgess AI, Poole DL, Mehta Z, Rothwell PM. Biomarkers and mortality after transient ischemic attack and minor ischemic stroke: population-based study. *Stroke*. 2015;46(3):659–666.
95. Shaafi S, Sharifipour E, Rahmanifar R, et al. Interleukin-6, a reliable prognostic factor for ischemic stroke. *Iran J Neurol*. 2014;13(2):70–76.
96. Smith CJ, Emsley HC, Gavin CM, et al. Peak plasma interleukin-6 and other peripheral markers of inflammation in the first week of ischaemic stroke correlate with brain infarct volume, stroke severity and long-term outcome. *BMC Neurol*. 2004;4:2.
97. Sotgiu S, Zanda B, Marchetti B, et al. Inflammatory biomarkers in blood of patients with acute brain ischemia. *Eur J Neurol*. 2006;13(5):505–513.
98. Montaner J, Rovira A, Molina CA, et al. Plasmatic level of neuroinflammatory markers predict the extent of diffusion-weighted image lesions in hyperacute stroke. *J Cereb Blood Flow Metab*. 2003;23(12):1403–1407.
99. Vila N, Castillo J, Davalos A, Chamorro A. Proinflammatory cytokines and early neurological worsening in ischemic stroke. *Stroke*. 2000;31(10):2325–2329.
100. Cojocaru IM, Cojocaru M, Miu G, Sapira V. Study of interleukin-6 production in Alzheimer's disease. *Rom J Intern Med*. 2011;49(1):55–58.
101. Zuliani G, Guerra G, Ranzini M, et al. High interleukin-6 plasma levels are associated with functional impairment in older patients with vascular dementia. *Int J Geriatr Psychiatry*. 2007;22(4):305–311.
102. Wada-Isoe K, Wakutani Y, Urakami K, Nakashima K. Elevated interleukin-6 levels in cerebrospinal fluid of vascular dementia patients. *Acta Neurol Scand*. 2004;110(2): 124–127.
103. Alvarez-Perez FJ, Castelo-Branco M, Alvarez-Sabin J. Usefulness of measurement of fibrinogen, D-dimer, D-dimer/fibrinogen ratio, C reactive protein and erythrocyte sedimentation rate to assess the pathophysiology and mechanism of ischaemic stroke. *J Neurol Neurosurg Psychiatry*. 2011;82(9):986–992.
104. Singh VK, Haria JM, Jain SK. C-reactive protein in ischemic stroke-an experimental study. *Int J Sci Stud*. 2014;2:25–27.

105. Curb JD, Abbott RD, Rodriguez BL, et al. C-reactive protein and the future risk of thromboembolic stroke in healthy men. *Circulation*. 2003;107(15):2016–2020.
106. Arenillas JF, Alvarez-Sabin J, Molina CA, et al. C-reactive protein predicts further ischemic events in first-ever transient ischemic attack or stroke patients with intracranial large-artery occlusive disease. *Stroke*. 2003;34(10):2463–2468.
107. Chaudhuri JR, Mridula KR, Umamahesh M, Swathi A, Balaraju B, Bandaru VC. High sensitivity C-reactive protein levels in acute ischemic stroke and subtypes: a study from a tertiary care center. *Iran J Neurol*. 2013;12(3):92–97.
108. Rost NS, Wolf PA, Kase CS, et al. Plasma concentration of C-reactive protein and risk of ischemic stroke and transient ischemic attack: the Framingham study. *Stroke*. 2001;32(11):2575–2579.
109. Di Napoli M, Schwaninger M, Cappelli R, et al. Evaluation of C-reactive protein measurement for assessing the risk and prognosis in ischemic stroke: a statement for health care professionals from the CRP Pooling Project members. *Stroke*. 2005;36(6): 1316–1329.
110. O'Bryant SE, Waring SC, Hobson V, et al. Decreased C-reactive protein levels in Alzheimer disease. *J Geriatr Psychiatry Neurol*. 2010;23(1):49–53.
111. Yarchoan M, Louneva N, Xie SX, et al. Association of plasma C-reactive protein levels with the diagnosis of Alzheimer's disease. *J Neurol Sci*. 2013;333(1–2):9–12.
112. Esquela-Kerscher A, Slack FJ. Oncomirs—microRNAs with a role in cancer. *Nat Rev Cancer*. 2006;6(4):259–269.
113. Ambros V. microRNAs: tiny regulators with great potential. *Cell*. 2001;107(7): 823–826.
114. Vemuganti R. The MicroRNAs and stroke: no need to be coded to be counted. *Transl Stroke Res*. 2010;1(3):158–160.
115. Li J, Liu Y, Xin X, et al. Evidence for positive selection on a number of MicroRNA regulatory interactions during recent human evolution. *PLoS Genet*. 2012;8(3): e1002578.
116. Chen X, Ba Y, Ma L, et al. Characterization of microRNAs in serum: a novel class of biomarkers for diagnosis of cancer and other diseases. *Cell Res*. 2008;18(10):997–1006.
117. Sano T, Reynolds JP, Jimenez-Mateos EM, Matsushima S, Taki W, Henshall DC. MicroRNA-34a upregulation during seizure-induced neuronal death. *Cell Death Dis*. 2012;3:e287.
118. Li P, Teng F, Gao F, Zhang M, Wu J, Zhang C. Identification of circulating microRNAs as potential biomarkers for detecting acute ischemic stroke. *Cell Mol Neurobiol*. 2015;35(3):433–447.
119. Kumar S, Reddy PH. Are circulating microRNAs peripheral biomarkers for Alzheimer's disease? *Biochim Biophys Acta*. 2016;1862(9):1617–1627.
120. Vijayan M, Reddy PH. Peripheral biomarkers of stroke: focus on circulatory microRNAs. *Biochim Biophys Acta*. 2016;1862(10):1984–1993.
121. Weidhaas J. Using microRNAs to understand cancer biology. *Lancet Oncol*. 2010;11(2):106–107.
122. Vemuganti R. All's well that transcribes well: non-coding RNAs and post-stroke brain damage. *Neurochem Int*. 2013;63(5):438–449.
123. Tan KS, Armugam A, Sepramaniam S, et al. Expression profile of MicroRNAs in young stroke patients. *PLoS One*. 2009;4(11):e7689
124. Tan JR, Tan KS, Koo YX, et al. Blood microRNAs in low or no risk ischemic stroke patients. *Int J Mol Sci*. 2013;14(1):2072–2084.
125. Jickling GC, Ander BP, Zhan X, Noblett D, Stamova B, Liu D. microRNA expression in peripheral blood cells following acute ischemic stroke and their predicted gene targets. *PLoS One*. 2014;9:e99283.

126. Wang W, Sun G, Zhang L, Shi L, Zeng Y. Circulating microRNAs as novel potential biomarkers for early diagnosis of acute stroke in humans. *J Stroke Cerebrovasc Dis.* 2014;23(10):2607–2613.
127. Long G, Wang F, Li H, et al. Circulating miR-30a, miR-126 and let-7b as biomarker for ischemic stroke in humans. *BMC Neurol.* 2013;13:178.
128. Sepramaniam S, Tan JR, Tan KS, et al. Circulating microRNAs as biomarkers of acute stroke. *Int J Mol Sci.* 2014;15(1):1418–1432.
129. Ragusa M, Bosco P, Tamburello L, et al. MiRNAs plasma profiles in vascular dementia: biomolecular data and biomedical implications. *Front Cell Neurosci.* 2016;10:51.
130. Absalon S, Kochanek DM, Raghavan V, Krichevsky AM. MiR-26b, upregulated in Alzheimer's disease, activates cell cycle entry, tau-phosphorylation, and apoptosis in postmitotic neurons. *J Neurosci.* 2013;33(37):14645–14659.
131. Hébert SS, Horré K, Nicolaï L, et al. Loss of microRNA cluster miR-29a/b-1 in sporadic Alzheimer's disease correlates with increased BACE1/beta-secretase expression. *Proc Natl Acad Sci USA.* 2008;105(17):6415–6420.
132. Weinberg RB, Mufson EJ, Counts SE. Evidence for a neuroprotective microRNA pathway in amnestic mild cognitive impairment. *Front Neurosci.* 2015;9:430.
133. Croce N, Gelfo F, Ciotti MT, et al. NPY modulates miR-30a-5p and BDNF in opposite direction in an in vitro model of Alzheimer disease: a possible role in neuroprotection? *Mol Cell Biochem.* 2013;376(1–2):189–195.
134. Tian N, Cao Z, Zhang Y. MiR-206 decreases brain-derived neurotrophic factor levels in a transgenic mouse model of Alzheimer's disease. *Neurosci Bull.* 2014;30(2):191–197.
135. Banzhaf-Strathmann J, Benito E, May S, et al. MicroRNA-125b induces tau hyperphosphorylation and cognitive deficits in Alzheimer's disease. *EMBO J.* 2014;33(15):1667–1680.
136. Ghanbari M, Ikram MA, de Looper HW, et al. Genome-wide identification of microRNA-related variants associated with risk of Alzheimer's disease. *Sci Rep.* 2016;6:28387.
137. Smith P, Al Hashimi A, Girard J, Delay C, Hébert SS. In vivo regulation of amyloid precursor protein neuronal splicing by microRNAs. *J Neurochem.* 2011;116(2): 240–247.
138. Pereira PA, Tomás JF, Queiroz JA, Figueiras AR, Sousa F. Recombinant pre-miR-29b for Alzheimer's disease therapeutics. *Sci Rep.* 2016;6:19946.
139. Zhang Y, Cheng L, Chen Y, Yang GY, Liu J, Zeng L. Clinical predictor and circulating microRNA profile expression in patients with early onset post-stroke depression. *J Affect Disord.* 2016;193:51–58.
140. Sorensen SS, Nygaard AB, Christensen T. miRNA expression profiles in cerebrospinal fluid and blood of patients with Alzheimer's disease and other types of dementia—an exploratory study. *Transl Neurodegener.* 2016;5:6.
141. Kim JM, Jung KH, Chu K, et al. Atherosclerosis-related circulating MicroRNAs as a predictor of stroke recurrence. *Transl Stroke Res.* 2015;6(3):191–197.
142. Dong H, Li J, Huang L, et al. Serum MicroRNA profiles serve as novel biomarkers for the diagnosis of Alzheimer's disease. *Dis Markers.* 2015;2015:625659.
143. Denk J, Boelmans K, Siegismund C, Lassner D, Arlt S, Jahn H. MicroRNA profiling of CSF reveals potential biomarkers to detect Alzheimer's disease. *PLoS One.* 2015;10(5), e0126423
144. Tan L, Yu JT, Tan MS, et al. Genome-wide serum microRNA expression profiling identifies serum biomarkers for Alzheimer's disease. *J Alzheimers Dis.* 2014;40(4): 1017–1027.
145. Sala Frigerio C, Lau P, Salta E, et al. Reduced expression of hsa-miR-27a-3p in CSF of patients with Alzheimer disease. *Neurology.* 2013;81(24):2103–2106.

146. Gan CS, Wang CW, Tan KS. Circulatory microRNA-145 expression is increased in cerebral ischemia. *Genet Mol Res*. 2012;11(1):147–152.
147. Alexandrov PN, Dua P, Hill JM, Bhattacharjee S, Zhao Y, Lukiw WJ. microRNA (miRNA) speciation in Alzheimer's disease (AD) cerebrospinal fluid (CSF) and extracellular fluid (ECF). *Int J Biochem Mol Biol*. 2012;3(4):365–373.
148. Zeng L, Liu J, Wang Y, et al. MicroRNA-210 as a novel blood biomarker in acute cerebral ischemia. *Front Biosci (Elite Ed)*. 2011;3:1265–1272.
149. Schipper HM, Maes OC, Chertkow HM, Wang E. MicroRNA expression in Alzheimer blood mononuclear cells. *Gene Regul Syst Bio*. 2007;1:263–274.
150. Schmitt A, Parlapani E, Bauer M, Heinsen H, Falkai P. Is brain banking of psychiatric cases valuable for neurobiological research? *Clinics (Sao Paulo)*. 2008;63(2):255–266.
151. Marian AJ. Modeling human disease phenotype in model organisms: "It's only a model!" *Circ Res*. 2011;109(4):356–359.
152. Pang XM, Liu JL, Li JP, et al. Fastigial nucleus stimulation regulates neuroprotection via induction of a novel microRNA, rno-miR-676-1, in middle cerebral artery occlusion rats. *J Neurochem*. 2015;133(6):926–934.
153. Peng Z, Li J, Li Y, et al. Downregulation of miR-181b in mouse brain following ischemic stroke induces neuroprotection against ischemic injury through targeting heat shock protein A5 and ubiquitin carboxyl-terminal hydrolase isozyme L1. *J Neurosci Res*. 2013;91(10):1349–1362.
154. Huang LG, Li JP, Pang XM, et al. MicroRNA-29c correlates with neuroprotection induced by FNS by targeting both Birc2 and Bak1 in rat brain after stroke. *CNS Neurosci Ther*. 2015;21(6):496–503.
155. Harraz MM, Eacker SM, Wang X, Dawson TM, Dawson VL. MicroRNA-223 is neuroprotective by targeting glutamate receptors. *Proc Natl Acad Sci USA*. 2012;109(46):18962–18967.
156. Liu P, Zhao H, Wang R, et al. MicroRNA-424 protects against focal cerebral ischemia and reperfusion injury in mice by suppressing oxidative stress. *Stroke*. 2015;46(2):513–519.
157. Vinciguerra A, Formisano L, Cerullo P, et al. MicroRNA-103-1 selectively downregulates brain NCX1 and its inhibition by anti-miRNA ameliorates stroke damage and neurological deficits. *Mol Ther*. 2014;22(10):1829–1838.
158. Ai J, Sun LH, Che H, et al. MicroRNA-195 protects against dementia induced by chronic brain hypoperfusion via its anti-amyloidogenic effect in rats. *J Neurosci*. 2013;33(9):3989–4001.
159. Moncini S, Lunghi M, Valmadre A, et al. The miR-15/107 family of microRNA genes regulates CDK5R1/p35 with implications for Alzheimer's disease pathogenesis. *Mol Neurobiol*. 2016. [Epub ahead of print], PMID: 27343180.
160. Sarkar S, Jun S, Rellick S, Quintana DD, Cavendish JZ, Simpkins JW. Expression of microRNA-34a in Alzheimer's disease brain targets genes linked to synaptic plasticity, energy metabolism, and resting state network activity. *Brain Res*. 1646;2016:139–151.

CHAPTER FIVE

MicroRNAs, Aging, Cellular Senescence, and Alzheimer's Disease

P.H. Reddy*,†,1, J. Williams*, F. Smith*, J.S. Bhatti*,‡, S. Kumar*, M. Vijayan*, R. Kandimalla*, C.S. Kuruva*, R. Wang*, M. Manczak*, X. Yin*, A.P. Reddy†

*Garrison Institute on Aging, Texas Tech University Health Sciences Center, Lubbock, TX, United States
†Texas Tech University Health Sciences Center, Lubbock, TX, United States
‡Department of Biotechnology, Sri Guru Gobind Singh College, Chandigarh, India
[1]Corresponding author: e-mail address: hemachandra.reddy@ttuhsc.edu

Contents

Progress in Molecular Biology and Translational Science, Volume 146
ISSN 1877-1173
http://dx.doi.org/10.1016/bs.pmbts.2016.12.009

Abstract

Aging is a normal process of living being. It has been reported that multiple cellular changes, including oxidative damage/mitochondrial dysfunction, telomere shortening, inflammation, may accelerate the aging process, leading to cellular senescence. These cellular changes induce age-related human diseases, including Alzheimer's, Parkinson's, multiple sclerosis, amyotrophic lateral sclerosis, cardiovascular, cancer, and skin diseases. Changes in somatic and germ-line DNA and epigenetics are reported to play large roles in accelerating the onset of human diseases. Cellular mechanisms of aging and age-related diseases are not completely understood. However, recent discoveries in molecular biology have revealed that microRNAs (miRNAs) are potential indicators of aging, cellular senescence, and Alzheimer's disease (AD). The purpose of our chapter is to highlight recent advancements in miRNAs and their involvement in cellular changes in aging, cellular senescence, and AD. This chapter also critically evaluates miRNA-based therapeutic drug targets for aging and age-related diseases, particularly Alzheimer's.

ABBREVIATIONS

ABCA1 adenosine triphosphate-binding cassette subfamily A member 1
AD Alzheimer's disease
ADAM10 a disintegrin and metalloproteinase domain-containing protein 10
ApoE apolipoprotein E
APP amyloid precursor protein
Aβ amyloid beta
BACE1 beta-site amyloid precursor protein cleaving enzyme 1
CAPE caffeic-acid phenethyl ester
CDK5R1 cyclin-dependent kinase 5, regulatory subunit 1
CSF cerebrospinal fluid
Drp1 dynamin-related protein 1
DUSP6 dual specificity phosphatase 6
ECF extracellular fluid
ERK extracellular signal-regulated kinase
FOXO3a forkhead box O3
GF growth factor
GluR1 glutamate receptor 1
GSK-3β glycogen synthase kinase-3 beta
HCN1 hyperpolarization-activated cyclic nucleotide-gated channel 1
IGF1 insulin-like growth factor 1
INDCs inflammatory neurological controls
IRS1 insulin receptor substrate 1
LRPAP1 low density lipoprotein receptor related protein associated protein 1
MAGL monoacylglycerol lipase
MANCOVA multivariate analysis of covariance
MAPT microtubule-associated protein tau
miRNA microRNA
mRNA messenger RNA

mTOR mechanistic target of rapamycin
NDUFC2 nicotinamide adenine dinucleotide ubiquinone oxidoreductase subunit C2
NF-κB nuclear factor-kappa B subunit
NINDCs noninflammatory controls
NR2A *N*-methyl-D-aspartate receptor subunit 2A
PI3K phosphatidylinositol-3-kinase
PIK3R2 phosphoinositide-3-kinase regulatory subunit 2
PPARγ peroxisome proliferator-activated receptor γ
PPP1CA protein phosphatase 1 catalytic subunit alpha
PS1 presenilin 1
PS2 presenilin 2
PTEN phosphatase and tensin homolog
Rb1 retinoblastoma 1
SAMP8 senescence-accelerated mouse prone 8
Sirt1 silent mating type information regulation 2 homolog
SORL1 sortilin-related receptor 1
SYT1 synaptotagmin 1
TG transgenic
TREM2 triggering receptor expressed on myeloid cells 2
VAMP2 vesicle-associated membrane protein 2

1. INTRODUCTION

Aging is the length of time during which a being or thing has existed and it is a natural process of life. While all cells are progressing towards death, many processes accelerate the aging process, leading to cellular senescence. Cellular changes, including oxidative damage/mitochondrial dysfunction, telomere shortening, changes in somatic and germ-line DNA, inflammation, may accelerate the aging process, leading to cellular senescence (Fig. 1). These cellular changes promote age-related conditions, including Alzheimer's, Parkinson's, multiple sclerosis, amyotrophic lateral sclerosis, cardiovascular, cancer, and skin diseases.

Cellular senescence involves permanent stoppage of growth and programmed cell death. During senescence, various tumor suppressors and death signals inhibit many of the normal functions of the cell. Common tumor suppressor systems such as p53, p21, and p16 are regulated by micro-RNA (miRNA) expression.[1–5] The role of miRNAs in modulating the expression of pathways leading to cellular senescence has led to an increased focus on their role in induced cell death.

With recent advancements in cell and molecular biology, researchers have found miRNAs as a potent form of genetic regulation in many species,

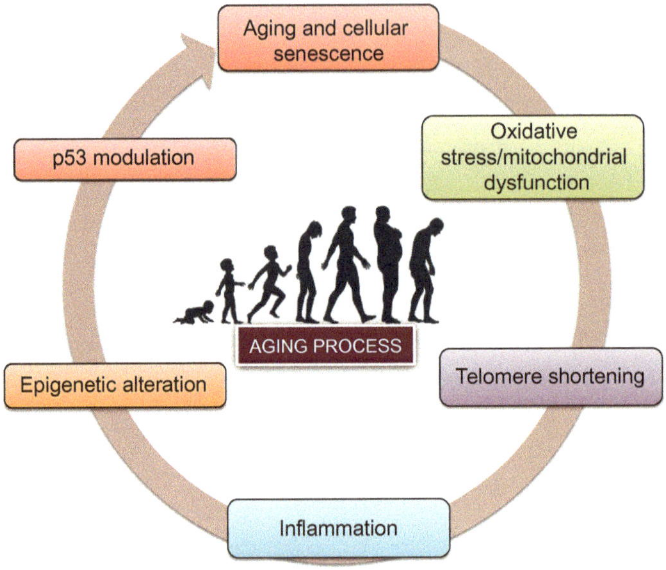

Fig. 1 Cellular processes in human aging.

including humans. By binding to the 3′-untranslated region (UTR) of messenger RNA (mRNA) of specific genes, miRNAs function to prevent the translation of specific genes.[6] Due to their role in silencing the expression of genes, miRNAs provide a promising target for research involving regulation of cellular processes. The purpose of this chapter is to highlight factors causing/promoting aging and cellular senescence and also age-related diseases with a particular focus on Alzheimer's disease (AD); this chapter also covers the involvement of miRNAs in aging, cellular senescence, and AD, and also critically evaluates miRNA-based therapeutic approaches for aging and Alzheimer's.

2. BIOGENESIS AND REGULATION OF miRNAs

miRNAs are a large family of conserved small (20–22 nucleotides), noncoding RNAs. miRNAs play a central role in the posttranscriptional regulation of gene expression.[7] In mammals, miRNAs are believed to control about 50% of all protein-coding genes.[7a] At present, over 2000 miRNAs have been identified (see details at http://www.mirbase.org). One-third of these miRNAs are found in the coding part of genes, and the remaining are in the intronic regions.

miRNA biogenesis is initiated in the nucleus with transcription of primary miRNA transcripts from miRNA-coding genes. Pri-miRNAs are converted to hairpin loop pre-miRNAs by enzymatic digestion with Drosha and DGCR8 proteins. Pre-miRNAs are transported to the cytoplasm by Exportin-5/Ran/GTP proteins where they are again digested by the cytoplasmic proteins Dicer and TRBP, resulting in the generation of the miRNA duplex. Duplex structure is unwinded by helicase, resulting in the generation of mature miRNA strands. Mature miRNAs form the RISC complex with the Ago2 protein and target the 3′-UTR of mRNA, and they modulate gene activity either by translation suppression or mRNA cleavage.[8] miRNAs can be found as a group or a family, harboring the same initial sequence and targeting a single gene. In some cases, a single miRNA can affect a large number of genes that are involved in the regulation of multiple cellular events and pathways.[9]

Research has revealed that miRNAs are differentially expressed in different cell types and tissues in mammals, including humans. miRNAs are believed to alter multiple cellular processes, including development, cell proliferation, replicative senescence, and aging.[10,11] Over 70% of reported miRNAs are expressed in the human brain. miRNAs found in the human brain include: miR-9, miR-7, miR-128, miR-125 a-b, miR-23, miR-132, miR-137, and miR-139. A recent deep RNA sequencing analysis has revealed a large number of miRNAs that are brain-specific, including: miR-134, miR-135, let-7g, miR-101, miR-181a-b, miR-191, miR-124, miR-let-7c, let-7a, miR-29a, and miR-107.[12,13] Most of these miRNAs are responsible for synaptic functions, neurotransmitter release, synapse formation, and neurite outgrowth. Expression levels of several miRNAs are altered in a diseased state, such as AD.

3. miRNA, AGING, AND CELLULAR SENESCENCE

A common theme seen with aging is the upregulation of miRNA expression, as is seen in the mouse liver.[14] MiR-34a has been shown to promote senescence in hepatocellular carcinoma cells via inhibition of the c-myc pathway, leading to inactivation of the hTERT pathway.[15] The inactivation of c-myc provides an example of a miRNA inhibiting the transcription of all RNAs, including miRNAs.

As aging progresses, cells are exposed to more stress. During stressful conditions, p53 is frequently upregulated. As previously discussed, a higher expression of p53 could potentially improve Drosha's ability to yield pre-miRNA, thus expediting the maturation process. One of the significant findings in a twin study on AD patients was an increase in DNA methylation in the cortex.[16] Additionally, DNA methylation in the temporal cortex has been shown to directly target the CpG islands of miRNAs.[17] By modulating the expression of regulatory miRNAs in the cortex, DNA methylation constitutes a process that can bring about an aging phenotype as seen in cells affected by AD.

The expression of glycogen synthase kinase-3 beta (GSK-3β) can be decreased via expression of miR-26a, leading to promotion of neuronal axon regeneration through gene regulation.[18] By inhibiting the proliferation of axons, miR-26 represents a case of gene regulation leading to accelerated aging. Consistent with rapid aging and neurodegeneration, increased expression of KSRP also resulted in a decrease in axonal outgrowth via downregulation of the growth factor GAP-43's mRNA.[18a] Decreasing expression of MeCP2 in glutamatergic neurons of mice led to a shorter lifespan and various neurological abnormalities.[19] All in all, regulation at the level of miRNA synthesis could have dire consequences on the lifespan of cells.

Photoaging has been shown to disrupt TGF-β signaling in skin cells, leading to failure to activate the Smad pathway.[20] When skin cells are unable to respond to the TGF-β signal, activation of Drosha by Smad could potentially be disrupted. These results suggest that the characteristic phenotype seen in photoaging could potentially be due to disruption of normal miRNA synthesis.

It has been shown that mice with a Dicer knockout in frontal cortex neurons displayed neurodegeneration.[21] An additional study showed that a Dicer knockout in adipose tissue effectively reversed the slowed rate of aging brought about by dietary restriction in mice.[22] The detrimental effects to longevity seen when Dicer, a key maturing step in the development of miRNAs, is knocked out indicates miRNAs could play a crucial role in many areas of development.

Much of the current knowledge of how miRNAs can induce senescence has come from research done in flies, worms, and mice. How well much of this research translates to humans is yet to be seen, but some of the findings in lower species have shown promising correlation in humans.

Drosophila melanogaster has been shown to develop in accordance with regulation of miRNAs. During overexpression of mirvana/miR-278 in *Drosophila* flies, tumors start to form in the developing eye, likely due to inhibition of apoptosis.[23] Additionally, mutation of miR-14 in *Drosophila* flies has been shown to decrease lifespan, indicating its role in preventing senescence.[24]

Some of the earliest recognition of miRNAs were in *Caenorhabditis elegans* worms. One study showed that the expression of lsy-6 miRNA dictates the left–right development of the nervous system of *C. elegans*.[25] The lifespan of *C. elegans* is also affected by the expression of miRNAs, with higher lin-4 miRNA expression correlating with decreased lin-14 expression and a longer lifespan.[26]

4. miRNAs, CELLULAR SENESCENCE, AND PATHWAYS

Many of the processes that carry out cellular senescence are regulated by miRNA expression. Via downregulation of transcripts of genes that typically promote cellular livelihood, miRNAs have become a central focus of research surrounding cellular senescence (Fig. 2).

4.1 miRNAs and Oxidative Stress

Oxidative stress is the condition in which the body is unable to completely detoxify ROS. Cells that experience oxidative stress may undergo senescence. In embryonic fibroblasts, from a mouse model that rapidly accumulates DNA damage from oxidative stress, miR-449a, miR-455, and miR-128 were found to be significantly downregulated.[27] These results mirrored expression found in kidneys of elderly mice. In a study, that used benzo-α-pyrene to induce oxidative stress in *C. elegans*, miR-1, miR-355, miR-50, miR-51, miR-58, miR-796, and miR-84 were found to have modified expression.[28] These miRNAs may be involved in the regulation of SKiN head-1, a transcription factor involved in antioxidant response element regulation, and gamma-glutamine cysteine synthase heavy chain.[28] In a mouse model, with oxidative liver injury induced by treatment with the anti-tuberculosis drug isoniazid, expression of miR-122 in tissue was significantly changed at days 3 and 5. In a study of miRNAs expressed in oxidatively stressed hippocampal neurons in senescence-accelerated mice, miR-329, miR-193b, miR-20a, miR-296, and miR-130b were found to be upregulated.[29] These miRNAs may be involved in a variety of processes, such as regulation of cell growth, apoptosis, signal transmission, and cancer development, especially through the mitogen-activated protein kinase signaling pathway. MiR-200c expression increases with oxidative stress,

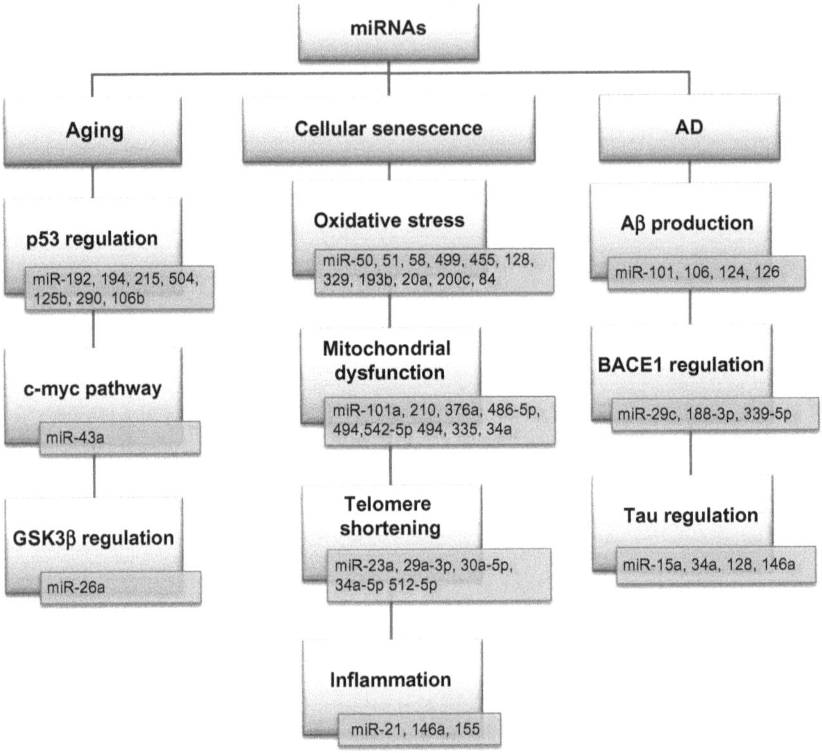

Fig. 2 Summary of microRNAs in aging, cellular senescence, and Alzheimer's disease.

leading to senescence and cell apoptosis.[30] Expression of miR-183 is upregulated in oxidative stress-induced senescence.[31] In addition, oxidative stress can be initiated by the altered expression of miRNAs. For example, miR-146a acts to downregulate the NOX4 subunit of NADPH oxidase. This miRNA is downregulated in senescence, leading to increased NADPH oxidase activity and further oxidative stress.[32] Details of miRNAs related to oxidative stress are given in Table 1.

4.2 miRNAs and Mitochondrial Dysfunction

Senescence is widely known to be induced by mitochondrial damage. MiRNAs can modulate this induction of senescence by controlling autophagy of the mitochondria. MiR-210, miR-376a*, miR-486-5p, miR-494, and miR-542-5p likely control autophagy through an mechanistic target of rapamycin (mTOR)-dependent mechanism.[56] On the other hand, miR-101 has been shown to inhibit autophagy by targeting a number

Table 1 Summary of MicroRNAs in Aging and Cellular Senescence

miRNA	Important Processes/Functions	Expression/Relationship	References
Oxidative stress/mitochondrial dysfunction			
miR-1	Oxidative stress/mitochondrial dysfunction; neurodegenerative disease; acute myocardial infarction; muscular aging	Decreased in response to oxidative stress. Potential biomarker for Parkinson's Disease. Elevated in the bloodstream of patients with acute myocardial infarction. Expression is downregulated in aging muscle of pig.	Wu *et al.*[28] Margis *et al.*[33] Wang *et al.*[34] Redshaw *et al.*[35]
miR-15b	Oxidative stress/mitochondrial dysfunction	Inhibits senescence-associated mitochondrial stress.	Lang *et al.*[36]
miR-20a	Oxidative stress/mitochondrial dysfunction; tumor suppression; neurodegenerative disease	Upregulated in response to oxidative stress. Downregulates p53 mRNA, leading to increased proliferation. Upregulated in the blood leukocytes of Parkinson's patients.	Zhang *et al.*[37] Poliseno *et al.*[38] Soreq *et al.*[39]
miR-21	Sarcopenia	Upregulated in response to muscle denervation. Involved in triggering atrophy.	Soares *et al.*[40]
miR-22	Oxidative stress/mitochondrial dysfunction; muscle progenitor cell differentiation	Targets mRNA of the ETC ATPase, decreasing mitochondrial efficiency. Upregulation promotes smooth muscle cell differentiation	Li *et al.*[41] Zhao *et al.*[42]
miR-23a	Sarcopenia	Targets the mRNA of MuRF1 and atrogin-1, two proatrophic proteins. MiR-23a's expression is decreased during muscle atrophy and sarcopenia.	Hudson *et al.*[43]
miR-24	Muscular aging	Expression is upregulated in aging muscle of pig.	Redshaw *et al.*[35]

Continued

Table 1 Summary of MicroRNAs in Aging and Cellular Senescence—cont'd

miRNA	Important Processes/Functions	Expression/Relationship	References
miR-34a	Oxidative stress/mitochondrial dysfunction; tumor suppression; telomere shortening; muscle progenitor cell differentiation	Promotes all processes listed. Inhibits c-myc function.	Yamakuchi *et al.*[5] Chang *et al.*[44] Bai *et al.*[45] Yang *et al.*[46] Xu *et al.*[15] Yu *et al.*[47]
miR-50	Oxidative stress/mitochondrial dysfunction	Altered in response to oxidative stress.	Wu *et al.*[28]
miR-51	Oxidative stress/mitochondrial dysfunction	Altered in response to oxidative stress.	Wu *et al.*[28]
miR-58	Oxidative stress/mitochondrial dysfunction	Altered in response to oxidative stress.	Wu *et al.*[28]
miR-84	Oxidative stress/mitochondrial dysfunction	Altered in response to oxidative stress.	Wu *et al.*[28]
miR-93	Oxidative stress/mitochondrial dysfunction; aging	Upregulated in the livers of aging mice; targets glutathione-*S*-transferases, which normally protect from oxidative stress. Downregulated in the livers of aging rats.	Maes *et al.*[14] Mimura *et al.*[48]
miR-122	Oxidative stress/mitochondrial dysfunction	Upregulated during oxidative stress.	Song and Lee[49]
miR-128	Oxidative stress/mitochondrial dysfunction	Downregulated during oxidative stress.	Nidadavolu *et al.*[27]
miR-130b	Oxidative stress/mitochondrial dysfunction	Upregulated in response to oxidative stress.	Zhang *et al.*[29]

miR-146a	Oxidative stress/mitochondrial dysfunction; neurodegenerative disease	Downregulated during senescence. Normally downregulates the NOX4 subunit of NADPH oxidase. Upregulated in the CSF/ECF of Alzheimer's patients.	Vasa-Nicotera *et al.*[32] Roggli *et al.*[50] Alexandrov *et al.*[51]
miR-183	Oxidative stress/mitochondrial dysfunction	Upregulated in response to oxidative stress.	Le *et al.*[52]
miR-193b	Oxidative stress/mitochondrial dysfunction	Upregulated in response to oxidative stress.	Zhang *et al.*[29]
miR-200c	Oxidative stress/mitochondrial dysfunction	Upregulated in response to oxidative stress.	Magenta *et al.*[30]
miR-206	Sarcopenia; muscular aging	Upregulated in response to muscle denervation. Involved in triggering atrophy. Also aids in maintaining the neuromuscular junction following nerve injury. Expression is downregulated in aging muscle of pig.	Soares *et al.*[40] Williams *et al.*[53] Redshaw *et al.*[35]
miR-214	Oxidative stress/mitochondrial dysfunction; osteoporosis	Upregulated in the livers of aging mice; targets glutathione-*S*-transferases, which normally protect from oxidative stress. Inhibits osteoblast activity.	Maes *et al.*[14] Wang *et al.*[54]
miR-296	Oxidative stress/mitochondrial dysfunction	Upregulated in response to oxidative stress.	Zhang *et al.*[29]
miR-329	Oxidative stress/mitochondrial dysfunction	Upregulated in response to oxidative stress.	Zhang *et al.*[29]
miR-335	Oxidative stress/mitochondrial dysfunction	Downregulates antioxidant enzymes in the mitochondria.	Bai *et al.*[45]

Continued

Table 1 Summary of MicroRNAs in Aging and Cellular Senescence—cont'd

miRNA	Important Processes/Functions	Expression/Relationship	References
miR-355	Oxidative stress/mitochondrial dysfunction	Altered in response to oxidative stress.	Wu *et al.*[28]
miR-449a	Oxidative stress/mitochondrial dysfunction	Downregulated in response to oxidative stress.	Nidadavolu *et al.*[27]
miR-455	Oxidative stress/mitochondrial dysfunction	Downregulated in response to oxidative stress.	Nidadavolu *et al.*[27]
miR-669c	Oxidative stress/mitochondrial dysfunction	Upregulated in the livers of aging mice; targets glutathione-*S*-transferases, which normally protect from oxidative stress.	Maes *et al.*[14]
miR-709	Oxidative stress/mitochondrial dysfunction	Upregulated in the livers of aging mice; targets glutathione-*S*-transferases, which normally protect from oxidative stress.	Maes *et al.*[4]
miR-796	Oxidative stress/mitochondrial dysfunction	Altered in response to oxidative stress.	Wu *et al.*[28]
miR-101a	Mitochondrial dysfunction	Targets mRNA of the ETC ATPase, decreasing mitochondrial efficiency.	Li *et al.*[55]
miR-210	Mitochondrial dysfunction and autophagy	Controls autophagy through mTOR-dependent mechanism and inhibits the mitochondrial ETC.	Faraonio *et al.*[56] Puissegur *et al.*[57]
miR-494	Mitochondrial dysfunction and autophagy	Controls autophagy of the mitochondria through mTOR-dependent mechanism. Likely controls ETC, cell cycle, and mitochondrial translation.	Faraonio *et al.*[3] Bandiera *et al.*[58]
miR-720	Mitochondrial dysfunction	Targets ETC ATPase mRNA.	Li *et al.*[55]
miR-721	Mitochondrial dysfunction	Targets ETC ATPase mRNA.	Li *et al.*[55]

Inflammation			
let-7	Inflammation	Activates toll-like receptors (TLRs), promoting inflammation.	Lehmann *et al.*[59]
miR-146	Inflammation	Inhibits inflammation. Potentially a negative feedback system.	Olivieri *et al.*[60] Taganov *et al.*[61]
Tumor suppression			
miR-19b	Tumor suppression	Downregulates p53 expression, leading to increased proliferation.	Fan *et al.*[62]
miR-21	Tumor suppression; osteosarcoma/osteoporosis	Targets p21 tumor suppressor mRNA, leading to increased proliferation. Biomarker for inflammation in pancreatic beta cells. Upregulated in osteosarcoma cells. Downregulated in osteoporotic mesenchymal stem cell.	Gomez-Cabello *et al.*[2] Roggli *et al.*[50] Ziyan *et al.*[63] Yang *et al.*[64]
miR-29	Tumor suppression; neurodegenerative disease	Activates p53, leading to tumor suppression. Potential blood biomarker for Parkinson's disease.	Ugalde *et al.*[4] Margis *et al.*[33]
miR-106b	Tumor suppression; aging	Inhibits tumor suppression by targeting p21 mRNA. Downregulated in the livers of aging rats.	Borgdorff *et al.*[1] Mimura *et al.*[48]
miR-125b	Tumor suppression; neurodegenerative disease	Inhibits tumor suppression by targeting p53 mRNA. Expression correlates with Alzheimer's disease.	Hu *et al.*[65] Le *et al.*[52] Tan *et al.*[66]
miR-192	Tumor suppression	Expression is induced by p53. Reduced in cancerous conditions.	Braun *et al.*[67]
miR-194	Tumor suppression	Expression is induced by p53. Reduced in cancerous conditions.	Braun *et al.*[67]

Continued

Table 1 Summary of MicroRNAs in Aging and Cellular Senescence—cont'd

miRNA	Important Processes/Functions	Expression/Relationship	References
miR-215	Tumor suppression	Expression is induced by p53. Reduced in cancerous conditions.	Braun *et al.*[67]
miR-290	Tumor suppression	Downregulates LRF, leading to increased p16 tumor suppressor expression.	Pitto *et al.*[68]
miR-504	Tumor suppression	Targets p53 mRNA, inhibiting tumor suppression.	Hu *et al.*[65]
Neurodegenerative disease			
miR-9	Neurodegenerative disease	Increased in the CSF and ECF of Alzheimer's patients.	Alexandrov *et al.*[51]
miR-16	Neurodegenerative disease	Downregulated in the blood leukocytes of Parkinson's patients.	Soreq *et al.*[39]
miR-16-2-3p	Neurodegenerative disease	Potential blood biomarker for Parkinson's disease.	Margis *et al.*[33]
miR-22-5p	Neurodegenerative disease	Potential blood biomarker for Parkinson's disease.	Margis *et al.*[33]
miR-26a	Neurodegenerative disease	Promotes axonal regeneration by inhibiting GSK-3β expression.	Jiang *et al.*[18]
miR-26a-2-3p	Neurodegenerative disease	Potential blood biomarker for Parkinson's disease	Margis *et al.*[33]
hsa-miR-27a-3p	Neurodegenerative disease	Expression is decreased in the CSF of Alzheimer's patients.	Sala *et al.*[69]
miR-30a	Neurodegenerative disease	Potential blood biomarker for Parkinson's disease.	Margis *et al.*[33]
miR-155	Neurodegenerative disease; rheumatoid arthritis	Upregulated in rheumatoid arthritis and the CSF/ECF of Alzheimer's patients.	Kurowska-Stolarska *et al.*[70] Alexandrov *et al.*[51]

miR-320	Neurodegenerative disease	Downregulated in the blood leukocytes of Parkinson's patients.	Soreq *et al.*[39]
miR-331-5p	Neurodegenerative disease	Upregulated in the plasma of Parkinson's patients.	Cardo *et al.*[71]
miR-450b-3p	Neurodegenerative disease	Upregulated in the plasma of Parkinson's patients.	Khoo *et al.*[72]
miR-505	Neurodegenerative disease	Upregulated in the plasma of Parkinson's patients.	Khoo *et al.*[72]
miR-626	Neurodegenerative disease	Upregulated in the plasma of Parkinson's patients.	Khoo *et al.*[72]
miR-1826	Neurodegenerative disease	Upregulated in the plasma of Parkinson's patients.	Khoo *et al.*[72]
Type II diabetes			
miR-15a	Type II diabetes	Downregulated in type II diabetes 5–10 years before the onset of the disease.	Zampetaki *et al.*[73]
miR-27a	Type II diabetes	Upregulated in type II diabetes; expression levels correlated with higher fasting glucose levels.	Karolina *et al.*[74]
miR-29b	Type II diabetes	Downregulated in type II diabetes 5–10 years before the onset of the disease.	Zampetaki *et al.*[73]
miR-126	Type II diabetes; inflammation; atherosclerosis	Inhibits the NF-κB pathway. Alters the expression of VCAM-1 in inflammation. When administered in apoptotic bodies to atherosclerosis patients, improved vasculature by increasing production of chemokine CXCL12. Downregulated in type II diabetes 5–10 years before the onset of the disease.	Harris *et al.*[75] Asgeirsdottir *et al.*[76] Qin *et al.*[77] Feng *et al.*[78] Zampetaki *et al.*[73]
miR-150	Type II diabetes	Upregulated in type II diabetes. Expression levels correlated with higher fasting glucose levels.	Karolina *et al.*[74]

Continued

Table 1 Summary of MicroRNAs in Aging and Cellular Senescence—cont'd

miRNA	Important Processes/Functions	Expression/Relationship	References
miR-193	Type II diabetes	Upregulated in type II diabetes. Expression levels correlated with higher fasting glucose levels.	Karolina *et al.*[74]
miR-223	Type II diabetes	Downregulated in type II diabetes 5–10 years before the onset of the disease.	Zampetaki *et al.*[73]
miR-320a	Type II diabetes	Upregulated in type II diabetes. Expression levels correlated with higher fasting glucose levels.	Karolina *et al.*[74]
miR-375	Type II diabetes	Upregulated in type II diabetes. Expression levels correlated with higher fasting glucose levels.	Karolina *et al.*[74]
Aging			
lin-4	Aging	Increases lifespan in *C. elegans*.	Boehm *et al.*[26]
miR-7a	Aging	Downregulated in the livers of aging rats.	Mimura *et al.*[48]
miR-14	Aging	Decreases lifespan in *Drosophila* flies.	Xu *et al.*[24]
miR-24-3p	Aging	Biomarker for aging in the saliva.	Machida *et al.*[79]
miR-29a	Aging	Upregulated in the livers of aging rats.	Mimura *et al.*[48]
miR-29c	Aging	Upregulated in the livers of aging rats.	Mimura *et al.*[48]
miR-148b-3p	Aging	Downregulated in the livers of aging rats.	Mimura *et al.*[48]
miR-185	Aging	Downregulated in the livers of aging rats.	Mimura *et al.*[48]
miR-195	Aging	Upregulated in the livers of aging rats.	Mimura *et al.*[48]
miR-301a/b	Aging	Downregulated in the livers of aging rats.	Mimura *et al.*[48]

miR-405a	Aging	Downregulated in the livers of aging rats.	Mimura *et al.*[48]
miR-497	Aging	Upregulated in the livers of aging rats.	Mimura *et al.*[48]
miR-539	Aging	Downregulated in the livers of aging rats.	Mimura *et al.*[48]
Telomere shortening			
miR-23	Telomere shortening; amino acid metabolism	Leads to shortened telomeres. Inhibits ATP production via amino acid metabolism.	Luo *et al.*[80] Guo *et al.*[81]
miR-143	Telomere shortening	Upregulated in cells lacking TERT.	Bonifacio *et al.*[82]
miR-145	Telomere shortening	Upregulated in cells lacking TERT.	Bonifacio *et al.*[82]
miR-512-5p	Telomere shortening	Targets hTERT mRNA.	Li *et al.*[83]
miR-101	Autophagy	Downregulates proautophagic genes.	Frankel *et al.*[84]
miR-204	Autophagy	Inhibits autophagy.	Mikhaylova *et al.*[85]
miR-376a*	Autophagy	Controls autophagy through mTOR-dependent mechanism.	Faraonio *et al.*[56]
miR-486-5p	Autophagy	Controls autophagy of the mitochondria through mTOR-dependent mechanism.	Faraonio *et al.*[56]
Osteoarthritis			
hsa-miR-149*	Osteoarthritis	Downregulated in the chondrocytes of osteoarthritic patients.	Diaz-Prado *et al.*[86]
hsa-mir-483-5p	Osteoarthritis	Upregulated in the chondrocytes of osteoarthritic patients.	Diaz-Prado *et al.*[86]
hsa-miR-576-5p	Osteoarthritis	Downregulated in the chondrocytes of osteoarthritic patients.	Diaz-Prado *et al.*[86]

Continued

Table 1 Summary of MicroRNAs in Aging and Cellular Senescence—cont'd

miRNA	Important Processes/Functions	Expression/Relationship	References
hsa-miR-582-3p	Osteoarthritis	Downregulated in the chondrocytes of osteoarthritic patients.	Diaz-Prado *et al.*[86]
hsa-miR-634	Osteoarthritis	Downregulated in the chondrocytes of osteoarthritic patients.	Diaz-Prado *et al.*[86]
hsa-miR-1227	Osteoarthritis	Downregulated in the chondrocytes of osteoarthritic patients.	Diaz-Prado *et al.*[86]
Development			
lsy-6	Development	Dictates left/right neuronal asymmetry in *C. elegans*.	Johnston and Hobert[25]
Acute myocardial infarction			
miR-133a	Acute myocardial infarction; osteoporosis	Elevated in the bloodstream of patients with acute myocardial infarction. Upregulated in the osteoclast precursors of osteoporotic women.	Wang *et al.*[34] Wang *et al.*[87]
miR-208a	Acute myocardial infarction	Elevated in the bloodstream of patients with acute myocardial infarction.	Wang *et al.*[34]
miR-499	Acute myocardial infarction	Elevated in the bloodstream of patients with acute myocardial infarction.	Wang *et al.*[34]
Tumor formation			
mirvana/miR-278	Tumor formation	Upregulation associated with tumor formation in *Drosophila* flies.	Nairz *et al.*[23]

of proautophagic proteins.[84] Evidence has been collected that miR-34a inhibits autophagy in *C. elegans* and probably in higher organisms as well.[46] MiR-204 has been shown to be significantly upregulated in proliferative endothelial cells, possibly functioning in autophagy inhibition.[85] The electron transport chain (ETC) is another target for miRNA action. MiR-210 is upregulated in senescence and has been shown to act by inhibiting translation of ETC protein machinery.[57] MiR-494 was also shown to induce senescence, likely by controlling ATP synthesis by the ETC, cell cycling, and mitochondrial translation.[58] Other miRNAs such as miR-335 and miR-34a increase ROS production, exerting their effect by downregulating the expression of antioxidative enzymes in the mitochondria.[45] MiR-23a/b decreases ATP production by targeting proteins involved in ATP synthesis via amino acid catabolism.[88] Expression of mitochondrial protein SIRT4 is upregulated by stress. MiR-15b acts as an inhibitor of SIRT4 and counteracts senescence-associated mitochondrial dysfunction.[36] Details of miRNAs related to mitochondrial dysfunction are given in Table 1.

4.3 miRNAs and p53

P53 is an important protein in the regulation of the cell cycle and preventing tumorigenesis. Its importance is so profound that it is often referred to as the guardian of the genome. P53, along with similar proteins p16 and p21, has a number of interactions and relationships with miRNAs. miR-34a is upregulated by p53 and may act by binding and deactivating silent mating type information regulation 2 homolog (SIRT1), an inhibitor of p53, causing a positive feedback loop.[5,44] Alternatively, it may act by suppressing expression of a different family of p53 inhibiting proteins.[89] MiR-192, miR-194, and miR-215 are also induced by p53, and can also cause cell cycle arrest.[67] The mRNA encoded by the p53 gene is directly targeted by miR-504 and miR-125b.[52,65] MiR-20a also inhibits p53 expression and, therefore, prevents cell death.[38] It acts by downregulating expression of LRF, which inhibits p19ARF, a repressor of p53 inhibitor MDM2.[38,90] This downregulation of LRF also allows for increased expression of miR-290, which leads to increased expression of the INK4A gene locus, particularly, the p16 tumor suppressor, in mouse fibroblast cells.[68] The miR-106b family of miRNAs target the 3′-UTR of the p21 transcript, preventing its translation and function as a tumor suppressor.[1] P53 was also shown to be a downstream target of miR-19b.[62] MiR-21 may target p21, as it has been shown to reverse the effects of a deletion of DGCR8, the Drosha auxiliary protein, which usually leads to senescence.[2] Details of miRNAs related to p53 are given in Table 1.

4.4 miRNAs and Telomerase Shortening

Telomeres are long repeating sequences of nucleotides at the end of linear chromosomes, such as those in humans. They shorten with age, and the process of shortening has been associated with miRNA expression and senescence. Telomeric repeat binding factor 1 and 2 (TRF1 and TRF2) are proteins essential for maintenance of telomeres. miR-23a has been shown to have the capacity to directly target the 3′-UTR of the TRF2 transcript.[80] Overexpression of miR-23a caused telomere dysfunction and more rapid onset of senescence. One study showed that a complex of natural yeast proteins caused upregulation of TRF2, while downregulating miR-29a-3p, miR-30a-5p, and miR-34a-5p in human fibroblasts.[91] Similar to miR-23a and TRF2, miR-155 has been shown to target TRF1 in human breast cancer cells, causing telomeres to be more fragile.[92] MiR-138 targets transcripts encoding telomerase reverse transcriptase (TERT),[93] a subunit of a protein called telomerase that lengthens telomeres by the addition of nucleotides. Human TERT has been identified as a target of miR-512-5p.[83] In a study that examined differences in miRNA expression in TERT-immortalized and nonimmortalized human foreskin fibroblasts, miR-143 and miR-145 levels were upregulated in senescent fibroblasts when compared to immortalized fibroblasts.[82] Details of miRNAs related to telomerase shortening are given in Table 1.

4.5 miRNAs and Inflammation

Inflammation is an immunological process characterized by redness, swelling, heat, and pain. Inflammation has been implicated in triggering senescence and tends to increase with age. MiR-21 has been demonstrated to be a biomarker for inflammation associated with aging, a process known as "inflamm-aging," as well as cardiovascular disease.[60] MiR-21 levels were also increased in the beta cells of the pancreas in response to exposure to inflammatory cytokines, leading to a decrease in expression of proteins involved in insulin secretion.[50] In addition, miR-21 decreases the expression of PDCD4, a known promoter of inflammatory activity.[94] miR-146a, like miR-21, is modulated in response to proinflammatory cytokines in the beta cells of the pancreas.[50] Additionally, miR-146 is known to be involved in regulation of transcripts in two important cellular signaling pathways in cells involved in vascular remodeling: the NF-κB and Toll-like receptor pathways.[60] Activation of these pathways leads to inflammation. MiR-146a and 146b have also been shown to downregulate proinflammatory cytokines IL-6 and IL-8 in human fibroblasts.[95] MiR-146 has even been suggested to act in a negative feedback loop, given that its expression increases as the inflammatory response progresses.[61] MiR-155 has been demonstrated to have higher expression levels in synovial fluid of patients with rheumatoid arthritis, an inflammatory process.[70] MiR-126 is involved in the inflammation process by altering expression of cell-adhesion molecules, like VCAM-1.[75–77] It also has a role in reducing the expression of IKBA, which inhibits the NF-κB pathway.[78] Details of miRNAs related to inflammation are given in Table 1.

5. miRNAs AND NEURODEGENERATIVE DISEASES

Well-studied miRNAs in human diseases include cancer,[96,97] cardiovascular diseases,[98] hypertension,[99] nephropathy,[100] stroke,[101] and neurodegenerative diseases such as schizophrenia,[102] Huntington's,[103] Parkinson's,[104] and AD.[7,105–107]

5.1 Alzheimer's Disease

AD is an age-related, multifactorial, progressive neurodegenerative disease, characterized by memory loss, multiple cognitive impairments, and changes in personality and behavior. Currently, 5.4 million Americans suffer from AD, and this number is expected to increase up to 16 million by 2050. With a growing aging population not only in the United States but also in the worldwide, AD has become a major health concern. AD has had a huge economic impact, with dementia health care costs alone reaching an estimated total of $818 billion worldwide in 2015.[108] Despite extensive research into AD, we still do not have drugs or agents that can delay or prevent AD progression, and we do not have detectable biomarkers for early AD diagnosis.

AD is associated with synaptic loss, mitochondrial dysfunction, amyloid beta (Aβ) production and accumulation, inflammatory responses, phosphorylated tau formation and accumulation, cell cycle deregulation, impaired cholinesterase transmission, deficits in neurotransmission and hormonal imbalance, neuronal loss, and an accumulation of senile plaques and neurofibrillary tangles in learning and memory regions of the brain[109–112] (Fig. 3). Synaptic pathology and mitochondrial damage have been identified as early events in AD pathogenesis.[113] An accumulation of Aβ and mislocalization of

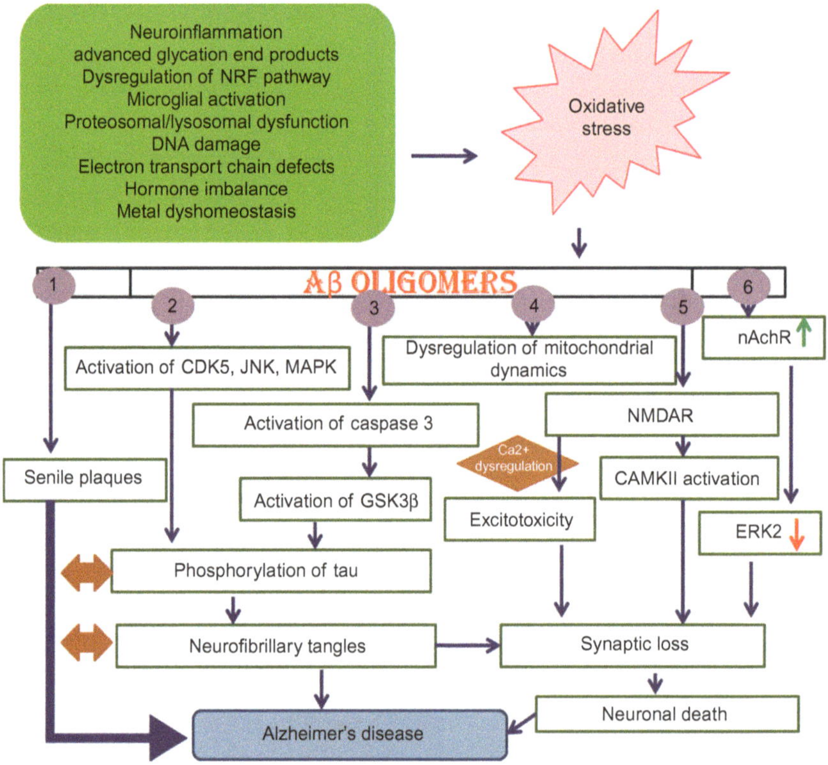

Fig. 3 Cellular changes in the progression and pathogenesis of Alzheimer's disease.

phosphorylated tau in synapses cause synaptic starvation and degeneration, and cognitive decline in AD patients. The precise cause underlying AD pathogenesis are not completely known or understood.

5.2 miRNAs and AD

miRNAs regulate genes that are responsible for Aβ production and phosphorylated tau, including amyloid precursor protein (APP), presenilin 1 (PS1), and presenilin 2 (PS2) (Table 2). Fig. 4 illustrates miRNA-based therapeutics in aging and AD.

5.3 The AD Brain and miRNAs

The progressive loss of synapses and neurons, reduced volume of the hippocampus, and reduction in size and weight are typical features of the AD brain. Similar to brains from humans with AD, brains from AD mice have most of these same features. The loss of synapses and synaptic damage

Table 2 miRNAs in Alzheimer's Diseases

miRNAs	Status in AD	Target Gene	Effect on Target Genes	References
Neuroprotective miRNAs in Alzheimer's disease				
miR-29a/b-1 cluster	Brain	*BACE1/β-secretase*	Upregulation	Hebert *et al.*[114]
miR-101	—	*APP*	Downregulation	Vilardo *et al.*[115]
miR-124	Brain↓	*BACE1*	Downregulation	Fang *et al.*[116]
miR-132/212	Brain↓	*PTEN*, *FOXO3a*, and *P300*	Downregulation	Wong *et al.*[117]
miR-34	—	*tau*	Downregulation	Dickson *et al.*[118]
miR-193b	Hippocampus↓	*APP*	Downregulation	Liu *et al.*[119]
miR-188-3p	Brain↓	*BACE1*	Downregulation	Zhang *et al.*[37]
miR-339-5p	Brain↓	*BACE1*	Downregulation	Long *et al.*[120]
miR-212/132 and miR-23a/b	Frontal cortex↓	*sirt1*	Upregulation	Weinberg *et al.*[121]
miR-219	Brain↓	*tau*	Downregulation	Santa-Maria *et al.*[122]
miR-16	Neuronal cells↓	*APP*	Downregulation	Zhang *et al.*[123]
Mir-29c	Peripheral blood↓	*BACE1*	Downregulation	Yang *et al.*[124]
miR-135b	Peripheral blood↓	*BACE1*	Downregulation	Zhang *et al.*[125]
miR-1229-3p	—	*SORL1*	Downregulation	Ghanbari *et al.*[126]
miR-15/107 family	Brain↓	*CDK5R1*	Downregulation	Moncini *et al.*[127]
miR-603	Hippocampus↑	*LRPAP1*	Downregulation	Zhang *et al.*[128]
miR-29	Brain↓	*hBACE1*	Downregulation	Pereira *et al.*[129]

Continued

Table 2 miRNAs in Alzheimer's Diseases—cont'd

miRNAs	Status in AD	Target Gene	Effect on Target Genes	References
Neurodegenerative miRNAs in Alzheimer's disease				
miR-26b	Brain cortex↑	*Rb1*	Upregulation	Absalon *et al.*[130]
miR-30a-5p	—	*BDNF*	Downregulation	Croce *et al.*[131]
miR-206	Brain↑	*BDNF*	Downregulation	Tian *et al.*[132]
miR-125	Brain↑	*DUSP6*, *PPP1CA*, and *Bcl-W*	Downregulation	Banzhaf-Strathmann *et al.*[133]
miR-33	—	*ABCA1*	Downregulation	Kim *et al.*[134]
miR-34a	Brain↑	*VAMP2*, *SYT1*, *HCN1*, *NR2A*, *GLUR1*, and *NDUFC2*,	Downregulation	Sarkar *et al.*[135]
miR-126	Brain↑	*IRS1* and *PIK3R2*	Downregulation	Kim *et al.*[136]

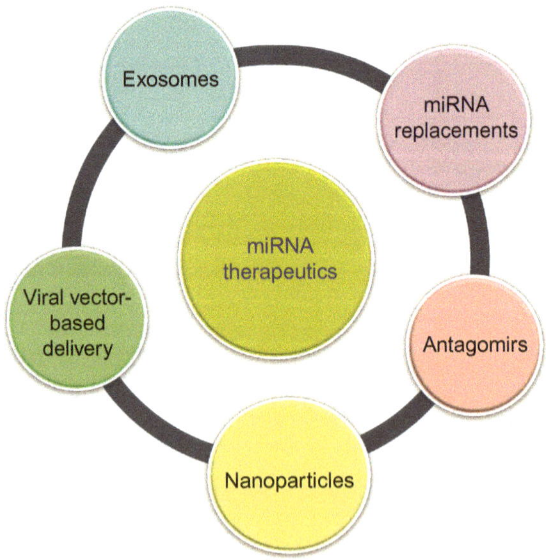

Fig. 4 miRNAs-based therapeutic strategies in Alzheimer's disease.

correlate the closest with cognitive decline and memory loss in AD patients and AD mice.[113] Studies revealed a reduction in miRNA expression in the AD brain, which in turn appears to correlate with a reduction and in Aβ production and reduced phosphorylated tau (Table 2). In contrast, several miRNAs are known to increase levels of Aβ, phosphorylated tau, and inflammation not only in the brains of humans with AD but also in the brains of mice with AD. Interestingly, the brains from Dicer knockout mice exhibited similar features found in the brains from humans and mice with AD, such as reduced brain size, enlarged ventricles, inflammation of brain, loss of synaptic branching and connectivity, and spine length.[137–139] Dicer knockout mice also showed oxidative stress, phosphorylated tau, and memory loss, and reduced levels of a large number of miRNAs,[137–139] conditions also found in the brains of humans and mice with AD. The similarities between the brains of humans with AD and Dicer knockout mice suggest that Dicer may play a large role in memory and cognition and that a progressive loss of Dicer may be linked to reduced learning and memory in persons with AD. Research is needed to investigate whether Dicer is linked to cognitive decline in AD.

5.4 Amyloid Beta and miRNAs

Aβ production and Aβ deposits in the brains of humans with AD are accompanied by alterations in the levels of many miRNAs from distinct miRNA classes. These miRNAs may be involved in AD pathogenesis, in particular the generation of Aβ.[140] Several miRNA classes, such as miR-101 and miR-106, target APP, resulting in an elevated generation of Aβ.[105,141–143] Interestingly, several nucleotide polymorphisms associated with AD are located in the miRNA-binding region of the APP mRNA, which can modulate protein expressions.[105] Fig. 5 summarizes the miRNA-based therapeutic targets.

Another class of miRNAs that are downregulated in AD is the neuron-specific miR-124.[144] The downregulated miR-124 leads to an overexpression of its targeted mRNA, polypyrimidine-tract binding protein 1 (a pre-RNA splicing regulator), in turn leading to the altered splicing of APP.[144] miR-124 also targets the Aβ cleaving enzyme 1 (BACE1), and the downregulation of miR-124 promotes the transformation of APP into Aβ, probably by activating BACE1.[116] It is not known whether there are reduced levels of other miRNAs in the brains from humans and mice with AD. Reduced levels of miR-9, miR-29a, miR-29b, and miR-107 may

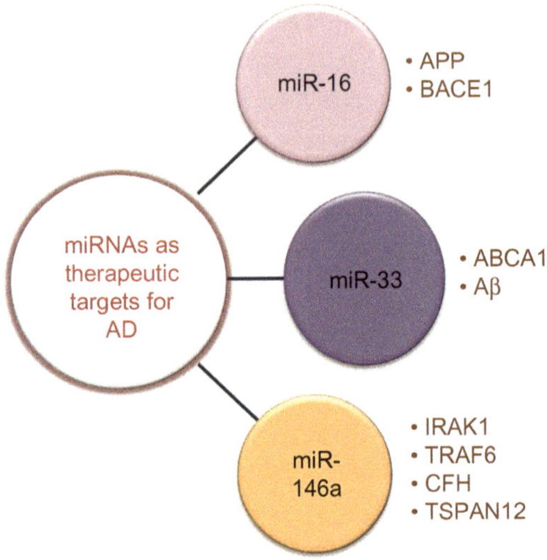

Fig. 5 miRNAs as potential therapeutics for Alzheimer's disease.

result in elevated BACE1 expression and an overproduction of Aβ known to characterize brains from humans and mice with AD.[142,145]

Studies in mutant AD mice suggest that miR-298 and miR-328 have similar roles.[146] Loss of miR-9, miR-29a, and miR-29b in AD, together with the loss of miR-137 and miR-181c disinhibits serine palmitoyltransferase, the rate-limiting enzyme of ceramide synthesis, leading to the mislocation of BACE1 in lipid rafts and augmenting the excessive processing of APP into Aβ.[147]

Increasing evidence suggests that miRNAs affect Aβ production. Several miRNAs increase Aβ levels and others reduce Aβ production (Table 2). On the other hand, Aβ itself reciprocally impacts the production of miRNAs, including miR-9, miR-106b, and let7.[148,149] Despite evidence in support of a role for reduced levels of miRNAs in AD, it remains unclear whether reduced miRNAs play a primary role in AD induction.

Kim and colleagues studied the effects of elevated levels of miRNA126 in dopamine neuronal cell survival in models of Parkinson's disease.[150] They showed that elevated levels of miR-126 increase the vulnerability of neurons to ubiquitous toxicity that is mediated by staurosporine or Aβ42. The neuroprotective factors IGF1, nerve growth factor, brain-derived neurotrophic factor, and soluble APP α could diminish, but not abrogate, the toxic effects of miR-126. MiR-126-overexpressing neurons from a Tg6799 familial mouse model of AD exhibited an increase in Aβ42 toxicity, but surprisingly, both Aβ42 and miR-126 promoted neurite sprouting. Pathway analysis revealed that the overexpression of miR-126 downregulated elements in the growth factor (GF)/phosphatidylinositol-3-kinase (PI3K)/AKT and extracellular signal-regulated kinase (ERK) signaling cascades, including AKT, GSK-3β, and ERK; the phosphorylation of tau; and the miR-126 targets insulin receptor substrate 1 (IRS1) and phosphoinositide-3-kinase regulatory subunit 2 (PIK3R2). In this same study, the inhibition of miR-126 was found to be neuroprotective against both STS and Aβ42 toxicity. Although the Kim study focused on Parkinson's disease, it provides evidence for a novel miR-126 mechanism in a neurodegenerative disease that is capable of regulating GF/PI3K signaling in neurons, suggesting that miR-126 may be an important mechanistic link between metabolic dysfunction and neurotoxicity in another neurodegenerative disease, namely AD.

5.5 BACE1 and miRNAs

Altered levels of miRNAs increase the production of Aβ and BACE1 activity. Yang and colleagues studied the expression levels of the miR-29 family in peripheral blood samples from patients with AD and age-matched controls.[124] They found a comparatively marked decrease in miR-29c expression and a significant increase in BACE1 expression in the samples from the AD patients. Correlation analysis revealed that miR-29c expression negatively correlated with the protein expression of BACE1 in the samples from the AD patients. Yang and colleagues also investigated the role of miR-29 on hippocampal neurons in vitro and in vivo. They found miR-29c upregulation promoted learning and memory behaviors in senescence-accelerated mouse prone 8 (SAMP8) mice by increasing the activity of the protein kinase A/cAMP response element-binding protein, which is involved in neuroprotection, suggesting that miR-29c may be a possible therapeutic target against AD.[124]

Zhang and colleagues studied the role of miR-188-3p that targets BACE1 in humans and mice with AD.[37] They found miR-188-3p to be significantly downregulated in the brains of AD humans and the AD transgenic (TG) mice, an APP mouse model of AD. The downregulated miR-188-3p was restored by the inhibition of monoacylglycerol lipase (MAGL). Overexpression of miR-188-3p in the hippocampus of the TG mice reduced BACE1, Aβ, and neuroinflammation, and prevented deterioration in hippocampal basal synaptic transmission, long-term potentiation, spatial learning, and memory. Loss of miR-188-3p function correlated with 2-AG-induced suppression of BACE1. Moreover, miR-188-3p expression was upregulated by 2-AG or peroxisome proliferator-activated receptor γ (PPARγ) agonists and suppressed by PPARγ antagonism or nuclear factor-kappa B subunit (NF-κB) activation. Reduction of Aβ and neuroinflammation by MAGL inhibition was occluded by PPARγ antagonism. In addition, BACE1 suppression by 2-AG and PPARγ activation was eliminated by the knockdown of NF-κB. The Zhang study revealed a novel molecular mechanism underlying improved synaptic and cognitive function in TG mice by 2-AG signalin—a mechanism that appears to upregulate miR-188-3p expression through the PPARγ and NF-κB signaling pathways, resulting in suppression of BACE1 expression and the consequent formation of Aβ.[37]

Long and colleagues identified miR-339-5p, a known miRNA, as a key contributor to the regulatory network.[120] Two distinct miR-339-5p target sites were predicted in the BACE1 3′-UTR by in silico analyses, and both were found to be linked to BACE1. Cotransfection of miR-339-5p with a BACE1 3′-UTR reporter construct resulted in significant reduction in reporter expression. Mutation of both target sites eliminated this effect. Delivery of the miR-339-5p mimic also significantly inhibited expression of the BACE1 protein in human glioblastoma cells and human primary brain cultures. Delivery of target protectors designed against the miR-339-5p BACE1 3′-UTR target sites in primary human brain cultures significantly elevated BACE1 expression. In addition, miR-339-5p levels were significantly reduced in brain specimens from AD patients compared to those from age-matched controls. Therefore, miR-339-5p appears to regulate BACE1 expression in human brain cells and to be dysregulated in at least a subset of AD patients, warranting the study of miR-339-5p as a novel drug target for patients with AD.[120]

Using AD mouse models, Boissonneault and colleagues studied the roles of miR-298 and miR-328 in Aβ production in mice with AD.[146] They observed a loss in correlation between BACE1 mRNA and protein levels in the hippocampus of the AD mouse model. These findings prompted an investigation of the regulatory role of the BACE1 3′-UTR element in AD progression and the possible involvement of specific miRNAs in cultured neuronal cells and fibroblastic cells from humans with AD. Using such experimental approaches, these researchers demonstrated that miR-298 and miR-328 recognize specific binding sites in the 3′-UTR of the BACE1 mRNA and exert regulatory effects on ACE1 protein expression in cultured neuronal cells.[146] These results may point to a molecular basis underlying BACE1 deregulation in AD and may offer new perspectives on AD.

Galimberti *et al.* studied the profiles of circulating miRNAs in serum and cerebrospinal fluids (CSF) from humans with AD and correlated them with profiles of AD patients.[151] Using a two-step analysis—microarray analysis followed by validation via real-time PCR—they found miR-23a to be downregulated in the serum from 22 AD patients compared to 18 non-inflammatory controls (NINDCs), 8 inflammatory neurological controls (INDCs), and 10 patients with frontotemporal dementia. Significant down-regulation of miR-125b and of miR-26b was also confirmed in the CSF from AD patients. These researchers found that cell-free miR-125b serum levels from AD patients are less than levels from INDCs and NINDCs with an accuracy of 82%.

Alexandrov and colleagues studied the effects of miRNA-34a on triggering receptor expressed on myeloid cells 2 (TREM2) mRNA 3′-UTR of TREM2 and found that miRNA-34a significantly downregulated TREM2 expression.[152] Mutations in TREM2 are known to cause rare, autosomal recessive forms of early onset dementia that present with and without bone cysts and fractures Aluminum-induced miRNA-34a upregulation and TREM2 downregulation were effectively quenched with the natural phenolic compound and the NF-κB inhibitor CAPE (2-phenylethyl-(2*E*)-3-(3,4-dihydroxyphenyl) acrylate; caffeic acid phenethyl ester). These results suggest that an epigenetic mechanism in AD involving an aluminum-triggered, NF-κB-sensitive, miRNA-34a-mediated downregulation of TREM2 expression may impair phagocytic responses that may ultimately contribute to an accumulation and aggregation of the Aβ42 peptide, amyloidogenesis, and inflammatory degeneration in the AD brain.[152]

Alexandrov *et al.* analyzed the relative amount of Aβ and miRNA in CSF from the neocortices of patients with AD and of age-matched controls, in short postmortem intervals (PMI < 2.1 h).[51] They found a decreased but nonsignificant abundance of Aβ42 in the CSF and extracellular fluid (ECF) of AD patients. The most abundant nucleic acids in the CSF and ECF from AD patients were miRNAs. This result led to additional studies of the speciation and inducibility. Fluorescent miRNA-array-based analysis indicated significant increases in miRNA-9, miRNA-125b, miRNA-146a, and miRNA-155 in the CSF and ECF of AD patients.[51] Primary human neuronal-glial cell cocultures stressed with AD-derived ECF also displayed an upregulation of these four miRNAs, an effect that was quenched using the anti-NF-κB agent caffeic-acid phenethyl ester. Increases in miRNAs were confirmed independently, using a highly sensitive LED-Northern dot blot assay. Several of these NF-κB-sensitive miRNAs are known to be upregulated in AD brain and are associated with the progressive spreading of inflammatory neurodegeneration. Results from these confirmation studies indicated that miRNA-9, miRNA-125b, miRNA-146a, and miRNA-155 are CSF and ECF abundant. NF-κB-sensitive proinflammatory miRNAs, and their enrichment in circulating CSF and ECF, suggest that miRNAs may be involved in the modulation or proliferation of miRNA-triggered pathogenic signaling throughout the central nervous system.[152]

Using SAMP8 mice and BALb/c mice, Liu *et al.* examined the posttranscriptional regulation mechanism of APP mediated by microribonucleic acids.[153] They found miR-16 to be one of the posttranscriptional regulators of APP in the SAMP8 mice. Overexpression of miR-16, both in vitro and in vivo, led to reduced APP expression. Further, miR-16 and APP displayed complementary expression patterns in the SAMP8 mice and BALb/c embryos. Taken together, these findings indicate that an abnormally low expression of miR-16 could lead to an accumulation of APP in AD mice and that APP may be a target for miR-16.[153]

Pogue and colleagues studied the effects of complement factor H on metal-sulfate-stressed human brain cells when miR-146a was modulated.[154] They found an NF-κB-sensitive, miRNA-146a-mediated modulation of CFH gene expression in AD neurons that may contribute to inflammatory responses in aluminum-stressed HN cells. This finding underscores the potential of just a nanomolar of aluminum that may be necessary to drive genotoxic mechanisms characteristic of neurodegenerative disease processes.

5.6 Alpha-Secretase and miRNAs

Aβ secretase is an enzyme in AD that cleaves the amyloid domain of APP and reduces the production of the Aβ peptide in neurons. There are numerous miRNAs that are involved in the increased activity of alpha-secretase in neurons, and there are other miRNAs responsible for reduced alpha-secretase activity, the latter group of which results in an increase in alpha-secretase production and a cascade of cellular changes in AD progression.

Interestingly, the loss of miR-107 disinhibits the alpha-secretase ADAM10 and favors the nonamyloidogenic pathway of APP processing. Compensatory effects from this disinhibitation shunt the cleavage of APP away from the generation of Aβ plaque toward the generation of soluble APP.[155]

5.7 CSF and miRNAs

Using open-array technology, Denk and colleagues studied the CSF of AD patients ($n=22$) and controls ($n=28$) to profile the expression of 1178 different miRNAs.[156] Using a Cq of 34 as cut-off, they identified positive signals from 441 miRNAs, but could not identify positive signals from 729 other miRNAs indicating that at least 37% of all miRNAs in the body are present in the brain. They found 74 downregulated miRNAs and 74 upregulated miRNAs with a 1.5-fold change threshold. By applying the new explorative "measure of relevance" method, they identified six reliable and nine informative biomarkers. Confirmatory multivariate analysis of covariance (MANCOVA) revealed reliable miR-100, miR-146a, and miR-1274a as differentially expressed in AD, an analysis that reached Bonferroni-corrected significance. MANCOVA also confirmed the differential expression of informative miR-103, miR-375, miR-505, miR-708, miR-4467, miR-219, miR-296, miR-766, and miR-3622b-3p. Discrimination analysis using a combination of miR-100, miR-103, and miR-375 detected AD in the CSF by positively classifying controls and AD patients with 96.4% and 95.5% accuracy, respectively. Using the ingenuity database, Denk *et al.* identified a set of AD-associated genes that these miRNAs targeted.[156] These targets included genes involved in the regulation of tau and amyloid pathways in AD, such as MAPT, BACE1, and mTOR.

5.8 Gamma-Secretase Complex and miRNAs

γ-Secretases are a group of widely expressed, intramembrane-cleaving proteases involved in many physiological processes associated with AD. Mutations in PS1 and PS2 lead to increased γ-secretase activity, which has been associated with the formation of Aβ in AD. In addition to PS1 and PS2, two other subunits—nicastrin and anterior-pharynx defective-1—were identified as essential cofactors. These four γ-secretase enzymes together generate an active and stable complex that cleaves APP at the end of the Aβ domain in APP. Inhibition of these enzymes redirects the amyloidogenic pathway toward the nonamyloidogenic pathway by reducing Aβ production. Similar to miRNAs that activate BACE1, several MiRNAs are believed to be involved in the increased production of γ-secretase.

Loss of PS function has been proposed to underlie memory impairment and neurodegeneration in AD pathogenesis.[157] Using brain tissue from the PS1 knockout mouse, Krichevsky and colleagues studied the miRNA profiles.[158] They found that the downregulation of miR-9 coincides with neurodegeneration in PS1 knockout mice. Other studies using zebrafish and mice found miR-9 to be an important regulator of neurogenesis.[159,160] Based on these results, miR-9, which is downregulated in the AD brain, may actively participate in maintaining neurons and in sustaining Aβ production. Further research is needed to determine the role of γ-secretase-linked miRNAs in AD.

5.9 Tau and miRNAs

The detrimental effects of miRNA changes in AD neurons might not be restricted to Aβ formation. For example, the loss of miR-15a favors the hyperphosphorylation of tau by disinhibiting ERK1.[161] Increased levels of miR-128 lead to decreased activity of Bcl2-associated athanogene, leading to a reduction in the removal of sarkosyl-insoluble tau, a reduction that is known to favor the formation of toxic tau inclusions.[162] Based on studies in an AD TG mouse model, putative increases in levels of miR-34a were found to suppress Bcl2 expression, which exacerbated neuronal loss by the recruitment of caspase-3 and apoptosis.[163] Increases in miR-206, in the temporal cortex of AD brains, led to the suppression of BNDF, which contributed to compromises in morphological and functional synaptic plasticity.[164]

Li and colleagues examined miRNA-146a levels in several human primary brain and retinal cell lines from the neocortex and hippocampus of patients in early-, moderate-, and late-stage AD, and of five different TG mouse models of AD (Tg2576, TgCRND8, PSAPP, 3xTg-AD, and 5xFAD).[55] Inducible expression of miRNA-146a was significantly upregulated in a primary coculture of human neuronal-glial cells that were stressed using interleukin 1-beta. This upregulation was quenched using specific NF-κB inhibitors that include curcumin. Expression of miRNA-146a correlated with senile plaque density and synaptic pathology in the Tg2576 and 5xFAD TG mouse models.[55]

5.10 ApoE4 and miRNAs

The apolipoprotein E4 (ApoE4) genotype is a major risk factor for late-onset sporadic AD. Alterations in miRNAs and cognitive decline are expected in humans with the ApoE4 genotype—mainly because the ApoE4 status increases the production of Aβ. Several lines of evidence support this notion. However, there are no published studies that have linked miRNAs and the ApoE4 genotype in AD patients.

Kim *et al.* studied miR-33 and its relationship to adenosine triphosphate-binding cassette subfamily A member 1 (ABCA1) and Aβ levels in the brain.[134] Overexpression of miR-33 impaired cellular cholesterol efflux and dramatically increased extracellular Aβ levels by promoting the secretion of Aβ and impairing the clearance of Aβ in neurons. In contrast, genetic deletion of mir-33 in mice dramatically increased ABCA1 levels and ApoE lipidation, but decreased endogenous Aβ levels in the cortex. Most importantly, pharmacological inhibition of miR-33 via antisense oligonucleotide specifically in the brain markedly decreased Aβ levels in the cortex of APP/PS1 mice, suggesting that miR-33 is a potential therapeutic strategy for AD. Additional research is needed to determine how ApoE4-linked miRNAs alter cellular changes in the AD brain, such as altering the production and accumulation of Aβ, the phosphorylation of tau, and the triggering of synaptic damage. Further research is needed to understand precise links between miRNAs and ApoE4 genotype association with Aβ levels in patients with AD and mouse models of AD.

5.11 Inflammation and miRNAs

Inflammatory responses are strongly associated with altered miRNA expressions in the AD brain. miRNA-155 is involved in diverse physiological and pathological mechanisms, such as inflammation and immunity. Recent studies indicate that miR-155 regulates T-cell functions during inflammation. Song and Lee investigated the role of miRNA-155 in AD, finding miRNA-155 to be a multifunctional miRNA in AD pathogenesis, with a distinct expression profile and links to T-cell functions.[49]

In addition, in studies of miR-125b and miR-146 in the human AD brain, levels of these miRNA were found to be elevated, which might aggravate neuroinflammation[165,166] and reduce complement factor H, which is associated with the neuronal release of mR-146a and miR-155 and inflammatory spreading in the AD brain.[165,167] Altered miR-106b levels impact the expression of transforming growth factor beta178.

In investigating the significance of miRNA release in the AD brain, Lehmann *et al.* focused on the miRNA let-7b.[59] They found that, following its release, let-7b activates the toll-like receptor 7, resulting in neuronal degeneration. They also found that loss of miR-29a disrupts the activity of another target gene, neuronal navigator 3, a protein that is involved in axonal guidance and is enriched in degenerating pyramidal neurons in AD.

5.12 Mitochondrial miRNAs and AD

Dysfunction of mitochondria and oxidative stress has been found to be involved in neurodegenerative diseases, including AD. Mitochondria are cytoplasmic organelles, and control/regulate cell survival and cell death. Mitochondria performs several key cellular functions, including ATP production, regulation of intracellular calcium, apoptotic cell death, sites of free radical production, and scavenging and activation of caspase family of proteases. Mitochondria are synthesized in cell soma, travel along axons and dendrites, and supply ATP for several synaptic functions, including synapse formation and outgrowth, neurotransmitter release and vesicle fusion. Mitochondria move from cell soma to nerve terminals via kinesin-based anterograde fashion and travel back to cell soma via dynein-based retrograde manner.

The human mitochondria carries 37 polypeptide genes in a 16.5 kb circular genome. The DNA of mitochondria has two strands: an outer strand enriched with guanine (heavy strand) and an inner strand enriched with cytosine (light strand). It also has a noncoding segment comprised of a displacement loop, a region of 1121 base pairs.

Studies have identified several mitochondrial miRNAs in the human mitochondria[168–174] (Fig. 6). More recently, Shinde and Bhandra have identified six pr-miRNAs and miRNAs from mitochondria genome, indicating miRNAs in the mitochondrial genome.[175] Barrey *et al.* identified 169 miRNAs in the mitochondrial genome that are believed to regulate polypeptide genes in the mitochondrial genome, oxidative phosphorylation, and ATP synthesis.[174]

miRNAs regulate mitochondrial structure. A recent study of miR-761 found that it is responsible for the downregulation of the mitochondrial fission factor and the suppression of mitochondrial fission machinery.[176] In studies of miR-30, Goud and Hua found miR-30a, -30b, and -30d to be highly expressed in myocardial cells that are exposed to hydrogen peroxide. The gene p53, a target of the miR-30 family, promotes dynamin-related protein 1 (Drp1) transcription while triggering apoptosis. Goud and Hua concluded that miR-30 regulates mitochondrial fission and apoptosis via the targeting of P53 and Drp1.[177]

miR-30 and miR-499 have been found to be involved in regulating mitochondrial dynamics via Drp1 through p53 and calcineurin.[87,178] Li *et al.* found that miR-30 family members inhibit mitochondrial fission and target p53, which is known to induce mitochondrial fission by transcriptionally upregulating Drp1 expression. miR-30 inhibits mitochondrial

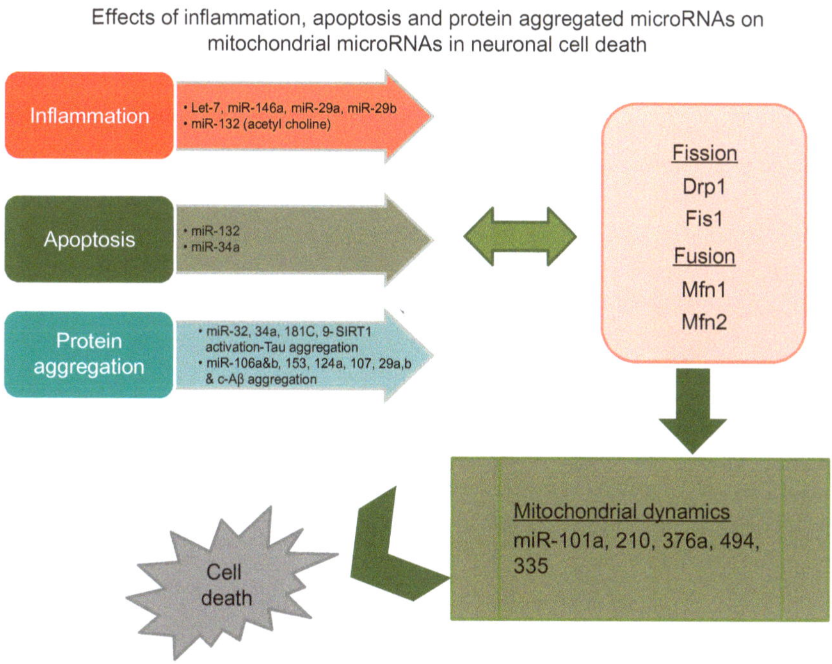

Fig. 6 Mitochondrial microRNAs in aging, cellular senescence, and Alzheimer's disease.

fission by suppressing the expression of p53 and its downstream target Drp1.[178] Regarding miR-499, Wang and colleagues found that it directly targets α and β isoforms of the calcineurin and inhibits cardiomyocyte apoptosis by suppressing calcineurin-mediated dephosphorylation of Drp1, which in turn decreases the translocation of Drp1 to mitochondria and Drp1-mediated activation of mitochondrial fission. Findings from these studies revealed that Drp1 is regulated by p53.[87,179] However, there are no published studies characterizing the role of miRNAs in mitochondrial dynamics either for fission activity or fusion in the AD. Additional research is needed to understand the role of miRNAs, particularly mitochondrial miRNAs, in mitochondrial dynamics and mitochondrial biogenesis in the disease process of AD.

6. CONCLUDING REMARKS

Multiple cellular changes, including oxidative damage, mitochondrial dysfunction, telomere shortening, and inflammation are reported to involve in aging process and cellular senescence. A large number of human diseases are associated with aging including Alzheimer's, Parkinson's, multiple sclerosis, amyotrophic lateral sclerosis, cardiovascular, cancer, and skin diseases. Changes in somatic and germ-line DNA and epigenetics are known to key role in accelerating the onset of human diseases. Among, age-related diseases, Alzheimer's continue to be a growing health concern that affects millions of persons worldwide. Although progress that has been made in AD research in terms of understanding the molecular basis of early onset familial AD and late-onset sporadic AD, we still do not have drugs or agents that can delay or prevent AD progression, and we still have not identified early detectable biomarkers for AD. A major breakthrough in developing such biomarkers is the discovery that miRNAs connect missing link between cellular changes and disease progression.

There are many questions about miRNAs in the AD brain that need to be answered, including which specific miRNAs are involved in cellular changes associated with AD pathogenesis and progression; which specific miRNAs are involved in cellular changes associated with other neurodegenerative diseases and aging; whether AD can be prevented, delayed, or stopped via strategic alterations of miRNA expression; and whether and how blood and imaging tests can be developed that focus on identifying miRNAs changes that correspond with AD onset and progression. Comprehensive epidemiological-based miRNA studies are urgently needed to inform the development of miRNA-based diagnostic tools.

ACKNOWLEDGMENTS

Work presented in this chapter is supported by NIH Grants—AG042178 and AG47812, the Garrison Family Foundation, and Sex and Gender Alzheimer's Association (SAGA) Grant (to P.H.R.). Present work is also supported by Alzheimer's Association New Investigator Research Grant 2016-NIRG-39787 and Center of Excellence for Translational Neuroscience and Therapeutics Grant number PN-CTNT20115-AR (to A.P.R.).

REFERENCES

1. Borgdorff V, Lleonart ME, Bishop CL, et al. Multiple microRNAs rescue from Ras-induced senescence by inhibiting p21(Waf1/Cip1). *Oncogene*. 2010;29(15):2262–2271.
2. Gomez-Cabello D, Adrados I, Gamarra D, et al. DGCR8-mediated disruption of miRNA biogenesis induces cellular senescence in primary fibroblasts. *Aging Cell*. 2013;12(5):923–931.
3. Overhoff MG, Garbe JC, Koh J, Stampfer MR, Beach DH, Bishop CL. Cellular senescence mediated by p16INK4A-coupled miRNA pathways. *Nucleic Acids Res*. 2014;42(3):1606–1618.

4. Ugalde AP, Ramsay AJ, de la Rosa J, et al. Aging and chronic DNA damage response activate a regulatory pathway involving miR-29 and p53. *EMBO J.* 2011;30(11): 2219–2232.
5. Yamakuchi M, Ferlito M, Lowenstein CJ. miR-34a repression of SIRT1 regulates apoptosis. *Proc Natl Acad Sci USA.* 2008;105(36):13421–13426.
6. Llave C, Xie Z, Kasschau KD, Carrington JC. Cleavage of Scarecrow-like mRNA targets directed by a class of Arabidopsis miRNA. *Science.* 2002;297(5589): 2053–2056.
7. Kumar S, Reddy PH. Are circulating microRNAs peripheral biomarkers for Alzheimer's disease? *Biochim Biophys Acta.* 2016;1862(9):1617–1627; (a) Maffioletti E, Tardito D, Gennarelli M, Bocchio-Chiavetto L. Micro spies from the brain to the periphery: new clues from studies on microRNAs in neuropsychiatric disorders. *Front Cell Neurosci.* 2014;8:75.
8. Peters L, Meister G. Argonaute proteins: mediators of RNA silencing. *Mol Cell.* 2007;26(5):611–623.
9. Baek D, Villen J, Shin C, Camargo FD, Gygi SP, Bartel DP. The impact of microRNAs on protein output. *Nature.* 2008;455(7209):64–71.
10. Noren Hooten N, Fitzpatrick M, Wood 3rd WH, et al. Age-related changes in microRNA levels in serum. *Aging (Albany NY).* 2013;5(10):725–740.
11. Bartel DP. MicroRNAs: genomics, biogenesis, mechanism, and function. *Cell.* 2004;116(2):281–297.
12. Shao NY, Hu HY, Yan Z, et al. Comprehensive survey of human brain microRNA by deep sequencing. *BMC Genomics.* 2010;11:409.
13. Adlakha YK, Saini N. Brain microRNAs and insights into biological functions and therapeutic potential of brain enriched miRNA-128. *Mol Cancer.* 2014;13:33.
14. Maes OC, An J, Sarojini H, Wang E. Murine microRNAs implicated in liver functions and aging process. *Mech Ageing Dev.* 2008;129(9):534–541.
15. Xu X, Chen W, Miao R, et al. miR-34a induces cellular senescence via modulation of telomerase activity in human hepatocellular carcinoma by targeting FoxM1/c-Myc pathway. *Oncotarget.* 2015;6(6):3988–4004.
16. Mastroeni D, McKee A, Grover A, Rogers J, Coleman PD. Epigenetic differences in cortical neurons from a pair of monozygotic twins discordant for Alzheimer's disease. *PLoS One.* 2009;4(8), e6617.
17. Villela D, Ramalho RF, Silva AR, et al. Differential DNA methylation of MicroRNA genes in temporal cortex from Alzheimer's disease individuals. *Neural Plast.* 2016;2016:2584940.
18. Jiang JJ, Liu CM, Zhang BY, et al. MicroRNA-26a supports mammalian axon regeneration in vivo by suppressing GSK3beta expression. *Cell Death Dis.* 2015;6, e1865; (a) Bird CW, Gardiner AS, Bolognani F. KSRP modulation of GAP-43 mRNA stability restricts axonal outgrowth in embryonic hippocampal neurons. *PLoS One.* 2013;8, e79255.
19. Meng X, Wang W, Lu H, et al. Manipulations of MeCP2 in glutamatergic neurons highlight their contributions to Rett and other neurological disorders. *Elife.* 2016;5. http://dx.doi.org/10.7554/eLife.14199, pii: e14199.
20. Han KH, Choi HR, Won CH, et al. Alteration of the TGF-beta/SMAD pathway in intrinsically and UV-induced skin aging. *Mech Ageing Dev.* 2005;126(5):560–567.
21. Cheng S, Zhang C, Xu C, Wang L, Zou X, Chen G. Age-dependent neuron loss is associated with impaired adult neurogenesis in forebrain neuron-specific Dicer conditional knockout mice. *Int J Biochem Cell Biol.* 2014;57:186–196.
22. Reis FC, Branquinho JL, Brandao BB, et al. Fat-specific Dicer deficiency accelerates aging and mitigates several effects of dietary restriction in mice. *Aging (Albany NY).* 2016;8(6):1201–1222.

23. Nairz K, Rottig C, Rintelen F, Zdobnov E, Moser M, Hafen E. Overgrowth caused by misexpression of a microRNA with dispensable wild-type function. *Dev Biol.* 2006;291(2):314–324.
24. Xu P, Vernooy SY, Guo M, Hay BA. The Drosophila microRNA Mir-14 suppresses cell death and is required for normal fat metabolism. *Curr Biol.* 2003;13(9): 790–795.
25. Johnston RJ, Hobert O. A microRNA controlling left/right neuronal asymmetry in Caenorhabditis elegans. *Nature.* 2003;426(6968):845–849.
26. Boehm M, Slack F. A developmental timing microRNA and its target regulate life span in C. elegans. *Science.* 2005;310(5756):1954–1957.
27. Nidadavolu LS, Niedernhofer LJ, Khan SA. Identification of microRNAs dysregulated in cellular senescence driven by endogenous genotoxic stress. *Aging (Albany NY).* 2013;5(6):460–473.
28. Wu H, Huang C, Taki FA, et al. Benzo-alpha-pyrene induced oxidative stress in Caenorhabditis elegans and the potential involvements of microRNA. *Chemosphere.* 2015;139:496–503.
29. Zhang R, Zhang Q, Niu J, et al. Screening of microRNAs associated with Alzheimer's disease using oxidative stress cell model and different strains of senescence accelerated mice. *J Neurol Sci.* 2014;338(1–2):57–64.
30. Magenta A, Cencioni C, Fasanaro P, et al. miR-200c is upregulated by oxidative stress and induces endothelial cell apoptosis and senescence via ZEB1 inhibition. *Cell Death Differ.* 2011;18(10):1628–1639.
31. Li G, Luna C, Qiu J, Epstein DL, Gonzalez P. Alterations in microRNA expression in stress-induced cellular senescence. *Mech Ageing Dev.* 2009;130(11–12):731–741.
32. Vasa-Nicotera M, Chen H, Tucci P, et al. miR-146a is modulated in human endothelial cell with aging. *Atherosclerosis.* 2011;217(2):326–330.
33. Margis R, Margis R, Rieder CR. Identification of blood microRNAs associated to Parkinson's disease. *J Biotechnol.* 2011;152(3):96–101.
34. Wang GK, Zhu JQ, Zhang JT, et al. Circulating microRNA: a novel potential biomarker for early diagnosis of acute myocardial infarction in humans. *Eur Heart J.* 2010;31(6):659–666.
35. Redshaw Z, Sweetman D, Loughna PT. The effects of age upon the expression of three miRNAs in muscle stem cells isolated from two different porcine skeletal muscles. *Differentiation.* 2014;88(4–5):117–123.
36. Lang A, Grether-Beck S, Singh M, et al. MicroRNA-15b regulates mitochondrial ROS production and the senescence-associated secretory phenotype through sirtuin 4/SIRT4. *Aging (Albany NY).* 2016;8(3):484–505.
37. Zhang J, Hu M, Teng Z, Tang YP, Chen C. Synaptic and cognitive improvements by inhibition of 2-AG metabolism are through upregulation of microRNA-188-3p in a mouse model of Alzheimer's disease. *J Neurosci.* 2014;34(45):14919–14933.
38. Poliseno L, Pitto L, Simili M, et al. The proto-oncogene LRF is under post-transcriptional control of MiR-20a: implications for senescence. *PLoS One.* 2008;3(7), e2542.
39. Soreq L, Salomonis N, Bronstein M, et al. Small RNA sequencing-microarray analyses in Parkinson leukocytes reveal deep brain stimulation-induced splicing changes that classify brain region transcriptomes. *Front Mol Neurosci.* 2013;6:10.
40. Soares RJ, Cagnin S, Chemello F, et al. Involvement of microRNAs in the regulation of muscle wasting during catabolic conditions. *J Biol Chem.* 2014;289(32): 21909–21925.
41. Li N, Bates DJ, An J, Terry DA, Wang E. Up-regulation of key microRNAs, and inverse down-regulation of their predicted oxidative phosphorylation target genes, during aging in mouse brain. *Neurobiol Aging.* 2011;32(5):944–955.

42. Zhao H, Wen G, Huang Y, et al. MicroRNA-22 regulates smooth muscle cell differentiation from stem cells by targeting methyl CpG-binding protein 2. *Arterioscler Thromb Vasc Biol.* 2015;35(4):918–929.
43. Hudson MB, Woodworth-Hobbs ME, Zheng B, et al. miR-23a is decreased during muscle atrophy by a mechanism that includes calcineurin signaling and exosome-mediated export. *Am J Physiol Cell Physiol.* 2014;306(6):C551–C558.
44. Chang TC, Wentzel EA, Kent OA, et al. Transactivation of miR-34a by p53 broadly influences gene expression and promotes apoptosis. *Mol Cell.* 2007;26(5):745–752.
45. Bai XY, Ma Y, Ding R, Fu B, Shi S, Chen XM. miR-335 and miR-34a Promote renal senescence by suppressing mitochondrial antioxidative enzymes. *J Am Soc Nephrol.* 2011;22(7):1252–1261.
46. Yang J, Chen D, He Y, et al. MiR-34 modulates Caenorhabditis elegans lifespan via repressing the autophagy gene atg9. *Age (Dordr).* 2013;35(1):11–22.
47. Yu X, Zhang L, Wen G, et al. Upregulated sirtuin 1 by miRNA-34a is required for smooth muscle cell differentiation from pluripotent stem cells. *Cell Death Differ.* 2015;22(7):1170–1180.
48. Mimura S, Iwama H, Kato K, et al. Profile of microRNAs associated with aging in rat liver. *Int J Mol Med.* 2014;34(4):1065–1072.
49. Song J, Lee JE. miR-155 is involved in Alzheimer's disease by regulating T lymphocyte function. *Front Aging Neurosci.* 2015;7:61.
50. Roggli E, Britan A, Gattesco S, et al. Involvement of microRNAs in the cytotoxic effects exerted by proinflammatory cytokines on pancreatic beta-cells. *Diabetes.* 2010;59(4):978–986.
51. Alexandrov PN, Dua P, Hill JM, Bhattacharjee S, Zhao Y, Lukiw WJ. microRNA (miRNA) speciation in Alzheimer's disease (AD) cerebrospinal fluid (CSF) and extracellular fluid (ECF). *Int J Biochem Mol Biol.* 2012;3(4):365–373.
52. Le MT, Teh C, Shyh-Chang N, et al. MicroRNA-125b is a novel negative regulator of p53. *Genes Dev.* 2009;23(7):862–876.
53. Williams AH, Valdez G, Moresi V, et al. MicroRNA-206 delays ALS progression and promotes regeneration of neuromuscular synapses in mice. *Science.* 2009;326(5959):1549–1554.
54. Wang X, Guo B, Li Q, et al. miR-214 targets ATF4 to inhibit bone formation. *Nat Med.* 2013;19(1):93–100.
55. Li YY, Cui JG, Hill JM, Bhattacharjee S, Zhao Y, Lukiw WJ. Increased expression of miRNA-146a in Alzheimer's disease transgenic mouse models. *Neurosci Lett.* 2011;487(1):94–98.
56. Faraonio R, Salerno P, Passaro F, et al. A set of miRNAs participates in the cellular senescence program in human diploid fibroblasts. *Cell Death Differ.* 2012;19(4):713–721.
57. Puissegur MP, Mazure NM, Bertero T, et al. miR-210 is overexpressed in late stages of lung cancer and mediates mitochondrial alterations associated with modulation of HIF-1 activity. *Cell Death Differ.* 2011;18(3):465–478.
58. Bandiera S, Ruberg S, Girard M, et al. Nuclear outsourcing of RNA interference components to human mitochondria. *PLoS One.* 2011;6(6), e20746.
59. Lehmann SM, Kruger C, Park B, et al. An unconventional role for miRNA: let-7 activates Toll-like receptor 7 and causes neurodegeneration. *Nat Neurosci.* 2012;15(6):827–835.
60. Olivieri F, Spazzafumo L, Santini G, et al. Age-related differences in the expression of circulating microRNAs: miR-21 as a new circulating marker of inflammaging. *Mech Ageing Dev.* 2012;133(11–12):675–685.
61. Taganov KD, Boldin MP, Chang KJ, Baltimore D. NF-kappaB-dependent induction of microRNA miR-146, an inhibitor targeted to signaling proteins of innate immune responses. *Proc Natl Acad Sci USA.* 2006;103(33):12481–12486.

62. Fan Y, Yin S, Hao Y, et al. miR-19b promotes tumor growth and metastasis via targeting TP53. *RNA*. 2014;20(6):765–772.
63. Ziyan W, Shuhua Y, Xiufang W, Xiaoyun L. MicroRNA-21 is involved in osteosarcoma cell invasion and migration. *Med Oncol*. 2011;28(4):1469–1474.
64. Yang N, Wang G, Hu C, et al. Tumor necrosis factor alpha suppresses the mesenchymal stem cell osteogenesis promoter miR-21 in estrogen deficiency-induced osteoporosis. *J Bone Miner Res*. 2013;28(3):559–573.
65. Hu W, Chan CS, Wu R, et al. Negative regulation of tumor suppressor p53 by microRNA miR-504. *Mol Cell*. 2010;38(5):689–699.
66. Tan L, Yu JT, Liu QY, et al. Circulating miR-125b as a biomarker of Alzheimer's disease. *J Neurol Sci*. 2014;336(1–2):52–56.
67. Braun CJ, Zhang X, Savelyeva I, et al. p53-Responsive micrornas 192 and 215 are capable of inducing cell cycle arrest. *Cancer Res*. 2008;68(24):10094–10104.
68. Pitto L, Rizzo M, Simili M, et al. miR-290 acts as a physiological effector of senescence in mouse embryo fibroblasts. *Physiol Genomics*. 2009;39(3):210–218.
69. Sala Frigerio C, Lau P, Salta E, et al. Reduced expression of hsa-miR-27a-3p in CSF of patients with Alzheimer disease. *Neurology*. 2013;81(24):2103–2106.
70. Kurowska-Stolarska M, Alivernini S, Ballantine LE, et al. MicroRNA-155 as a proinflammatory regulator in clinical and experimental arthritis. *Proc Natl Acad Sci USA*. 2011;108(27):11193–11198.
71. Cardo LF, Coto E, de Mena L, et al. Profile of microRNAs in the plasma of Parkinson's disease patients and healthy controls. *J Neurol*. 2013;260(5):1420–1422.
72. Khoo SK, Petillo D, Kang UJ, et al. Plasma-based circulating MicroRNA biomarkers for Parkinson's disease. *J Parkinson's Dis*. 2012;2(4):321–331.
73. Zampetaki A, Kiechl S, Drozdov I, et al. Plasma microRNA profiling reveals loss of endothelial miR-126 and other microRNAs in type 2 diabetes. *Circ Res*. 2010; 107(6):810–817.
74. Karolina DS, Tavintharan S, Armugam A, et al. Circulating miRNA profiles in patients with metabolic syndrome. *J Clin Endocrinol Metab*. 2012;97(12):E2271–E2276.
75. Harris TA, Yamakuchi M, Ferlito M, Mendell JT, Lowenstein CJ. MicroRNA-126 regulates endothelial expression of vascular cell adhesion molecule 1. *Proc Natl Acad Sci USA*. 2008;105(5):1516–1521.
76. Asgeirsdottir SA, van Solingen C, Kurniati NF, et al. MicroRNA-126 contributes to renal microvascular heterogeneity of VCAM-1 protein expression in acute inflammation. *Am J Physiol Renal Physiol*. 2012;302(12):F1630–F1639.
77. Qin B, Yang H, Xiao B. Role of microRNAs in endothelial inflammation and senescence. *Mol Biol Rep*. 2012;39(4):4509–4518.
78. Feng X, Wang H, Ye S, et al. Up-regulation of microRNA-126 may contribute to pathogenesis of ulcerative colitis via regulating NF-kappaB inhibitor IkappaBalpha. *PLoS One*. 2012;7(12), e52782.
79. Machida T, Tomofuji T, Ekuni D, et al. MicroRNAs in salivary exosome as potential biomarkers of aging. *Int J Mol Sci*. 2015;16(9):21294–21309.
80. Luo Z, Feng X, Wang H, et al. Mir-23a induces telomere dysfunction and cellular senescence by inhibiting TRF2 expression. *Aging Cell*. 2015;14(3):391–399.
81. Guo N, Parry EM, Li LS, et al. Short telomeres compromise beta-cell signaling and survival. *PLoS One*. 2011;6(3), e17858.
82. Bonifacio LN, Jarstfer MB. MiRNA profile associated with replicative senescence, extended cell culture, and ectopic telomerase expression in human foreskin fibroblasts. *PLoS One*. 2010;5(9). http://dx.doi.org/10.1371/journal.pone.0012519, pii: e12519.
83. Li J, Lei H, Xu Y, Tao ZZ. miR-512-5p suppresses tumor growth by targeting hTERT in telomerase positive head and neck squamous cell carcinoma in vitro and in vivo. *PLoS One*. 2015;10(8), e0135265.

84. Frankel LB, Wen J, Lees M, et al. microRNA-101 is a potent inhibitor of autophagy. *EMBO J.* 2011;30(22):4628–4641.
85. Mikhaylova O, Stratton Y, Hall D, et al. VHL-regulated MiR-204 suppresses tumor growth through inhibition of LC3B-mediated autophagy in renal clear cell carcinoma. *Cancer Cell.* 2012;21(4):532–546.
86. Diaz-Prado S, Cicione C, Muinos-Lopez E, et al. Characterization of microRNA expression profiles in normal and osteoarthritic human chondrocytes. *BMC Musculoskelet Disord.* 2012;13:144.
87. Wang JX, Jiao JQ, Li Q, et al. miR-499 regulates mitochondrial dynamics by targeting calcineurin and dynamin-related protein-1. *Nat Med.* 2011;17(1):71–78.
88. Gao P, Tchernyshyov I, Chang TC, et al. c-Myc suppression of miR-23a/b enhances mitochondrial glutaminase expression and glutamine metabolism. *Nature.* 2009;458(7239):762–765.
89. Tazawa H, Tsuchiya N, Izumiya M, Nakagama H. Tumor-suppressive miR-34a induces senescence-like growth arrest through modulation of the E2F pathway in human colon cancer cells. *Proc Natl Acad Sci USA.* 2007;104(39):15472–15477.
90. Pomerantz J, Schreiber-Agus N, Liegeois NJ, et al. The Ink4a tumor suppressor gene product, p19Arf, interacts with MDM2 and neutralizes MDM2's inhibition of p53. *Cell.* 1998;92(6):713–723.
91. Yan X, Serre C, Bergeron L, et al. Modulation of a specific pattern of microRNAs, including miR-29a, miR-30a and miR-34a, in cultured human skin fibroblasts, in response to the application of a biofunctional ingredient that protects against cellular senescence in vitro. *J Cosmet Dermatol Sci Appl.* 2015;5(4):332–342.
92. Dinami R, Ercolani C, Petti E, et al. miR-155 drives telomere fragility in human breast cancer by targeting TRF1. *Cancer Res.* 2014;74(15):4145–4156.
93. Mitomo S, Maesawa C, Ogasawara S, et al. Downregulation of miR-138 is associated with overexpression of human telomerase reverse transcriptase protein in human anaplastic thyroid carcinoma cell lines. *Cancer Sci.* 2008;99(2):280–286.
94. Sheedy FJ, Palsson-McDermott E, Hennessy EJ, et al. Negative regulation of TLR4 via targeting of the proinflammatory tumor suppressor PDCD4 by the microRNA miR-21. *Nat Immunol.* 2010;11(2):141–147.
95. Bhaumik D, Scott GK, Schokrpur S, et al. MicroRNAs miR-146a/b negatively modulate the senescence-associated inflammatory mediators IL-6 and IL-8. *Aging (Albany NY).* 2009;1(4):402–411.
96. Nicolas FE, Lopez-Martinez AF. MicroRNAs in human diseases. *Recent Pat DNA Gene Seq.* 2010;4(3):142–154.
97. Bao B, Ali S, Kong D, et al. Anti-tumor activity of a novel compound-CDF is mediated by regulating miR-21, miR-200, and PTEN in pancreatic cancer. *PLoS One.* 2011;6(3), e17850.
98. Quiat D, Olson EN. MicroRNAs in cardiovascular disease: from pathogenesis to prevention and treatment. *J Clin Invest.* 2013;123(1):11–18.
99. Wei C, Henderson H, Spradley C, et al. Circulating miRNAs as potential marker for pulmonary hypertension. *PLoS One.* 2013;8(5)e64396.
100. Papagregoriou G, Erguler K, Dweep H, et al. A miR-1207-5p binding site polymorphism abolishes regulation of HBEGF and is associated with disease severity in CFHR5 nephropathy. *PLoS One.* 2012;7(2), e31021.
101. Koutsis G, Siasos G, Spengos K. The emerging role of microRNA in stroke. *Curr Top Med Chem.* 2013;13(13):1573–1588.
102. Beveridge NJ, Cairns MJ. MicroRNA dysregulation in schizophrenia. *Neurobiol Dis.* 2012;46(2):263–271.
103. Diez-Planelles C, Sanchez-Lozano P, Crespo MC, et al. Circulating microRNAs in Huntington's disease: emerging mediators in metabolic impairment. *Pharmacol Res.* 2016;108:102–110.

104. Mouradian MM. MicroRNAs in Parkinson's disease. *Neurobiol Dis.* 2012;46(2):279–284.
105. Delay C, Calon F, Mathews P, Hebert SS. Alzheimer-specific variants in the 3'UTR of amyloid precursor protein affect microRNA function. *Mol Neurodegener.* 2011;6:70.
106. Nicolas G, Wallon D, Goupil C, et al. Mutation in the 3'untranslated region of APP as a genetic determinant of cerebral amyloid angiopathy. *Eur J Hum Genet.* 2016;24(1):92–98.
107. Hernandez-Rapp J, Rainone S, Hebert SS. MicroRNAs underlying memory deficits in neurodegenerative disorders. *Prog Neuropsychopharmacol Biol Psychiatry.* 2016;73:79–86.
108. World Alzheimer Report 2015. https://www.alz.co.uk/research/WorldAlzheimerReport2015.pdf.
109. Selkoe DJ. Alzheimer's disease: genes, proteins, and therapy. *Physiol Rev.* 2001;81(2):741–766.
110. Mattson MP. Pathways towards and away from Alzheimer's disease. *Nature.* 2004;430(7000):631–639.
111. LaFerla FM, Green KN, Oddo S. Intracellular amyloid-beta in Alzheimer's disease. *Nat Rev Neurosci.* 2007;8(7):499–509.
112. Reddy PH, Manczak M, Mao P, Calkins MJ, Reddy AP, Shirendeb U. Amyloid-beta and mitochondria in aging and Alzheimer's disease: implications for synaptic damage and cognitive decline. *J Alzheimers Dis.* 2010;20(suppl 2):S499–S512.
113. Reddy PH, Tripathi R, Troung Q, et al. Abnormal mitochondrial dynamics and synaptic degeneration as early events in Alzheimer's disease: implications to mitochondria-targeted antioxidant therapeutics. *Biochim Biophys Acta.* 2012;1822(5):639–649.
114. Hebert SS, Horre K, Nicolai L, et al. Loss of microRNA cluster miR-29a/b-1 in sporadic Alzheimer's disease correlates with increased BACE1/beta-secretase expression. *Proc Natl Acad Sci USA.* 2008;105(17):6415–6420.
115. Vilardo E, Barbato C, Ciotti M, Cogoni C, Ruberti F. MicroRNA-101 regulates amyloid precursor protein expression in hippocampal neurons. *J Biol Chem.* 2010;285(24):18344–18351.
116. Fang M, Wang J, Zhang X, et al. The miR-124 regulates the expression of BACE1/beta-secretase correlated with cell death in Alzheimer's disease. *Toxicol Lett.* 2012;209(1):94–105.
117. Wong HK, Veremeyko T, Patel N, et al. De-repression of FOXO3a death axis by microRNA-132 and -212 causes neuronal apoptosis in Alzheimer's disease. *Hum Mol Genet.* 2013;22(15):3077–3092.
118. Dickson JR, Kruse C, Montagna DR, Finsen B, Wolfe MS. Alternative polyadenylation and miR-34 family members regulate tau expression. *J Neurochem.* 2013;127(6):739–749.
119. Liu CG, Song J, Zhang YQ, Wang PC. MicroRNA-193b is a regulator of amyloid precursor protein in the blood and cerebrospinal fluid derived exosomal microRNA-193b is a biomarker of Alzheimer's disease. *Mol Med Rep.* 2014;10(5):2395–2400.
120. Long JM, Ray B, Lahiri DK. MicroRNA-339-5p down-regulates protein expression of beta-site amyloid precursor protein-cleaving enzyme 1 (BACE1) in human primary brain cultures and is reduced in brain tissue specimens of Alzheimer disease subjects. *J Biol Chem.* 2014;289(8):5184–5198.
121. Weinberg RB, Mufson EJ, Counts SE. Evidence for a neuroprotective microRNA pathway in amnestic mild cognitive impairment. *Front Neurosci.* 2015;9:430.
122. Santa-Maria I, Alaniz ME, Renwick N, et al. Dysregulation of microRNA-219 promotes neurodegeneration through post-transcriptional regulation of tau. *J Clin Invest.* 2015;125(2):681–686.

123. Zhang B, Chen CF, Wang AH, Lin QF. MiR-16 regulates cell death in Alzheimer's disease by targeting amyloid precursor protein. *Eur Rev Med Pharmacol Sci*. 2015;19(21):4020–4027.
124. Yang G, Song Y, Zhou X, et al. MicroRNA-29c targets beta-site amyloid precursor protein-cleaving enzyme 1 and has a neuroprotective role in vitro and in vivo. *Mol Med Rep*. 2015;12(2):3081–3088.
125. Zhang Y, Xing H, Guo S, Zheng Z, Wang H, Xu D. MicroRNA-135b has a neuroprotective role via targeting of beta-site APP-cleaving enzyme 1. *Exp Ther Med*. 2016;12(2):809–814.
126. Ghanbari M, Ikram MA, de Looper HW, et al. Genome-wide identification of microRNA-related variants associated with risk of Alzheimer's disease. *Sci Rep*. 2016;6:28387.
127. Moncini S, Lunghi M, Valmadre A, et al. The miR-15/107 family of microRNA genes regulates CDK5R1/p35 with implications for Alzheimer's disease pathogenesis. *Mol Neurobiol*. 2016, PMID: 27343180, [Epub ahead of print].
128. Zhang C, Lu J, Liu B, Cui Q, Wang Y. Primate-specific miR-603 is implicated in the risk and pathogenesis of Alzheimer's disease. *Aging (Albany NY)*. 2016;8(2):272–290.
129. Pereira PA, Tomas JF, Queiroz JA, Figueiras AR, Sousa F. Recombinant pre-miR-29b for Alzheimer's disease therapeutics. *Sci Rep*. 2016;6:19946.
130. Absalon S, Kochanek DM, Raghavan V, Krichevsky AM. MiR-26b, upregulated in Alzheimer's disease, activates cell cycle entry, tau-phosphorylation, and apoptosis in postmitotic neurons. *J Neurosci*. 2013;33(37):14645–14659.
131. Croce N, Gelfo F, Ciotti MT, et al. NPY modulates miR-30a-5p and BDNF in opposite direction in an in vitro model of Alzheimer disease: a possible role in neuroprotection? *Mol Cell Biochem*. 2013;376(1–2):189–195.
132. Tian N, Cao Z, Zhang Y. MiR-206 decreases brain-derived neurotrophic factor levels in a transgenic mouse model of Alzheimer's disease. *Neurosci Bull*. 2014;30(2):191–197.
133. Banzhaf-Strathmann J, Benito E, May S, et al. MicroRNA-125b induces tau hyperphosphorylation and cognitive deficits in Alzheimer's disease. *EMBO J*. 2014;33(15):1667–1680.
134. Kim J, Yoon H, Horie T, et al. MicroRNA-33 regulates ApoE lipidation and amyloid-beta metabolism in the brain. *J Neurosci*. 2015;35(44):14717–14726.
135. Sarkar S, Jun S, Rellick S, Quintana DD, Cavendish JZ, Simpkins JW. Expression of microRNA-34a in Alzheimer's disease brain targets genes linked to synaptic plasticity, energy metabolism, and resting state network activity. *Brain Res*. 1646;2016:139–151.
136. Kim W, Noh H, Lee Y, et al. MiR-126 regulates growth factor activities and vulnerability to toxic insult in neurons. *Mol Neurobiol*. 2016;53(1):95–108.
137. Hebert LE, Bienias JL, Aggarwal NT, et al. Change in risk of Alzheimer disease over time. *Neurology*. 2010;75(9):786–791.
138. Davis TH, Cuellar TL, Koch SM, et al. Conditional loss of Dicer disrupts cellular and tissue morphogenesis in the cortex and hippocampus. *J Neurosci*. 2008;28(17):4322–4330.
139. Kawase-Koga Y, Low R, Otaegi G, et al. RNAase-III enzyme Dicer maintains signaling pathways for differentiation and survival in mouse cortical neural stem cells. *J Cell Sci*. 2010;123(pt 4):586–594.
140. Reddy PH, Tonk S, Kumar S, et al. A critical evaluation of neuroprotective and neurodegenerative MicroRNAs in Alzheimer's disease. *Biochem Biophys Res Commun*. 2016. http://dx.doi.org/10.1016/j.bbrc.2016.08.067, pii: S0006-291X(16)31324-9.
141. Schonrock N, Gotz J. Decoding the non-coding RNAs in Alzheimer's disease. *Cell Mol Life Sci*. 2012;69(21):3543–3559.
142. Hebert SS, Horre K, Nicolai L, et al. MicroRNA regulation of Alzheimer's amyloid precursor protein expression. *Neurobiol Dis*. 2009;33(3):422–428.

143. Patel N, Hoang D, Miller N, et al. MicroRNAs can regulate human APP levels. *Mol Neurodegener.* 2008;3:10.
144. Smith P, Al Hashimi A, Girard J, Delay C, Hebert SS. In vivo regulation of amyloid precursor protein neuronal splicing by microRNAs. *J Neurochem.* 2011;116(2):240–247.
145. Wang WX, Huang Q, Hu Y, Stromberg AJ, Nelson PT. Patterns of microRNA expression in normal and early Alzheimer's disease human temporal cortex: white matter versus gray matter. *Acta Neuropathol.* 2011;121(2):193–205.
146. Boissonneault V, Plante I, Rivest S, Provost P. MicroRNA-298 and microRNA-328 regulate expression of mouse beta-amyloid precursor protein-converting enzyme 1. *J Biol Chem.* 2009;284(4):1971–1981.
147. Schonrock N, Humphreys DT, Preiss T, Gotz J. Target gene repression mediated by miRNAs miR-181c and miR-9 both of which are down-regulated by amyloid-beta. *J Mol Neurosci.* 2012;46(2):324–335.
148. Schonrock N, Ke YD, Humphreys D, et al. Neuronal microRNA deregulation in response to Alzheimer's disease amyloid-beta. *PLoS One.* 2010;5(6)e11070.
149. Wang H, Liu J, Zong Y, et al. miR-106b aberrantly expressed in a double transgenic mouse model for Alzheimer's disease targets TGF-beta type II receptor. *Brain Res.* 2010;1357:166–174.
150. Kim W, Lee Y, McKenna ND, et al. miR-126 contributes to Parkinson's disease by dysregulating the insulin-like growth factor/phosphoinositide 3-kinase signaling. *Neurobiol Aging.* 2014;35(7):1712–1721.
151. Galimberti D, Villa C, Fenoglio C, et al. Circulating miRNAs as potential biomarkers in Alzheimer's disease. *J Alzheimers Dis.* 2014;42(4):1261–1267.
152. Alexandrov PN, Zhao Y, Jones BM, Bhattacharjee S, Lukiw WJ. Expression of the phagocytosis-essential protein TREM2 is down-regulated by an aluminum-induced miRNA-34a in a murine microglial cell line. *J Inorg Biochem.* 2013;128:267–269.
153. Liu W, Liu C, Zhu J, et al. MicroRNA-16 targets amyloid precursor protein to potentially modulate Alzheimer's-associated pathogenesis in SAMP8 mice. *Neurobiol Aging.* 2012;33(3):522–534.
154. Pogue AI, Li YY, Cui JG, et al. Characterization of an NF-kappaB-regulated, miRNA-146a-mediated down-regulation of complement factor H (CFH) in metal-sulfate-stressed human brain cells. *J Inorg Biochem.* 2009;103(11):1591–1595.
155. Augustin R, Endres K, Reinhardt S, et al. Computational identification and experimental validation of microRNAs binding to the Alzheimer-related gene ADAM10. *BMC Med Genet.* 2012;13:35.
156. Denk J, Boelmans K, Siegismund C, Lassner D, Arlt S, Jahn H. MicroRNA profiling of CSF reveals potential biomarkers to detect Alzheimer's disease. *PLoS One.* 2015;10(5), e0126423.
157. Shen J, Kelleher 3rd RJ. The presenilin hypothesis of Alzheimer's disease: evidence for a loss-of-function pathogenic mechanism. *Proc Natl Acad Sci USA.* 2007;104(2): 403–409.
158. Krichevsky AM, King KS, Donahue CP, Khrapko K, Kosik KS. A microRNA array reveals extensive regulation of microRNAs during brain development. *RNA.* 2003;9(10):1274–1281.
159. Jing L, Jia Y, Lu J, et al. MicroRNA-9 promotes differentiation of mouse bone mesenchymal stem cells into neurons by Notch signaling. *Neuroreport.* 2011;22(5):206–211.
160. Bonev B, Pisco A, Papalopulu N. MicroRNA-9 reveals regional diversity of neural progenitors along the anterior-posterior axis. *Dev Cell.* 2011;20(1):19–32.
161. Hebert SS, Papadopoulou AS, Smith P, et al. Genetic ablation of Dicer in adult forebrain neurons results in abnormal tau hyperphosphorylation and neurodegeneration. *Hum Mol Genet.* 2010;19(20):3959–3969.

162. Carrettiero DC, Hernandez I, Neveu P, Papagiannakopoulos T, Kosik KS. The cochaperone BAG2 sweeps paired helical filament- insoluble tau from the microtubule. *J Neurosci*. 2009;29(7):2151–2161.
163. Wang X, Liu P, Zhu H, et al. miR-34a, a microRNA up-regulated in a double transgenic mouse model of Alzheimer's disease, inhibits bcl2 translation. *Brain Res Bull*. 2009;80(4–5):268–273.
164. Lee ST, Chu K, Jung KH, et al. miR-206 regulates brain-derived neurotrophic factor in Alzheimer disease model. *Ann Neurol*. 2012;72(2):269–277.
165. Lukiw WJ, Alexandrov PN. Regulation of complement factor H (CFH) by multiple miRNAs in Alzheimer's disease (AD) brain. *Mol Neurobiol*. 2012;46(1):11–19.
166. Li Z, Gu X, Fang Y, Xiang J, Chen Z. microRNA expression profiles in human colorectal cancers with brain metastases. *Oncol Lett*. 2012;3(2):346–350.
167. Redis RS, Calin S, Yang Y, You MJ, Calin GA. Cell-to-cell miRNA transfer: from body homeostasis to therapy. *Pharmacol Ther*. 2012;136(2):169–174.
168. Chan SY, Zhang YY, Hemann C, Mahoney CE, Zweier JL, Loscalzo J. MicroRNA-210 controls mitochondrial metabolism during hypoxia by repressing the iron-sulfur cluster assembly proteins ISCU1/2. *Cell Metab*. 2009;10(4):273–284.
169. Huang L, Mollet S, Souquere S, et al. Mitochondria associate with P-bodies and modulate microRNA-mediated RNA interference. *J Biol Chem*. 2011;286(27): 24219–24230.
170. Favaro E, Ramachandran A, McCormick R, et al. MicroRNA-210 regulates mitochondrial free radical response to hypoxia and Krebs cycle in cancer cells by targeting iron sulfur cluster protein ISCU. *PLoS One*. 2010;5(4), e10345.
171. Kren BT, Wong PY, Sarver A, Zhang X, Zeng Y, Steer CJ. MicroRNAs identified in highly purified liver-derived mitochondria may play a role in apoptosis. *RNA Biol*. 2009;6(1):65–72.
172. Chen Z, Li Y, Zhang H, Huang P, Luthra R. Hypoxia-regulated microRNA-210 modulates mitochondrial function and decreases ISCU and COX10 expression. *Oncogene*. 2010;29(30):4362–4368.
173. Lung B, Zemann A, Madej MJ, et al. Identification of small non-coding RNAs from mitochondria and chloroplasts. *Nucleic Acids Res*. 2006;34(14):3842–3852.
174. Barrey E, Saint-Auret G, Bonnamy B, Damas D, Boyer O, Gidrol X. Pre-microRNA and mature microRNA in human mitochondria. *PLoS One*. 2011;6(5), e20220.
175. Shinde S, Bhadra U. A complex genome-microRNA interplay in human mitochondria. *Biomed Res Int*. 2015;2015:206382.
176. Long B, Wang K, Li N, et al. miR-761 regulates the mitochondrial network by targeting mitochondrial fission factor. *Free Radic Biol Med*. 2013;65:371–379.
177. Goud RM, Hua ZA. Role of microRNA in the regulation of mitochondrial functions. *Sci Lett*. 2015;3(2):83–88.
178. Li J, Donath S, Li Y, Qin D, Prabhakar BS, Li P. miR-30 regulates mitochondrial fission through targeting p53 and the dynamin-related protein-1 pathway. *PLoS Genet*. 2010;6(1), e1000795.
179. Kandimalla R, Reddy PH. Multiple faces of dynamin-related protein 1 and its role in Alzheimer's disease pathogenesis. *Biochim Biophys Acta*. 2016;1862(4):814–828.

CHAPTER SIX

Mitochondria-Targeted Molecules as Potential Drugs to Treat Patients With Alzheimer's Disease

A.P. Reddy[†,1], **P.H. Reddy**[*,†]
*Garrison Institute on Aging, Texas Tech University Health Sciences Center, Lubbock, TX, United States
†Texas Tech University Health Sciences Center, Lubbock, TX, United States
[1]Corresponding author: e-mail address: Arubala.reddy@ttuhsc.edu

Contents

Abstract

Alzheimer's disease (AD) is the most common multifactorial mental illness affecting the elderly population in the world. Its prevalence increases as person ages. There is no known drug or agent that can delay or prevent the AD and its progression. Extensive research has revealed that multiple cellular pathways involved, including amyloid beta production, mitochondrial structural and functional changes, hyperphosphorylation of Tau and NFT formation, inflammatory responses, and neuronal loss in AD pathogenesis. Amyloid beta-induced synaptic damage, mitochondrial abnormalities, and phosphorylated Tau are major areas of present research investigations. Synaptic pathology and mitochondrial oxidative damage are early events in disease process. In this chapter, a systematic literature survey has been conducted and presented a summary of antioxidants used in (1) AD mouse models, (2) elderly populations, and (3) randomized clinical

Progress in Molecular Biology and Translational Science, Volume 146
ISSN 1877-1173
http://dx.doi.org/10.1016/bs.pmbts.2016.12.010

trials in AD patients. This chapter highlights the recent progress in developing and testing mitochondria-targeted molecules using AD cell cultures and AD mouse models. This chapter also discusses recent research on AD pathogenesis and therapeutics, focusing on mitochondria-targeted molecules as potential therapeutic targets to delay or prevent AD progression.

ABBREVIATIONS

ABAD amyloid beta-induced alcohol dehydrogenase
ApoE4 apolipoprotein epsilon 4 genotype
APP amyloid precursor protein
ATP adenosine triphosphate
Aβ amyloid beta
CD2AP CD2-associated protein
ETC electron transport chain
MCAT mitochondria-targeted catalase
MitoQ mitochondria-quinone
NFTs neurofibrillary tangles
OXPHOS oxidative phosphorylation
PET positron emission tomography
PS1 presenilin 1
PS2 presenilin 2
ROS reactive oxygen species
SS31 peptide Szeto–Schiller peptide
VDAC1 voltage-dependent anion channel protein 1

1. INTRODUCTION

Alzheimer's disease (AD) is a progressive, heterogeneous, age-dependent, neurodegenerative disorder, characterized by the loss of memory, impairment of multiple cognitive functions, and changes in the personality and behavior.[1–3] Currently, 36 million people older than 65 years are living with AD-related dementia worldwide, with numbers in this age group expecting to double to 66 million by 2030 and increase to 115 million by 2050. According to 2015 estimates from the World Alzheimer Report, worldwide dementia is currently costing $818 billion annually.[4]

Pathological and morphological examination of autopsied brains from patients with AD revealed that AD is mainly associated with (1) intracellular neurofibrillary tangles (NFTs), (2) extracellular amyloid beta (Aβ) plaques, (3) synaptic damage, loss of synapses, and loss of synaptic proteins,

(4) proliferation of reactive astrocytes and activated microglia, (5) defects and alterations in cholinergic neurons, (6) an age-dependent imbalance in hormones, and (7) structural and functional changes in mitochondria.[1–3,5–13] Among these changes, synaptic damage and the loss of synapses and mitochondrial oxidative damage are widely recognized as early events in the pathogenesis and progression of AD.[8,14–16] Also, the loss of synapses and synaptic damage are the best correlates of cognitive decline found in AD patients.[11,12,17]

Histopathological studies have revealed that neuronal damage is initiated in layer 2 of the entorhinal cortex and then spreads to the hippocampus, temporal cortex, frontoparietal cortex, and finally to subcortical nuclei.[18–20] During this degenerative process, the entorhinal cortex and the hippocampal dentate gyrus become progressively disconnected, and subsequently, neuronal connections between pre- and postsynaptic neurons within hippocampal regions become detached.[21] This degenerative process then spreads to the neocortex, leading to neuronal disconnections in layer 3 of the cortex. These degenerative events occur in the brain regions that control and regulate learning, memory, and other cognitive functions in AD patients.

AD occurs in two forms: (1) early-onset "familial" AD involves genetic mutations and (2) late-onset "sporadic" AD. Genetic mutations in amyloid precursor protein (APP), presenilin 1 (PS1), and presenilin 2 cause a small proportion of familial AD.[22–24] Patients with Down syndrome carry an extra chromosome 21 that harbors the APP gene, and they have been reported to develop AD pathology and dementia. Genetic polymorphisms in multiple genes—including the apolipoprotein E gene with the E4 genotype, sortilin-related receptor 1, clusterin, complement component receptor 1, CD2AP, CD33, EPHA1, and MS4A4/MS4A6E—are involved in sporadic AD.[25–30] In addition to these genetic factors, lifestyle activities (diet, exposure to toxic environments, including chemicals) and oxidative mitochondrial DNA damage are major contributing factors found to affect the onset of sporadic AD. Above all, aging has been found to be the #1 risk factor affecting the onset of familial AD and sporadic AD.

AD with its concomitant decrease in cognitive function has become a major health problem worldwide, especially with populations reaching 85 years and older in greater numbers than ever before. Therapeutic interventions are urgently needed to minimize the effects of AD on cognition. The purposes of this chapter are to highlight recent developments in AD research; to present a summary of antioxidants used in AD mouse models, elderly populations, and randomized clinical trials investigating AD treatments; and to report on mitochondria-targeted molecules that hold promise as therapeutic approaches.

2. Aβ AND ALZHEIMER'S DISEASE

Aβ, a 4-kDa peptide, is a major component of Aβ plaques found in AD brains. Recent molecular, cellular, and animal model studies have provided evidence that Aβ—a product of APP due to the cleavage of β and γ secretases—is a key factor in AD development and progression.[3,15] The formation and subsequent accumulation of Aβ peptide in the brains of AD patients is a progressive, sequential process. Aβ exists in multiple forms. In AD, Aβ exists in various forms, and in any of its forms, it aggregates and accumulates in different subcellular organelles of neurons.[3,31] The most prevalent forms of Aβ are Aβ40 and Aβ42, with Aβ42 being found to be the more highly toxic. Aβ42 is known to aggregate into accumulations of different sizes, ranging from monomeric to multimeric Aβ, and to participate in the formation of multimeric, diffusible, soluble aggregations, protofibrils, insoluble fibrils, and Aβ deposits.[32]

The continuous production and reduced clearance of Aβ in neurons may lead to a cascade of events in the AD process.[2,15] This is primarily due to the increased accumulation of Aβ in subcellular compartments of cell including—Golgi apparatus, lysosomes, endoplasmic reticulum, and mitochondria, leading to disruption of these subcellular organelles.[15] It is also important to note that an age-dependent, decreased production of Aβ-degrading enzymes—neprilysin, insulin-degrading enzyme, and others—has been found to be contributing factors in the accumulation of Aβ in AD neurons.[3] These findings, taken together, suggest that factors that are involved in increased production and decreased clearance of Aβ are altered with aging.

Recent research on Aβ, using mouse models of AD, including the APP/PS1 and 3XAD.Tg mice, has found that intraneuronal Aβ facilitates tau pathology. Further, Aβ deposits have been found to be associated with activated microglia and astrocytes, and to trigger an inflammatory response.[33] In addition, Aβ has been found to enter mitochondria, to interact with mitochondrial matrix proteins, CypD proteins (an Aβ-induced alcohol dehydrogenase [ABAD]), to disrupt the electron transport chain (ETC), to generate reactive oxygen species (ROS) and free radicals derived from molecular oxygen in the mitochondria, and to inhibit the generation of cellular energy (ATP).[34,35]

Recent research has also revealed that Aβ interacts with the mitochondrial fission protein Drp1 and enhances the enzymatic activity of GTPase Drp1, causing increased mitochondrial fragmentation and synaptic damage in AD-affected neurons.[36] More recently, in AD neurons, Aβ has been found to interact with VDAC1, a mitochondrial outer-membrane protein, and to interfere with permeability transition pore gating in mitochondria, leading to low ATP production, mitochondrial dysfunction, and defects in mitochondrial oxidative phosphorylation (OXPHOS).[37]

Overall, increased production and low clearance of Aβ in AD neurons has been found to cause a cascade of cellular changes, leading to malfunctioning of multiple subcellular organelles, and mitochondrial and synaptic functions.

3. PHOSPHORYLATED TAU AND ALZHEIMER'S DISEASE

Phosphorylated Tau and NFTs are a second major pathological hallmark in AD pathogenesis. Although they are secondary events, they play a significant role in damaging neurons structurally and functionally. Growing evidence suggests that phosphorylated Tau is involved in AD pathogenesis by impairing axonal transport of proteins, vesicles, and subcellular organelles, including mitochondria in AD neurons.[33]

Tau is a major microtubule-associated protein, abundantly present in the central nervous system and is predominantly expressed in neuronal axons. Tau performs several important functions in neurons, including the stabilization of microtubules, the promotion of neurite outgrowth and of membrane interactions, and facilitation of enzyme anchoring and of axonal transport of organelles to nerve terminals.[38–40] However, in AD neurons, Tau is hyperphosphorylated, accumulates in neurons, and forms paired helical filaments, resulting in an inability for Tau to bind with microtubules, which ultimately leads to the impairment of organelle axonal transport to nerve terminals, and causes synaptic degeneration[38–42]

Extensive research using brain tissues from transgenic mouse models of Tau, APP/PS1, and 3XAD.Tg revealed that overexpressed normal Tau and/or overexpressed mutant Tau in neurons become hyperphosphorylated, causing oxidative stress, mitochondrial dysfunction, synaptic deprivation, and neuronal damage.[33] In support of phosphorylated Tau involvement in mitochondrial dysfunction and synaptic damage in AD, several research groups reported oxidative damage, defective mitochondrial activities, disrupted calcium homeostasis, and defective mitochondrial function in 3xTg-AD mice[43–47] and APP/PS1,[48] both of which are mouse models that produce hyperphosphorylated Tau.

Using postmortem brain tissues from AD patients at different stages of disease progression, including AD patients at late-stage disease progression who exhibited cognitive decline; from control subjects without AD; and from AβPP, AβPPxPS1, and 3xTg-AD mice, we studied the interaction between Aβ and phosphorylated Tau.[49] Using immunohistological and double-immunofluorescence analyses and AD postmortem brains, we also studied the localization of monomeric and oligomeric Aβ with phosphorylated Tau. We found monomeric and oligomeric Aβ interacting with phosphorylated Tau in neurons affected by AD. Further, these interactions progressively increased as AD progressed. Double-labeling analysis of monomeric and oligomeric Aβ and phosphorylated Tau revealed colocalization of monomeric and oligomeric Aβ with phosphorylated Tau, confirming that Aβ and Tau increasingly interact as AD progresses. Based on these findings, we concluded that phosphorylated abnormally interacts with Aβ, and this interaction damages neuronal structure and function, particularly synapses, leading to cognitive decline in AD patients.[49]

Overall, these studies provided strong evidence that hyperphosphorylated Tau in brain tissue from AD patients is involved in cellular changes related primarily to mitochondrial dysfunction and synaptic damage.

4. SYNAPTIC DAMAGE AND ALZHEIMER'S DISEASE

In healthy, intact synapses, synaptic terminals function actively to transmit signals between neurons and to process information.[15,17] However, in elderly individuals and in AD patients and in elderly patients,[50,51] intact synaptic terminals exhibited changes that are responsible for cognitive decline.

In a study of synaptic loss in the cerebellum (unaffected in AD) and hippocampus (affected in AD), researchers found no significant differences in the synapse-to-neuron ratio in samples taken from the cerebellum of adult persons without AD, nonelderly patients with AD, and elderly patients with AD. However, the synapse-to-neuron ratio in samples from the hippocampus decreased more than 50% in adults without AD and in elderly patients with AD. These observations suggest that the loss of synapses is confined to affected brain region in AD.[17]

Several morphological and ultrastructural studies revealed a 25%–30% decrease in synapses in the cortex of AD patients and a 15%–35% decrease in synapses per cortical neuron in AD patients, suggesting that the loss of synapses in AD patients may more robustly correlate with cognitive decline than the number of Aβ plaques and NFTs.[11,12] Further, recent quantification studies of synaptic proteins in AD patients and nondemented healthy control adults revealed decreased levels of the presynaptic vesicle proteins synaptophysin, synaptotagmin, and Rab3a; the postsynaptic proteins synaptopodin, neurogranin, and PSD95; and the synaptic membrane proteins GAP43 and synaptobrevin in the AD patients.[52–54]

These results suggest that membrane-bound, presynaptic and postsynaptic proteins may be critically involved in AD progression[52–54] and that the loss of synapses and synaptic proteins may be confined to AD-affected regions of the brain. It has been proposed that soluble Aβ, localized at synaptic terminals, may be responsible for this loss of synapses and synaptic proteins.[52–54]

The loss of synapses and synaptic proteins that occurs before neuronal death in patients with AD appears to be accompanied by axonal degeneration and defective axonal transport of mitochondria.[55–57] To explore the loss of synapses in AD neurons, we conducted an investigation of mRNA changes in synaptic genes of AD neurons.[55] This investigation of mRNA levels of synaptic genes in APP hippocampal neurons revealed significantly reduced synaptic genes, suggesting that neurons that produce Aβ may also be deficient in synaptic mRNA. We also found significantly reduced mitochondrial anterograde axonal transport in AD neurons.[55,56] Further, many research groups found Aβ accumulated at synapses in AD neurons.[8,55–59]

These results strongly suggest synaptic degeneration and synaptic functional failure primarily due to defective mitochondria and mitochondria malfunction in AD neurons. This research provides compelling evidence that Aβ-induced synaptic and mitochondrial damage plays a large role in cognitive decline in AD.

5. DECREASED GLUCOSE METABOLISM AND ALZHEIMER'S DISEASE

The development of positron emission tomography (PET) methodologies has made it possible to study brain imaging and to provide information related to cerebral energy metabolism that can be correlated to cognitive behavior. PET images of brain specimens from elderly individuals without

AD and elderly AD patients showed large decreases in glucose metabolism in cortico-temporal–parietal regions, in contrast to PET images of brains from healthy aging individuals, which did not show such decreases.[60–66] The results from these studies using PET methodologies suggest that the impairment of energy metabolism is involved in AD development and progression.

To determine the relationship between metabolic decline and cognitive decline, using PET methodologies, Small *et al.* investigated cerebral metabolic rates in the brains of elderly persons at risk for AD, as determined by their ApoE genotype E4.[63] They found that a single copy of the ApoE-4 allele was associated with metabolic decline in the inferior parietal, lateral temporal, and posterior cingulate metabolism. Further, they found that the presence of the ApoE-4 allele was a predictor of cognitive decline in elderly persons. Overall, findings from this study suggest that the combination of cerebral metabolic rates and genetic risk factors (such as the ApoE4 genotype) may provide a means to detect preclinical AD and to monitor cognitive decline during experimental treatments.

Recently, in a longitudinal study, Jagust and colleagues[66] investigated the connection between glucose metabolism and cognitive decline in the medial temporal lobe volumes in brains from 60 healthy elderly adults.[66] Over a 3.8-year period, they took PET scans of [18F]-fluorodeoxyglucose and structural magnetic resonance images of these brains, to determine each elderly person's global cognition. They also administered tests, including the Modified Mini-Mental State Examination (MMSE) and delayed recall tests. They quantified baseline brain volumes and glucose metabolism, and determined how these measurements correlated with scores from the cognitive tests. Baseline PET scans showed that brain volume did not correspond to a decline in the MMSE scores. However, regions in the left and right angular gyrus, left mid-temporal gyrus, and left mid-frontal gyrus correlated with the rate of change in MMSE scores ($P < 0.001$). The volume of the medial temporal brain was found to correspond to memory decline, suggesting that, for healthy elderly persons, the decline in temporal and parietal glucose metabolism may correspond to a decline in global cognitive function and that preclinical symptoms of AD pathology may be found in the medial temporal brain, the left and right angular gyrus, the left mid-temporal gyrus, and the left mid-frontal gyrus.

These findings, taken together, suggest that decreased glucose uptake may be associated with low ATP production in AD neurons. The metabolism of glucose into energy may occur in combination with oxygen in humans. It is believed that oxygen used in oxidative metabolism may be found in the mitochondria. It is also possible that, due to decreased glucose uptake, fewer mitochondria are transported to synapses, resulting in low synaptic ATP and synaptic damage and, ultimately, to cognitive decline in patients with AD.

6. MITOCHONDRIA AND ROS IN AGING AND ALZHEIMER'S DISEASE

Mitochondria are essential cytoplasmic organelles that are critical for cell survival and cell death. Structurally, mitochondria are compartmentalized with two lipid membranes—an outer and an inner membrane. The outer membrane is highly porous and allows the flow of small molecules and metabolites into its inner-membrane space. The inner membrane covers the matrix, which contains beta-oxidation and tricarboxylic acid. The inner membrane is highly nonporous and restricts ionic flow into the mitochondrial matrix. The ETC is localized within the inner membrane and participates in OXPHOS and in the production of essential cellular ATP.[6,14,67] The ROS that is produced in mitochondria is a physiological by-product of the ETC, created by electron leaks. During the transfer of electrons to molecular oxygen, 1%–5% of electrons lose their way and participate in the formation of superoxide radicals ($O_2^{\bullet -}$) at multiple sites in the mitochondria: in complexes I and III, components of the tricarboxylic acid cycle, and the outer mitochondrial membrane. The $O_2^{\bullet -}$ produced in the mitochondria activates the mitochondrial permeability transition pore and destroys mitochondrial cells by apoptosis.[14,67–69]

Increased production of ROS has been documented in the aging process. This increase may be primarily due to an increased accumulation of mtDNA mutations, which in turn could damage mitochondrial structures and functions, consequently altering enzymatic activities, disrupting mitochondrial pore gating, altering mitochondrial calcium levels, and lowering ATP; and ultimately leading to cellular senescence.[14,15] An age-dependent and a mitochondrial damaged DNA-induced, increased production of ROS is known to activate β and γ secretases in APP molecules and to produce Aβ in sporadic AD neurons. These peptides may further enter mitochondria, and induce and increase the production of free radicals, leading to increase levels of Aβ levels. Thus, mitochondrial Aβ may ultimately result in a cascade of events: (1) Aβ could interact with mitochondrial proteins, (2) cause defective axonal transport of mitochondria, (3) supply low mitochondrial ATP to synapses, (4) cause synaptic degeneration, and (5) ultimately lead to neuronal damage and dysfunction.

In studies of Aβ-induced mitochondrial function in cell cultures and Aβ transgenic mice, several groups found that Aβ produced by APP, PS1, and PS2 genetic mutations participate in enhancing ROS production, mitochondrial dysfunction, and neuronal damage in familial AD neurons.[1,10] Mutations of APP, PS1, and PS2 genes induce ROS production, mitochondrial dysfunction, and neuronal damage early in the familial AD process, but in sporadic AD, aging induces ROS production similar to genetic mutations, but takes more time to trigger events in the sporadic AD process.[1–3,5]

Overall, increased mitochondrial ROS production is a key event in neuronal damage in both familial AD and sporadic AD. The lowering of mitochondrial ROS may be a potential therapeutic approach to aging and AD.

7. Aβ AND PHOSPHORYLATED TAU IN MITOCHONDRIA

Biochemical, cell biology, and immunohistochemical analyses, and transmission electron microscopy studies have revealed that mutant APP, Aβ, and N-terminal ApoE4 fragments are associated with mitochondrial membranes and the mitochondrial matrix.[34,35,46,67,70–74] Further, these mutant proteins disrupt OXPHOS, induce ROS production, and cause mitochondrial dysfunction in AD neurons.

Recently, we studied the relationship between Aβ and VDAC1.[37] We found Aβ interacting with VDAC1 in postmortem brains from AD patients and APP transgenic mice, and we found that these interactions increased as AD progressed and as mitochondrial dysfunction increased.

Recent biochemical studies of mitochondria and phosphorylated Tau revealed that in the N-terminal, truncated Tau interacts with mitochondria.[72–74] Studies found that toxic effects resulting from the interaction of two N-terminal Tau fragments (NH2, 1–25aa and NH2, 26–44aa) and mitochondria.[74] They found OXPHOS defects in the NH2, 26–66aa fragments, but not in the NH2, 1–25aa fragments. Recent study has found that 20–22-kDa N-terminal Tau fragments were enriched in synaptosomal mitochondria in AD brains, and that the increase in Tau correlated with pathological structural changes and functional impairment in synapses.[72]

Recently, we investigated the relationship between phosphorylated Tau and VDAC1.[37] Using postmortem brain tissues from AD patients and from 3XTg.AD mice, we found that phosphorylated Tau increased as AD progressed. We also performed co-IP analysis using the phosphorylated Tau antibody, and immunoblotting analysis using the VDAC1 antibody and protein lysates of cortical tissues from control subjects; from patients with mild, definite, and severe AD; and from 3XTg.AD mice. We found that phosphorylated Tau interacted with VDAC1 in the cortical tissues of the 3XTg.AD mice and of AD patients at all three stages of disease progression. We found less interaction between phosphorylated Tau and VDAC1 in the control subjects.[37] These results suggest that phosphorylated Tau may have a role in blocking mitochondrial pores and/or in causing OXPHOS defects.

Overall, these studies suggest that both Aβ and phosphorylated Tau induce ROS production and cause OXPHOS defects and mitochondrial dysfunction in AD.

8. NATURAL ANTIOXIDANTS AND MITOCHONDRIAL THERAPEUTIC APPROACHES TO ALZHEIMER'S DISEASE

As discussed earlier, oxidative stress and mitochondrial dysfunction have been largely implicated in AD progression and pathogenesis. Two approaches that have been used to study the effects of antioxidants—natural antioxidants and mitochondria-targeted molecules—may be important in treating elderly individuals and patients with AD.

As shown in Table 1, several transgenic mouse lines, including overexpressed mutant APP, mutant APP + PS1, mutant APP + PS1, and mutant tau, have been used to study the effects of vitamin E, vitamin C, ginkgo biloba, melatonin, *N*-acetyl-L-cysteine, alpha-lipoic acid, R-lipoic acid, CoQ10, ferulic acid, pyrrolyl-alpha-nitronyl nitroxide, zeolite supplementation, ApoE mimetic peptide Ac-hE18A-NH2, and curcumin.[76–100] The major outcomes from these studies are reduced Aβ levels, ameliorated cognitive decline, reduced phosphorylated Tau, reduced mitochondrial dysfunction, reduced microglial activation, and enhanced synaptic activity. These results indicate that antioxidant treatment is beneficial in reducing and/or preventing disease progression in mice that carry AD mutations.

In some studies, a combination of exercise, alpha-lipoic acid, melatonin, and R-lipoic acid and *N*-acetyl-L-cysteine were administered and/or supplemented in the diets of AD mouse models to study their effects on cognitive behavior and AD pathology.[84,86,92] Results also indicated beneficial

Table 1 Antioxidant Use in Transgenic Mouse Models of Alzheimer's Disease

Transgenic Line	Antioxidant	Treatment Period	Major Findings	Reference
Tg2576	Ginkgo biloba	6 months	Blocked age-dependent spatial cognition; no change in Aβ levels	Stackman *et al.*[75]
Tg2576	Vitamin E	4 weeks	Reduced learning deficits; reduced lipid peroxidation levels	Conte *et al.*[76]
Tg2576	Vitamin E	8 months to mid groups 6 months to older groups	Reduced Aβ and lipid peroxidation; reduced lipid peroxidation	Sung *et al.*[77]
Tau	Vitamin E	Not available	Delayed development of tau pathology, which correlated with improvement in the health and attenuation of motor weakness	Nakashima *et al.*[78]
Tg2576	Melatonin	Not available	Melatonin partially inhibited the expected time-dependent elevation of Aβ; reduced abnormal nitration of proteins; increased survival	Matsubara *et al.*[79]
Tg2576	Melatonin	4 months	Melatonin alleviated learning and memory deficits; decreased choline acetyltransferase activity in frontal cortex and hippocampus of APP mice; acetyltransferase activity increased by melatonin supplement in frontal cortex and hippocampus	Feng *et al.*[80]
Tg2576	Melatonin	4 months	No change in Aβ levels	Quinn *et al.*[81]

PS1 L235P	CoQ10	60 days	Attenuated Aβ pathology; reduced MDA levels; upregulated SOD activity	Yang *et al.*[82]
APP	R-lipoic acid	10 months	Reduced oxidative modifications; no change in Aβ levels	Siedlak *et al.*[83]
ApoE4	R-lipoic acid Acetyl-L-carnitine	Not available	Improved spatial and temporal memory	Shenk *et al.*[84]
APP/PS1	Vitamin C	Not available	Improved cognitive behavior; no change in amyloid pathology	Harrison *et al.*[85]
NSE/APP	Lipoic acid exercise	16 weeks	No change in Aβ; ameliorated spatial learning and memory deficits	Cho *et al.*[86]
APP/PS1-Knockin	*N*-Acetyl-L-cysteine	3 months	Reduced lipid peroxidation, oxidative stress, and glutathione peroxidase	Huang *et al.*[87]
APP/PS1	Melatonin	1 month	Increased mitochondrial function	Dragicevic *et al.*[88]
APP	Vitamin C	6 months	Reduced Aβ oligomers; reduced tau phosphorylation; reduced oxidative stress	Murakami *et al.*[89]
Tg19959	CoQ10	5 months	Improved cognitive behavior; reduced Aβ pathology	Dumont *et al.*[90]
3XAD.Tg	MitoQ	5 months	Prevented cognitive decline, oxidative stress, Aβ accumulation, synaptic loss, and caspase activation	McManus *et al.*[91]

Continued

Table 1 Antioxidant Use in Transgenic Mouse Models of Alzheimer's Disease—cont'd

Transgenic Line	Antioxidant	Treatment Period	Major Findings	Reference
3XAD.Tg	Melatonin + exercise	6 months	Protected against cognitive impairment, oxidative stress, and mtDNA changes	Garcia-Mesa *et al.*[92]
3XAD.Tg	Catalase mimetic	5 months	Protected against oxidative stress, DNA, and protein oxidation; reduced Aβ and tau phosphorylation	Clausen *et al.*[93]
APP/PS1	Melatonin	Long term	Reduced hippocampal protein oxidation; improved cognitive behavior	Bano Otalora *et al.*[94]
APP/PS1	Ferulic acid	6 months	Enhanced novel object recognition; reduced amyloid deposition and inflammation	Yan *et al.*[95]
APP/PS1	Pyrrolyl alpha nitronyl nitroxide	1 month	Improved spatial learning and memory; reduced astrocyte activation, Aβ pathology, and tau phosphorylation	Shi *et al.*[96]
APP/PS1	Zeolite supplementation	5 months	Increased endogenous SOD; reduced Aβ levels and plaque load	Montinaro *et al.*[97]
APP/PS1	ApoE mimetic peptide Ac-hE18A-NH2	6 weeks	Reduced oxidative stress and ApoE secretion; inhibited Aβ plaque deposition	Handattu *et al.*[98]
APP/PS1	Curcumin	3 months	Reduced Aβ 40 and 42, and Aβ-derived diffusible ligands; increased Aβ degrading enzymes	Wang *et al.*[99]

effects of combined therapies in terms of improving cognitive behavior and reducing AD pathology.

The positive findings from these studies are promising and warrant perspective studies (antioxidant treatment for elderly individuals without AD) and clinical trials (antioxidants treatment for patients with AD).

9. HUMAN CLINICAL TRIALS AND PERSPECTIVE STUDIES ON ALZHEIMER'S DISEASE

Based on positive outcomes from studies of antioxidant treatment for AD mouse models, several perspective studies have been conducted to test the effects of natural antioxidants (including vitamin E, vitamin C, combination of vitamin E and vitamin C, antioxidant supplement, combination of vitamin E, vitamin C, and beta carotene) on elderly populations.[100–110] As shown in Table 2, 6 out of 15 perspective studies showed positive outcomes in terms of improved cognitive function and/or reduced risk of AD in elderly individuals.[101–105,107] A closer examination of the six positive studies revealed that a high intake of vitamin E (one study) and/or a combination of vitamin E and vitamin C (four studies)[103–105,107] resulted in the greatest beneficial effects.

In addition, in placebo-controlled, double-blind, randomized clinical trials, *Melissa officinalis* and *Salvia officinalis* were administered to patients with mild and moderate AD, and their cognitive functions also significantly improved.[111]

As shown in Table 3, multiple clinical studies have been conducted to determine the effects of vitamin E, vitamin C, and a combination of vitamin E and vitamin C in patients in advanced stages of AD.[114–117] In only one of four studies,[117] disease progression was delayed.

Overall, elderly individuals whose diets were supplemented with a combination of vitamins E and C exhibited positive effects in terms of cognitive function and delayed risk for AD but not when vitamin E and vitamin C were administered individually. These results suggest possible problems: (1) in the experimental design, including the testing antioxidant treatment on AD patients in late-stage disease progression, (2) intake and/or dose of antioxidants, and (3) natural antioxidants might not reach the sites of free radical production. It is yet to be tested whether therapies as indicated in Table 3 can delay disease progression in AD patients who are not yet in advanced stages of disease progression.

Table 2 Antioxidant Administration and Supplementation in Diet and Measured Cognitive Functions in Studies With Elderly Populations

Study Population	Antioxidant and Population Size	Treatment Period	Major Findings	Reference
Rural Elderly, Southwestern Pennsylvania	Antioxidant supplement; $n=1059$	2 years	Cognitive decline not reduced by antioxidant supplement	Mendelsen *et al.*[100]
Elderly Women, Nurse Health Study	Vitamin E, vitamin C; $n=14,968$	5 years	Improved cognitive functions	Grodstein *et al.*[101]
Chicago Health Aging Study Population	Vitamin E; $n=815$	7 years	Vitamin E intake via food associated with reduced risk for AD	Morris *et al.*[102]
Japanese American Male Population	Vitamin E, vitamin C; $n=3385$	2 years	Improved cognitive functions with diet supplemented with vitamin E and/or C supplementation	Masaki *et al.*[103]
Netherland Population	Antioxidant supplement; $n=5395$	6 years	Dietary high intake of vitamins E and C associated with lower risk of AD	Engelhart *et al.*[104]
Canadian Population	Combined vitamin C, vitamin E; $n=894$	5 years	Combined vitamins C and E improved cognitive functions	Maxwell *et al.*[105]
Washington Heights-Inwood Columbia Population	Vitamin E; $n=4023$	4 years	No positive effects with vitamin E in diet	Luschinger *et al.*[106]
Cache County Population	Vitamin E, vitamin C; $n=4740$	4 years	Reduced prevalence and incidence of AD associated with combined vitamins E and C supplements	Zandi *et al.*[107]
Honolulu-Asia Aging Population	Vitamin E; $n=2459$	30 years	No effect shown by vitamin E and vitamin C in diet	Laurin *et al.*[108]

Table 2 Antioxidant Administration and Supplementation in Diet and Measured Cognitive Functions in Studies With Elderly Populations—cont'd

Study Population	Antioxidant and Population Size	Treatment Period	Major Findings	Reference
Duke Establishment Population	Vitamin E, vitamin C; *n*=616	10 years	No positive effects from dietary intake of vitamin E and vitamin C	Fillenbaum *et al.*[109]
Group Health Cooperative Study Population	Vitamin E, vitamin C, vitamins E and C combined; *n*=2969	5.5 years	No positive effect from vitamin E, vitamin C, and combined vitamins E and C	Gray *et al.*[110]

Table 3 Randomized Antioxidant Clinical Trials Using Patients With AD

Study Population	Antioxidant and Population Size	Treatment Period	Major Findings	Reference
Alzheimer's Disease Cooperative Study Population	Vitamin E 2000 IU, donepezil 10 mg/day; *n*=769	3 years	MCI patients unaffected by vitamin E; donepezil associated with lower rate of AD progression in first 12 months	Petersen *et al.*[112]
Age-related Eye Disease Study Population	Vitamin C 500 mg, vitamin E 400 IU, beta carotene 15 mg/daily; *n*=2166	6.9 years	No significant positive effects in treated individuals	Yaffe *et al.*[113]
Women's Health Study Population	Vitamin E 600 IU, supplementation on alternate days; *n*=38,876	6 years	No effect on cognitive function	Kang *et al.*[114]
Alzheimer's Disease Cooperative Study Population	Vitamin E 2000 IU/day; *n*=341	2 years	Disease progression slowed by vitamin E	Sano *et al.*[115]

10. THERAPIES FOR ALZHEIMER'S DISEASE USING MITOCHONDRIA-TARGETED MOLECULES

In the last decade, tremendous progress has been made in developing and testing molecules designed to deliver treatments to mitochondria in neurodegenerative diseases in order to reduce ROS and mitochondrial dysfunction. These molecules are (1) triphenylphosphonium lipophilic cation-based molecules (e.g., MitoQ, MitoVitE, MitoPBN, and Mito-α-lipoic acid)[67,116,118]; (2) small-cell permeable tetra peptide molecules (SS31 and SS20)[67,118–121]; and (3) choline esters of glutathione and *N*-acetyl-L-cysteine.[67]

Triphenylphosphonium lipophilic cation-attached molecules have been studied extensively in cell and experimental animal models for AD and other neurodegenerative diseases.[67,68,117,118] These molecules have been developed using positively charged triphenylphosphonium lipophilic cation attached to antioxidants, such as mitoquinone, vitamin E, alpha-lipoic acid, and PBN. These positively charged molecules are dragged to cell plasma membranes due to charge difference and enter several hundred mitochondria via the mitochondrial matrix due to a charge difference between the molecule (positive charge) and the mitochondrial matrix (negative charge). These molecules scavenge free radicals in the mitochondria and induce mitochondria to function normally.

Among these mitochondrial-targeted molecules, MitoQ has been extensively studied in cell and animal models.[6,67,68,117,118] McManus and colleagues[91] studied the effects of MitoQ on mitochondria, using cortical neurons in cell cultures and in 3xTg.AD mice. They found that MitoQ attenuated Aβ-induced neurotoxicity in cortical neurons, prevented increased production of free radicals, and prevented loss of the mitochondrial membrane potential. They also treated young female 3xTg.AD mice with MitoQ for 5 months and studied the effects of MitoQ on AD progression. MitoQ reduced cognitive decline in these mice, and studies of their brain specimens after treatment revealed lower levels of oxidative stress, lower Aβ accumulation, increased astrocytes and microglia, increased synaptic loss, and increased caspase activation. They concluded that mitochondria-targeted therapeutics in diseases involving oxidative stress and metabolic failure, such as AD, may be able to slow disease progression.

We investigated the effects of MitoQ and SS31, and the antiaging agent resveratrol on neurons from a mouse model (Tg2576 line) of AD and on mouse neuroblastoma (N2a) cells incubated with the Aβ peptide.[120] Using electron and confocal microscopy, gene expression analysis, and biochemical methods, we studied mitochondrial structure and function, and neurite outgrowth in N2a cells treated with MitoQ, SS31, and resveratrol, and then we incubated the cells with Aβ. In the N2a cells that were incubated only with Aβ, we found increased expressions of mitochondrial fission genes and decreased expressions of fusion genes, and also decreased expressions of peroxiredoxins. Electron microscopy of the N2a cells that were incubated with Aβ revealed a significantly increased number of mitochondria, indicating that Aβ fragments mitochondria. Biochemical analysis revealed defective mitochondrial function in Aβ-treated N2a cells. Neurite outgrowth in Aβ-incubated N2a cells was significantly decreased, indicating that Aβ affects neurite outgrowth. However, in the N2a cells that were treated with MitoQ, SS31, and resveratrol, and then incubated with Aβ, abnormal expressions of peroxiredoxins and mitochondrial structural genes were prevented, and mitochondrial function was normal. Further, intact mitochondria were present and neurite outgrowth was significantly increased. In primary neurons from AβPP transgenic mice that were treated with MitoQ and SS31, neurite outgrowth was significantly increased and cyclophilin D expression was significantly decreased.[118] These findings suggest that MitoQ and SS31 prevent Aβ toxicity, and they warrant the study of MitoQ and SS31 as potential drugs to treat patients with AD.

10.1 Cell-Permeable Tetra Peptides to Defective Mitochondria in AD Patients

Szeto and Schiller have developed four cell-permeable tetra peptides and small peptide molecules that have been reported to protect mitochondria from oxidative damage. These peptides and molecules are (1) SS-01 H-Tyr-D-Arg-Phe-Lys-NH2, (2) SS-02 H-Dmt-D-Arg-Phe-Lys-NH2, (3) SS-31 H-D-Arg-Dmt-Lys-Phe-NH2, and (4) SS-20 H-Phe-D-Arg-Phe-Lys-NH2.[67,118,119,122,123] The structural motif of these SS peptides centers on alternating aromatic residues and basic amino acids, and the SS peptides have a sequence motif that allows them to target mitochondria. They scavenge H_2O_2 and ONOO −, and they inhibit lipid peroxidation. Their antioxidant action can be attributed to the tyrosine, or dimethyl-tyrosine (Dmt), residue, which is known to scavenge oxyradicals. Among these oxyradicals, SS31 has been studied extensively.[67,118,119,122,123]

Using primary neurons from a well-characterized AβPP mouse model (Tg2576 mouse line), we studied mitochondrial activity, including axonal transport of mitochondria, and mitochondrial dynamics, morphology, and function.[55] We also studied the nature of Aβ-induced synaptic alterations and cell death in primary neurons from Tg2576 mice, to determine whether the mitochondria-targeted antioxidant SS31 is capable of mitigating the effects of oligomeric Aβ. We found significantly decreased anterograde mitochondrial movement, increased mitochondrial fission and decreased fusion, abnormal mitochondrial and synaptic proteins, and defective mitochondrial function in primary neurons from AβPP mice compared to wild-type neurons. Transmission electron microscopy revealed a large number of small mitochondria and structurally damaged mitochondria, with broken cristae in AβPP primary neurons.[55] We also found an increased accumulation of oligomeric Aβ and increased apoptotic neuronal death in the primary neurons from the AβPP mice relative to the wild-type neurons. Our results revealed an accumulation of intraneuronal oligomeric Aβ, leading to mitochondrial and synaptic deficiencies, ultimately causing neurodegeneration in AβPP cultures. Most importantly, we found that the mitochondria-targeted antioxidant SS31 restored mitochondrial transport and synaptic viability, and decreased the percentage of defective mitochondria, indicating that SS31 protects mitochondria and synapses from Aβ toxicity.[55]

11. EVIDENCE SUPPORTING NEURONAL FUNCTION IN MCAT MICE

Schriner and colleagues tested the hypotheses that mitochondria-targeted catalase reduces mitochondria-derived ROS, enhances mitochondrial function, boosts cell survival, delays the aging phenotype, and extends life span.[120] They generated three different lines of AD transgenic mice (mitochondria-targeted catalase, nuclear-targeted catalase, and peroxisomal-targeted catalase). They then assessed all three lines of mice for morbidity and life span from birth to terminal stage of disease progression. They found mitochondria-targeted catalase mice survived 5.5 months longer than their wild-type counterparts, indicating that overexpressed catalase in mitochondria scavenges ROS, boosts mitochondrial function, and extends the life span of cells. However, nuclear- and peroxisomal-targeted catalase mice did not exhibit any significant change in mitochondrial function or in life span compared to their nontransgenic wild-type mice. These results further support the hypothesis that mitochondria-derived ROS is critical for senescence.

We recently sought to determine the relationship between mitochondria-derived ROS and the production of Aβ in AD. We crossed MCAT mice and APP transgenic mice, producing four lines of mice: MCAT, AβPP, MCAT/AβPP, and wild type.[121] We then followed these mice from birth to terminal stage of disease progression, for morbidity and life span. We also studied their APP, soluble APPα, the C-terminal fragments CTF99 and CTF83, monomeric and oligomeric Aβ, Aβ deposits, and the beta site amyloid precursor protein cleaving enzyme 1 (BACE1) at three different ages (6, 12, and 24 months), corresponding to three different stages of disease progression. Using quantitative reverse transcriptase polymerase chain reaction and immunostaining analyses, we studied the expression of catalase, BACE1, the AD markers, synaptophysin, APP, neprilysin, insulin-degrading enzyme, and transthyretin in the MCAT, AβPP, MCAT/AβPP, and wild-type mice. Using high-pressure liquid chromatography analysis of 8-hydroxy-2-deoxyguanosine, we measured oxidative DNA damage in the cerebral cortical tissues from all four lines of mice. We found that the AβPP transgenic mice carrying the human MCAT gene lived 5 months longer than did the AβPP mice without the human MCAT gene. We also found that overexpression of MCAT in the brain sections from the MCAT/AβPP transgenic mice significantly correlated with reduced levels of full-length APP, CTF99, BACE1, Aβ levels (40 and 42), Aβ deposits, and oxidative DNA damage. Further, we found significantly increased levels of soluble APPα and CTF83 in the MCAT/AβPP mice, relative to the AβPP mice. These data provide direct evidence that oxidative stress plays a primary role in AD etiopathology and that in MCAT mice that express Aβ, MCAT prevents abnormal APP processing, reduces Aβ levels, and enhances Aβ-degrading enzymes at different stages of disease progression. These findings indicate that mitochondria-targeted molecules, such as MitoQ and SS31, may be an effective therapeutic approach to treat patients with AD.

Overall, findings from these studies suggest that mitochondria-targeted molecules protect neurons from age- and AD mutant protein-induced free radicals and mitochondrial dysfunction, maintain healthy neuronal function in elderly individuals and AD patients, and extend healthy cell survival.

12. CONCLUSIONS AND FUTURE STUDIES

Several years of research into AD pathogenesis have revealed the involvement of several cellular pathways, including Aβ production, mitochondrial structural and functional changes, hyperphosphorylation of Tau, NFT formation, inflammatory response, and neuronal loss in AD pathogenesis. Several years of research into AD have also revealed synaptic pathology and mitochondrial damage early in AD progression. The dominant involvement of mitochondrial Aβ in AD suggests that therapeutics targeting mitochondria may be effective in delaying disease progression.

Studies of natural antioxidants as possible therapeutics to mouse models of AD have produced positive results. Further, several studies have been conducted in elderly individuals using natural antioxidants such as vitamin A, vitamin C, beta carotene, antioxidant supplements and a combination of vitamin A and vitamin C, and most importantly, combination treatments/diet supplements of vitamin A and vitamin C provided positive findings. However, clinical trials using advanced stage of AD patients did not provide positive findings. There are multiple reasons for the failure of antioxidant AD clinical trials in AD patients: (1) clinical trials started in AD patients with advanced stages of disease progression, (2) combination administration/diet supplementation of vitamin A and vitamin C and/or antioxidant cocktail is still not optimized, and (3) high intake of antioxidants is not appropriately optimized so far.

Recent studies of experimental and transgenic mouse models have revealed ameliorated cognitive decline and reduced AD pathology and strongly suggested to start mitochondria-targeted molecules, such as MitoQ and SS31, as therapeutic targets for the treatment in elderly individuals and AD patients. Further, it is important to start natural antioxidants and mitochondria-targeted molecules very early in disease process; in other words treatments must be started before clinical symptoms develop and these molecules provided have no adverse effects.

ACKNOWLEDGMENTS

Work presented in this chapter is supported by NIH grants—AG042178 and AG47812, the Garrison Family Foundation, and Sex and Gender Alzheimer's Association (SAGA) grant (to P.H.R.). Present work is also supported by Alzheimer's Association New Investigator Research Grant 2016-NIRG-39787 and Center of Excellence for Translational Neuroscience and Therapeutics grant number PN-CTNT20115-AR (to A.P.R.).

REFERENCES

1. Mattson MP. Pathways towards and away from Alzheimer's disease. *Nature*. 2004;430:631–639.
2. LaFerla FM, Green KN, Oddo S. Intracellular amyloid-beta in Alzheimer's disease. *Nat Rev Neurosci*. 2007;8:499–509.
3. Reddy PH, Manczak M, Mao P, Calkins MJ, Reddy AP, Shirendeb U. Amyloid-beta and mitochondria in aging and Alzheimer's disease: implications for synaptic damage and cognitive decline. *J Alzheimers Dis*. 2010;20(suppl 2):S499–S512.
4. World Alzheimer Report: The Global Impact of Dementia, Analysis of Prevalence, Incidence, Costs and Trends. https://www.alz.co.uk/research/WorldAlzheimerReport2015.pdf, 2015.
5. Nunomura A, Perry G, Aliev G, et al. Oxidative damage is the earliest event in Alzheimer disease. *J Neuropathol Exp Neurol*. 2001;60:759–767.
6. Reddy PH, Tripathi R, Troung Q, et al. Abnormal mitochondrial dynamics and synaptic degeneration as early events in Alzheimer's disease: implications to mitochondria-targeted antioxidant therapeutics. *Biochim Biophys Acta*. 2012;1822:639–649.
7. Zhu X, Perry G, Smith MA, Wang X. Abnormal mitochondrial dynamics in the pathogenesis of Alzheimer's disease. *J Alzheimers Dis*. 2013;33(suppl 1):S253–S262.
8. Du H, Guo L, Yan S, Sosunov AA, McKhann GM, Yan SS. Early deficits in synaptic mitochondria in an Alzheimer's disease mouse model. *Proc Natl Acad Sci USA*. 2010;107:18670–18675.
9. McGeer PL, McGeer EG. The inflammatory response system of brain: implications for therapy of Alzheimer and other neurodegenerative diseases. *Brain Res Brain Res Rev*. 1995;21:195–218.
10. Manczak M, Mao P, Nakamura K, Bebbington C, Park B, Reddy PH. Neutralization of granulocyte macrophage colony-stimulating factor decreases amyloid beta 1-42 and suppresses microglial activity in a transgenic mouse model of Alzheimer's disease. *Hum Mol Genet*. 2009;18:3876–3893.
11. Terry RD, Masliah E, Salmon DP, et al. Physical basis of cognitive alterations in Alzheimer's disease: synapse loss is the major correlate of cognitive impairment. *Ann Neurol*. 1991;30:572–580.
12. DeKosky ST, Scheff SW, Styrene SD. Structural correlates of cognition in dementia: quantification and assessment of synapse change. *Neurodegeneration*. 1996;5:417–421.
13. Swerdlow RH. Brain aging, Alzheimer's disease, and mitochondria. *Biochim Biophys Acta*. 2011;1812:1630–1639.
14. Reddy PH. Amyloid precursor protein-mediated free radicals and oxidative damage: implications for the development and progression of *Alzheimer's disease. J Neurochem*. 2006;96:1–13.
15. Reddy PH, Beal MF. Amyloid beta, mitochondrial dysfunction and synaptic damage: implications for cognitive decline in aging and Alzheimer's disease. *Trends Mol Med*. 2008;14:45–53.
16. Tampellini D, Gouras GK. Synapses, synaptic activity and intraneuronal abeta in Alzheimer's disease. *Front Aging Neurosci*. 2010;2. http://dx.doi.org/10.3389/fnagi.2010.00013, pii: 13.
17. Bertoni-Freddari C, Fattoretti P, Casoli T, Caselli U, Meier-Ruge W. Deterioration threshold of synaptic morphology in aging and senile dementia of Alzheimer's type. *Anal Quant Cytol Histol*. 1996;18:209–213.
18. Hyman BT, Van Hoesen GW, Kromer LJ, Damasio AR. Perforant pathway changes and the memory impairment of Alzheimer's disease. *Ann Neurol*. 1986;20:472–481.
19. Berg L, McKeel Jr DW, Miller JP, Baty J, Morris JC. Neuropathological indexes of Alzheimer's disease in demented and nondemented persons aged 80 years and older. *Arch Neurol*. 1993;50:349–358.

20. Braak H, Braak E. Neuropathological staging of Alzheimer-related changes. *Acta Neuropathol.* 1991;82:239–259.
21. Samuel W, Masliah E, Hill LR, Butters N, Terry R. Hippocampal connectivity and Alzheimer's dementia: effects of synapse loss and tangle frequency in a two-component model. *Neurology.* 1994;44:2081–2088.
22. Goate A, Chartier-Harlin MC, Mullan M, et al. Segregation of a missense mutation in the amyloid precursor protein gene with familial Alzheimer's disease. *Nature.* 1991;349:704–706.
23. Levy-Lahad E, Wasco W, Poorkaj P, et al. Candidate gene for the chromosome 1 familial Alzheimer's disease locus. *Science.* 1995;269:973–977.
24. Schellenberg GD, Bird TD, Wijsman EM, et al. Genetic linkage evidence for a familial Alzheimer's disease locus on chromosome 14. *Science.* 1992;258:668–671.
25. Harold D, Abraham R, Hollingworth P, et al. Genome-wide association study identifies variants at CLU and PICALM associated with Alzheimer's disease. *Nat Genet.* 2009;41:1088–1093.
26. Hollingworth P, Harold D, Sims R, et al. Common variants at ABCA7, MS4A6A/MS4A4E, EPHA1, CD33 and CD2AP are associated with Alzheimer's disease. *Nat Genet.* 2011;43:429–435.
27. Lambert JC, Heath S, Even G, et al. Genome-wide association study identifies variants at CLU and CR1 associated with Alzheimer's disease. *Nat Genet.* 2009;41:1094–1099.
28. Naj AC, Jun G, Beecham GW, et al. Common variants at MS4A4/MS4A6E, CD2AP, CD33 and EPHA1 are associated with late-onset Alzheimer's disease. *Nat Genet.* 2011;43:436–441.
29. Rogaeva E, Meng Y, Lee JH, et al. The neuronal sortilin-related receptor SORL1 is genetically associated with Alzheimer disease. *Nat Genet.* 2007;39(2):168–177.
30. Strittmatter WJ, Weisgraber KH, Huang DY, et al. Binding of human apolipoprotein E to synthetic amyloid beta peptide: isoform-specific effects and implications for late-onset Alzheimer disease. *Proc Natl Acad Sci USA.* 1993;90:8098–8102.
31. Broersen K, Rousseau F, Schymkowitz J. The culprit behind amyloid beta peptide related neurotoxicity in Alzheimer's disease: oligomer size or conformation? *Alzheimers Res Ther.* 2010;2:12.
32. Walsh DM, Selkoe DJ. A beta oligomers—a decade of discovery. *J Neurochem.* 2007;101:1172–1184.
33. Reddy PH. Abnormal tau, mitochondrial dysfunction, impaired axonal transport of mitochondria, and synaptic deprivation in Alzheimer's disease. *Brain Res.* 2011;1415:136–148.
34. Lustbader JW, Cirilli M, Lin C, et al. ABAD directly links Abeta to mitochondrial toxicity in Alzheimer's disease. *Science.* 2004;304:448–452.
35. Du H, Guo L, Fang F, et al. Cyclophilin D deficiency attenuates mitochondrial and neuronal perturbation and ameliorates learning and memory in Alzheimer's disease. *Nat Med.* 2008;14:1097–1105.
36. Manczak M, Calkins MJ, Reddy PH. Impaired mitochondrial dynamics and abnormal interaction of amyloid beta with mitochondrial protein Drp1 in neurons from patients with Alzheimer's disease: implications for neuronal damage. *Hum Mol Genet.* 2011;20:2495–2509.
37. Manczak M, Reddy PH. Abnormal interaction of VDAC1 with amyloid beta and phosphorylated tau causes mitochondrial dysfunction in Alzheimer's disease. *Hum Mol Genet.* 2012;21:5131–5146.
38. Avila J, Lucas JJ, Perez M, Hernandez F. Role of tau protein in both physiological and pathological conditions. *Physiol Rev.* 2004;84:361–384.
39. Iqbal K, Alonso Adel C, Chen S, et al. Tau pathology in Alzheimer disease and other tauopathies. *Biochim Biophys Acta.* 2005;1739:198–210.

40. Lee VM, Goedert M, Trojanowski JQ. Neurodegenerative tauopathies. *Annu Rev Neurosci*. 2011;24:1121–1159.
41. Mocanu MM, Nissen A, Eckermann K, et al. The potential for beta-structure in the repeat domain of tau protein determines aggregation, synaptic decay, neuronal loss, and coassembly with endogenous Tau in inducible mouse models of tauopathy. *J Neurosci*. 2008;28:737–748.
42. Zempel H, Mandelkow EM. Linking amyloid-β and tau: amyloid-β induced synaptic dysfunction via local wreckage of the neuronal cytoskeleton. *Neurodegener Dis*. 2012;10:64–72.
43. Drago D, Cavaliere A, Mascetra N, et al. Aluminum modulates effects of beta amyloid (1-42) on neuronal calcium homeostasis and mitochondria functioning and is altered in a triple transgenic mouse model of Alzheimer's disease. *Rejuvenation Res*. 2008;11:861–871.
44. Resende R, Moreira PI, Proença T, et al. Brain oxidative stress in a triple-transgenic mouse model of Alzheimer disease. *Free Radic Biol Med*. 2008;44:2051–2057.
45. Sensi SL, Rapposelli IG, Frazzini V, Mascetra N. Altered oxidant-mediated intraneuronal zinc mobilization in a triple transgenic mouse model of Alzheimer's disease. *Exp Gerontol*. 2008;43:488–492.
46. Yao J, Irwin RW, Zhao L, Nilsen J, Hamilton RT, Brinton RD. Mitochondrial bioenergetic deficit precedes Alzheimer's pathology in female mouse model of Alzheimer's disease. *Proc Natl Acad Sci USA*. 2009;106:14670–14675.
47. David DC, Hauptmann S, Scherping I, et al. Proteomic and functional analyses reveal a mitochondrial dysfunction in P301L tau transgenic mice. *J Biol Chem*. 2005;280: 23802–23814.
48. Rhein V, Song X, Wiesner A, et al. Amyloid-beta and tau synergistically impair the oxidative phosphorylation system in triple transgenic Alzheimer's disease mice. *Proc Natl Acad Sci USA*. 2009;106:20057–20062.
49. Manczak M, Reddy PH. Abnormal interaction of oligomeric amyloid-β with phosphorylated tau: implications to synaptic dysfunction and neuronal damage. *J Alzheimers Dis*. 2013;36:285–295.
50. Scheff SW, Anderson KJ, DeKosky ST. Strain comparison of synaptic density in hippocampal CA1 of aged rats. *Neurobiol Aging*. 1985;6:29–34.
51. Scheff SW, Scott SA, DeKosky ST. Quantitation of synaptic density in the septal nuclei of young and aged Fischer 344 rats. *Neurobiol Aging*. 1991;12:3–12.
52. Gylys KH, Fein JA, Yang F, Wiley DJ, Miller CA, Cole GM. Synaptic changes in Alzheimer's disease: increased amyloid-beta and gliosis in surviving terminals is accompanied by decreased PSD-95 fluorescence. *Am J Pathol*. 2004;165:1809–1817.
53. Reddy PH, Mani G, Park BS, et al. Differential loss of synaptic proteins in Alzheimer's disease: implications for synaptic dysfunction. *J Alzheimers Dis*. 2005;7:103–117.
54. Almeida CG, Tampellini D, Takahashi RH, et al. Beta-amyloid accumulation in APP mutant neurons reduces PSD-95 and GluR1 in synapses. *Neurobiol Dis*. 2005;20:187–198.
55. Calkins MJ, Manczak M, Mao P, Shirendeb U, Reddy PH. Impaired mitochondrial biogenesis, defective axonal transport of mitochondria, abnormal mitochondrial dynamics and synaptic degeneration in a mouse model of Alzheimer's disease. *Hum Mol Genet*. 2011;20:4515–4529.
56. Calkins MJ, Reddy PH. Amyloid beta impairs mitochondrial anterograde transport and degenerates synapses in Alzheimer's disease neurons. *Biochim Biophys Acta*. 2011;1812:507–513.
57. Wang X, Perry G, Smith MA, Zhu X. Amyloid-beta-derived diffusible ligands cause impaired axonal transport of mitochondria in neurons. *Neurodegener Dis*. 2010;7:56–59.

58. Wang X, Su B, Siedlak SL, et al. Amyloid-beta overproduction causes abnormal mitochondrial dynamics via differential modulation of mitochondrial fission/fusion proteins. *Proc Natl Acad Sci USA*. 2008;105:19318–19323.
59. Wang X, Su B, Lee HG, et al. Impaired balance of mitochondrial fission and fusion in Alzheimer's disease. *J Neurosci*. 2009;29:9090–9103.
60. Benson DF, Kuhl DE, Hawkins RA, Phelps ME, Cummings JL, Tsai SY. The fluorodeoxyglucose 18F scan in Alzheimer's disease and multi-infarct dementia. *Arch Neurol*. 1983;40:711–714.
61. Kennedy AM, Frackowiak RS, Newman SK, et al. Deficits in cerebral glucose metabolism demonstrated by positron emission tomography in individuals at risk of familial Alzheimer's disease. *Neurosci Lett*. 1915;186:17–20.
62. Vander Borght T, Minoshima S, Giordani B, et al. Cerebral metabolic differences in Parkinson's and Alzheimer's diseases matched for dementia severity. *J Nucl Med*. 1997;38:797–802.
63. Small GW, Mazziotta JC, Collins MT, et al. Apolipoprotein E type 4 allele and cerebral glucose metabolism in relatives at risk for familial Alzheimer disease. *JAMA*. 1995;273:942–947.
64. Small SA, Nava AS, Perera GM, Delapaz R, Stern Y. Evaluating the function of hippocampal subregions with high-resolution MRI in Alzheimer's disease and aging. *Microsc Res Tech*. 2000;51:101–108.
65. Small GW, Ercoli LM, Silverman DH, et al. Cerebral metabolic and cognitive decline in persons at genetic risk for Alzheimer's disease. *Proc Natl Acad Sci USA*. 2000;97:6037–6042.
66. Jagust W, Gitcho A, Sun F, Kuczynski B, Mungas D, Haan M. Brain imaging evidence of preclinical Alzheimer's disease in normal aging. *Ann Neurol*. 2006;59:673–681.
67. Reddy PH. Mitochondrial medicine for aging and neurodegenerative diseases. *Neuromolecular Med*. 2008;10:291–315.
68. Reddy PH. Mitochondrial dysfunction in aging and Alzheimer's disease: strategies to protect neurons. *Antioxid Redox Signal*. 2007;9:1647–1658.
69. Reddy PH. Role of mitochondria in neurodegenerative diseases: mitochondria as a therapeutic target in Alzheimer's disease. *CNS Spectr*. 2009;14:8–13. discussion 16–8.
70. Caspersen C, Wang N, Yao J, et al. Mitochondrial Abeta: a potential focal point for neuronal metabolic dysfunction in Alzheimer's disease. *FASEB J*. 2005;19:2040–2041.
71. Manczak M, Anekonda TS, Henson E, Park BS, Quinn J, Reddy PH. Mitochondria are a direct site of A beta accumulation in Alzheimer's disease neurons: implications for free radical generation and oxidative damage in disease progression. *Hum Mol Genet*. 2006;15:1437–1449.
72. Amadoro G, Corsetti V, Atlante A, et al. Interaction between NH(2)-tau fragment and Aβ in Alzheimer's disease mitochondria contributes to the synaptic deterioration. *Neurobiol Aging*. 2012;33. 833.e1-25.
73. Atlante A, Amadoro G, Bobba A, et al. A peptide containing residues 26-44 of tau protein impairs mitochondrial oxidative phosphorylation acting at the level of the adenine nucleotide translocator. *Biochim Biophys Acta*. 2008;1777:1289–1300.
74. Quintanilla RA, Matthews-Roberson TA, Dolan PJ, Johnson GV. Caspase-cleaved tau expression induces mitochondrial dysfunction in immortalized cortical neurons: implications for the pathogenesis of Alzheimer disease. *J Biol Chem*. 2009;284:18754–18766.
75. Stackman RW, Eckenstein F, Frei B, Kulhanek D, Nowlin J, Quinn JF. Prevention of age-related spatial memory deficits in a transgenic mouse model of Alzheimer's disease by chronic Ginkgo biloba treatment. *Exp Neurol*. 2003;184:510–520.
76. Conte V, Uryu K, Fujimoto S, et al. Vitamin E reduces amyloidosis and improves cognitive function in Tg2576 mice following repetitive concussive brain injury. *J Neurochem*. 2004;90:758–764.

77. Sung S, Yao Y, Uryu K, et al. Early vitamin E supplementation in young but not aged mice reduces Abeta levels and amyloid deposition in a transgenic model of Alzheimer's disease. *FASEB J*. 2004;18:323–325.
78. Nakashima H, Ishihara T, Yokota O, et al. Effects of alpha-tocopherol on an animal model of tauopathies. *Free Radic Biol Med*. 2004;37:176–186.
79. Matsubara E, Bryant-Thomas T, Pacheco Quinto J, et al. Melatonin increases survival and inhibits oxidative and amyloid pathology in a transgenic model of Alzheimer's disease. *J Neurochem*. 2003;85:1101–1108.
80. Feng Z, Chang Y, Cheng Y, et al. Melatonin alleviates behavioral deficits associated with apoptosis and cholinergic system dysfunction in the APP 695 transgenic mouse model of Alzheimer's disease. *J Pineal Res*. 2004;37:129–136.
81. Quinn J, Kulhanek D, Nowlin J, et al. Chronic melatonin therapy fails to alter amyloid burden or oxidative damage in old Tg2576 mice: implications for clinical trials. *Brain Res*. 2005;1037:209–213.
82. Yang X, Dai G, Li G, Yang ES. Coenzyme Q10 reduces beta-amyloid plaque in an APP/PS1 transgenic mouse model of Alzheimer's disease. *J Mol Neurosci*. 2010;41:110–113.
83. Siedlak SL, Casadesus G, Webber KM, et al. Chronic antioxidant therapy reduces oxidative stress in a mouse model of Alzheimer's disease. *Free Radic Res*. 2009;43:156–164.
84. Shenk JC, Liu J, Fischbach K, et al. The effect of acetyl-L-carnitine and R-alpha-lipoic acid treatment in ApoE4 mouse as a model of human Alzheimer's disease. *J Neurol Sci*. 2009;283:199–206.
85. Harrison FE, Hosseini AH, McDonald MP, May JM. Vitamin C reduces spatial learning deficits in middle-aged and very old APP/PSEN1 transgenic and wild-type mice. *Pharmacol Biochem Behav*. 2009;93:443–450.
86. Cho JY, Um HS, Kang EB, et al. The combination of exercise training and alpha-lipoic acid treatment has therapeutic effects on the pathogenic phenotypes of Alzheimer's disease in NSE/APPsw-transgenic mice. *Int J Mol Med*. 2010;25:337–346.
87. Huang Q, Aluise CD, Joshi G, et al. Potential in vivo amelioration by N-acetyl-L-cysteine of oxidative stress in brain in human double mutant APP/PS-1 knock-in mice: toward therapeutic modulation of mild cognitive impairment. *J Neurosci Res*. 2010;88:2618–2629.
88. Dragicevic N, Copes N, O'Neal-Moffitt G, et al. Melatonin treatment restores mitochondrial function in Alzheimer's mice: a mitochondrial protective role of melatonin membrane receptor signaling. *J Pineal Res*. 2011;51:75–86.
89. Murakami K, Murata N, Ozawa Y, et al. Vitamin C restores behavioral deficits and amyloid-β oligomerization without affecting plaque formation in a mouse model of Alzheimer's disease. *J Alzheimers Dis*. 2011;26:7–18.
90. Dumont M, Kipiani K, Yu F, et al. Coenzyme Q10 decreases amyloid pathology and improves behavior in a transgenic mouse model of Alzheimer's disease. *J Alzheimers Dis*. 2011;27:211–223.
91. McManus MJ, Murphy MP, Franklin JL. The mitochondria-targeted antioxidant MitoQ prevents loss of spatial memory retention and early neuropathology in a transgenic mouse model of Alzheimer's disease. *J Neurosci*. 2011;31:15703–15715.
92. García-Mesa Y, Giménez-Llort L, López LC, et al. Melatonin plus physical exercise are highly neuroprotective in the 3xTg-AD mouse. *Neurobiol Aging*. 2012;33. 1124.e13-29.
93. Clausen A, Xu X, Bi X, Baudry M. Effects of the superoxide dismutase/catalase mimetic EUK-207 in a mouse model of Alzheimer's disease: protection against and interruption of progression of amyloid and tau pathology and cognitive decline. *J Alzheimers Dis*. 2012;30:183–208.
94. Baño Otalora B, Popovic N, Gambini J, et al. Circadian system functionality, hippocampal oxidative stress, and spatial memory in the APPswe/PS1dE9 transgenic model of Alzheimer disease: effects of melatonin or ramelteon. *Chronobiol Int*. 2012;29:822–834.

95. Yan JJ, Jung JS, Kim TK, et al. Protective effects of ferulic acid in amyloid precursor protein plus presenilin-1 transgenic mouse model of Alzheimer disease. *Biol Pharm Bull.* 2013;36:140–143.
96. Shi TY, Zhao DQ, Wang HB, et al. A new chiral pyrrolyl α-nitronyl nitroxide radical attenuates β-amyloid deposition and rescues memory deficits in a mouse model of Alzheimer disease. *Neurotherapeutics.* 2013;10:340–353.
97. Montinaro M, Uberti D, Maccarinelli G, Bonini SA, Ferrari-Toninelli G, Memo M. Dietary zeolite supplementation reduces oxidative damage and plaque generation in the brain of an Alzheimer's disease mouse model. *Life Sci.* 2013;92:903–910.
98. Handattu SP, Monroe CE, Nayyar G, et al. In vivo and in vitro effects of an apolipoprotein e mimetic peptide on amyloid-β pathology. *J Alzheimers Dis.* 2013;36:335–347.
99. Wang P, Su C, Li R, et al. Mechanisms and effects of curcumin on spatial learning and memory improvement in APPswe/PS1dE9 mice. *J Neurosci Res.* 2014;92:218–231.
100. Mendelsohn AB, Belle SH, Stoehr GP, Ganguli M. Use of antioxidant supplements and its association with cognitive function in a rural elderly cohort: the MoVIES Project. Monongahela Valley Independent Elders Survey. *Am J Epidemiol.* 1998;148:38–44.
101. Grodstein F, Chen J, Willett WC. High-dose antioxidant supplements and cognitive function in community-dwelling elderly women. *Am J Clin Nutr.* 2003;77:975–984.
102. Morris MC, Evans DA, Bienias JL, et al. Dietary intake of antioxidant nutrients and the risk of incident Alzheimer disease in a biracial community study. *JAMA.* 2002;287:3230–3237.
103. Masaki KH, Losonczy KG, Izmirlian G, et al. Association of vitamin E and C supplement use with cognitive function and dementia in elderly men. *Neurology.* 2000;54:1265–1272.
104. Engelhart MJ, Geerlings MI, Ruitenberg A, et al. Dietary intake of antioxidants and risk of Alzheimer disease. *JAMA.* 2002;287:3223–3229.
105. Maxwell CJ, Hicks MS, Hogan DB, Basran J, Ebly EM. Supplemental use of antioxidant vitamins and subsequent risk of cognitive decline and dementia. *Dement Geriatr Cogn Disord.* 2005;20:45–51.
106. Luchsinger JA, Tang MX, Shea S, Mayeux R. Antioxidant vitamin intake and risk of Alzheimer disease. *Arch Neurol.* 2003;60:203–208.
107. Zandi PP, Anthony JC, Khachaturian AS, et al. Reduced risk of Alzheimer disease in users of antioxidant vitamin supplements: the Cache County Study. *Arch Neurol.* 2004;61:82–88.
108. Laurin D, Masaki KH, Foley DJ, White LR, Launer LJ. Midlife dietary intake of antioxidants and risk of late-life incident dementia: the Honolulu-Asia Aging Study. *Am J Epidemiol.* 2004;159(10):959–967.
109. Fillenbaum GG, Kuchibhatla MN, Hanlon JT, et al. Dementia and Alzheimer's disease in community-dwelling elders taking vitamin C and/or vitamin E. *Ann Pharmacother.* 2005;39:2009–2014.
110. Gray SL, Anderson ML, Crane PK, et al. Antioxidant vitamin supplement use and risk of dementia or Alzheimer's disease in older adults. *J Am Geriatr Soc.* 2008;56:291–295.
111. Akhondzadeh S, Noroozian M, Mohammadi M, Ohadinia S, Jamshidi AH, Khani M. Salvia officinalis extract in the treatment of patients with mild to moderate Alzheimer's disease: a double blind, randomized and placebo-controlled trial. *J Clin Pharm Ther.* 2003;28:53–59.
112. Petersen RC, Thomas RG, Grundman M, et al. Vitamin E and donepezil for the treatment of mild cognitive impairment. *N Engl J Med.* 2005;352:2379–2388.
113. Yaffe K, Clemons TE, McBee WL, Lindblad AS. Age-Related Eye Disease Study Research Group. Impact of antioxidants, zinc, and copper on cognition in the elderly: a randomized, controlled trial. *Neurology.* 2004;63:1705–1707.

114. Kang JH, Cook N, Manson J, Buring JE, Grodstein F. A randomized trial of vitamin E supplementation and cognitive function in women. *Arch Intern Med.* 2006;166: 2462–2468.
115. Sano M, Ernesto C, Thomas RG, et al. A controlled trial of selegiline, alpha-tocopherol, or both as treatment for Alzheimer's disease. The Alzheimer's Disease Cooperative Study. *N Engl J Med.* 1997;336:1216–1222.
116. Smith RA, Hartley RC, Murphy MP. Mitochondria-targeted small molecule therapeutics and probes. *Antioxid Redox Signal.* 2011;15:3021–3038.
117. Szeto HH. Mitochondria-targeted peptide antioxidants: novel neuroprotective agents. *AAPS J.* 2006;8:E521–E531.
118. Manczak M, Mao P, Calkins MJ, et al. Mitochondria-targeted antioxidants protect against amyloid-beta toxicity in Alzheimer's disease neurons. *J Alzheimers Dis.* 2010; 20(suppl 2):S609–S631.
119. Jin H, Kanthasamy A, Ghosh A, Anantharamaiah V, Kalyanaraman B, Kanthasamy AG. Mitochondria-targeted antioxidants for treatment of Parkinson's disease: preclinical and clinical outcomes. *Biochim Biophys Acta.* 2014;1842(8):1282–1294.
120. Schriner SE, Linford NJ, Martin GM, et al. Extension of murine life span by overexpression of catalase targeted to mitochondria. *Science.* 2005;308:1909–1911.
121. Mao P, Manczak M, Calkins MJ, et al. Mitochondria-targeted catalase reduces abnormal APP processing, amyloid β production and BACE1 in a mouse model of Alzheimer's disease: implications for neuroprotection and lifespan extension. *Hum Mol Genet.* 2012;21:2973–2990.
122. Szeto HH. Development of mitochondria-targeted aromatic-cationic peptides for neurodegenerative diseases. *Ann N Y Acad Sci.* 2008;1147:112–121.
123. Szeto HH, Schiller PW. Novel therapies targeting inner mitochondrial membrane—from discovery to clinical development. *Pharm Res.* 2011;28:2669–2679.

CHAPTER SEVEN

Mitochondrial-Targeted Catalase: Extended Longevity and the Roles in Various Disease Models

D.-F. Dai, Y.-A. Chiao, G.M. Martin, D.J. Marcinek, N. Basisty, E.K. Quarles, P.S. Rabinovitch[1]
University of Washington, Seattle, WA, United States
[1]Corresponding author: e-mail address: PeterR@medicine.washington.edu

Contents

Abstract

The free-radical theory of aging was proposed more than 50 years ago. As one of the most popular mechanisms explaining the aging process, it has been extensively studied in several model organisms. However, the results remain controversial. The mitochondrial version of free-radical theory of aging proposes that mitochondria are both the primary sources of reactive oxygen species (ROS) and the primary targets of ROS-induced damage. One critical ROS is hydrogen peroxide, which is naturally degraded by catalase in peroxisomes or glutathione peroxidase within mitochondria. Our laboratory developed mice-overexpressing catalase targeted to mitochondria (mCAT), peroxisomes (pCAT), or the nucleus (nCAT) in order to investigate the role

Progress in Molecular Biology and Translational Science, Volume 146
ISSN 1877-1173
http://dx.doi.org/10.1016/bs.pmbts.2016.12.015

of hydrogen peroxide in different subcellular compartments in aging and age-related diseases. The mCAT mice have demonstrated the largest effects on life span and healthspan extension. This chapter will discuss the mCAT phenotype and review studies using mCAT to investigate the roles of mitochondrial oxidative stresses in various disease models, including metabolic syndrome and atherosclerosis, cardiac aging, heart failure, skeletal muscle pathology, sensory defect, neurodegenerative diseases, and cancer. As ROS has been increasingly recognized as essential signaling molecules that may be beneficial in hormesis, stress response and immunity, the potential pleiotropic, or adverse effects of mCAT are also discussed. Finally, the development of small-molecule mitochondrial-targeted therapeutic approaches is reviewed.

1. INTRODUCTION

The potential connections between free radicals, particularly reactive oxygen species (ROS) and aging, have a long, complicated, and often controversial history. It is within this context that the transgenic mouse with catalase targeted to mitochondria (mCAT) has served as an elegant tool to dissect the role of mitochondrial ROS (mtROS) in healthspan and disease.

Harman first proposed the free-radical theory of aging in 1956, in which he suggested that free-radical-induced accumulation of macromolecular damage was a driving force in aging and a primary determinant of life span.[1] This theory was attractive in its simplicity and became highly tested. Initially, broad confirmation of the increased prevalence of ROS-mediated damage to macromolecules with age was demonstrated. Subsequent attempts to prove a more causal role by increasing or decreasing the antioxidant capacity of animals, however, had conflicting, although usually negative results.[2,3] Numerous attempts to apply antioxidant supplementation were undertaken based on the free-radical theory, almost all of which have had negative results.[4,5] These results, and others, led Harman to modify his original theory to specify mitochondria as both the primary sources of ROS and the primary targets of ROS damage.[6] Thus, the mitochondrial free-radical theory postulates a central role for mitochondria, both in generating ROS from the electron transport chain during production of energy (ATP) and in the numerous feedback loops that could be envisioned within mitochondria, in which redox state and ROS might create a "vicious cycle." Mutations or deletions in mtDNA can result in damaged proteins, including the subset of respiratory chain (RC) proteins that are encoded by mtDNA; damage to these proteins is hypothesized to lead to greater electron leakage and ROS production from the RC, as well as changes in the mitochondrial redox

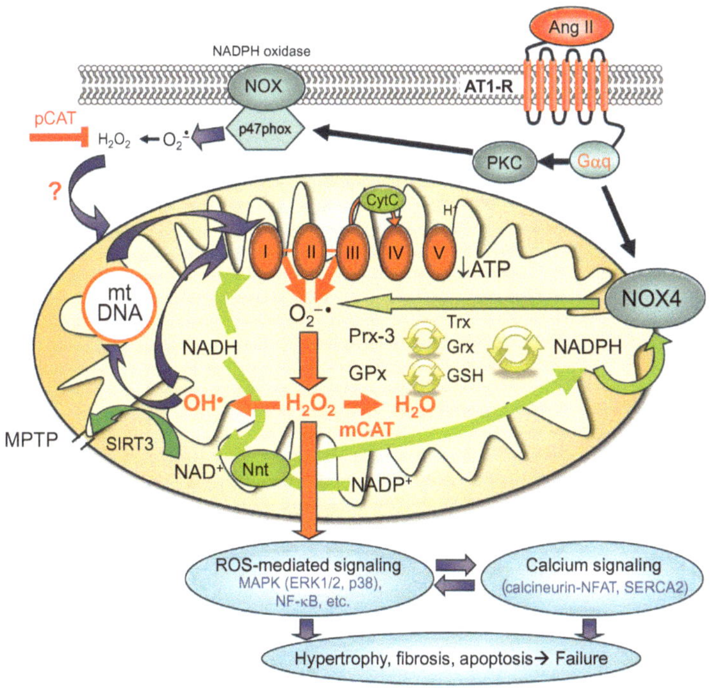

Fig. 1 Mitochondrial ROS and ROS scavenging, the vicious cycle of ROS-induced mtDNA damage, and ROS-induced ROS signaling. *Adapted from Dai DF, Rabinovitch PS, Ungvari Z. Mitochondria and cardiovascular aging.* Circ Res. *2012;110(8):1109–1124.*

balance, including glutathione and NAD(P)H redox pairs (see Fig. 1 and following sections). Importantly, these reductants are utilized to regenerate glutathione peroxidase (GPx) and peroxiredoxin (PRx), the primary intrinsic mitochondrial antioxidant enzymes that detoxify hydrogen peroxide, preventing it's conversion to the highly damaging hydroxyl radical. Furthermore, the reductive potential of NADPH is in balance with that of NADH, via the activity of nicotinamide nucleotide transferase (NNT); the balance of NAD/NADH regulates sirtuin histone deacetylases, including mitochondrial SIRT3.[7] This is a further example of a redox cycle in which ROS can have diverse metabolic consequences.

The mitochondrial version of the free-radical theory suggested that past failures to validate the general theory might be explained by failure of experimental interventions to target mtROS and that, conversely, antioxidants that were targeted to mitochondria might contribute to healthspan extension. It is within this context that transgenic overexpression of catalase has played a valuable role. Comparison of overexpression targeted to peroxisomes (the endogenously targeted location, pCAT) could be compared to mitochondrial-targeted expression (mCAT) to directly assess the relative importance of antioxidant potential in cytoplasmic vs mitochondrial compartments. Moreover, unlike other antioxidant mechanisms, catalase does not directly consume ATP or alter glutathione or NAD(P)H redox balance as it detoxifies hydrogen peroxide. It is can thus be utilized as a direct and targeted strategy to examine the effects of reduced ROS.

In recent years it has become more apparent that not all of the effects of ROS are necessarily detrimental, and in fact the theory of mitochondrial hormesis (mitohormesis) hypothesizes that low levels of ROS may trigger adaptive responses that improve overall stress resistance. Some of this effect may be through increased endogenous antioxidant defense, which may reduce chronic oxidative damage and subsequently improve healthspan.[8] Thus, an ideal antioxidant therapy might be one that prevents oxidative damage induced under pathological conditions without interfering with ROS needed for physiological ROS signaling or hormesis. The preponderance of positive effects of mCAT in aging and disease models, described subsequently, might suggest that it is able to function in such a capacity. Key to this may be that the K_m of the catalytic activity of catalase for conversion of hydrogen peroxide to water is >10 mM, so that this activity is less likely to be effective at the lower intracellular H_2O_2 concentrations that may be involved in signaling or hormesis.[9,10]

2. LIFE SPAN AND HEALTHSPAN EXTENSION IN MICE-OVEREXPRESSING CATALASE

Catalase is a heme-containing tetrameric protein, the activities of which are highest in the liver, kidney, and red blood cells and lower in the brain, heart, and skeletal muscle. In normal physiology, catalase functions in the peroxisome to breakdown the H_2O_2 generated by peroxisomal β-oxidation of long-chain fatty acids. If the intracellular H_2O_2 reaches sufficiently high levels, catalase in peroxisome can also break down H_2O_2 diffusing from the other sources. As mentioned earlier, catalase converts two H_2O_2 molecules to water and oxygen only at relatively high H_2O_2 levels. At lower levels, when a second H_2O_2 may be in limited supply, catalase may act as a peroxidase and oxidize a variety of substrates[11]; thus, catalase overexpression may not interfere with low level of H_2O_2 acting as signaling molecules. In mitochondria, excessive H_2O_2 is usually degraded by GPx using reduced glutathione (GSH), as part of complex redox enzymes maintaining GSH/GSSG and NAD(P)H/NAD(P) balance, which implies that GPx degradation of H_2O_2 is dependent on the availability of GSH, or the overall redox status (Fig. 1). Alternative detoxification of H_2O_2 by PRx also relies on NAD(P)H as its ultimate electron donor. The advantage of mitochondrial catalase in this context includes degradation of H_2O_2 in a manner independent of the availability of electron donors, which may prevent a potential "vicious cycle" within mitochondria.

Since both nucleus and mitochondria are susceptible to oxidative damage, the objectives of the original studies were to understand the role of oxidative damage to nuclear DNA and mitochondria during aging. There are no specific antioxidants in the nucleus, despite the fact that oxidative damage to the DNA can lead to gene mutations, genomic instability, cancer, and other phenotypes of aging. This led to the design of targeting catalase to the nucleus (nCAT) or to mitochondria (mCAT), in addition to overexpressing the wild-type peroxisomal catalase (pCAT). All three mouse models of catalase overexpression utilized the same beta-actin promoter and CMV enhancer element which resulted in similar tissue distribution of catalase expression and activities.[12] All mice in the longevity cohorts of pCAT, mCAT, and nCAT were allowed to age until death, to determine the relevance of H_2O_2 levels in life span and healthspan.[13] Censoring because of ethical euthanasia only occurred in the most severe end-of-life pathologic conditions. Of the three mouse longevity cohorts of catalase overexpression, the mCAT enhanced both maximal and median life span extension by ~20%. The pCAT showed a modest and overall nonsignificant extension of life span, and contradictory to our initial hypothesis, the nCAT had no positive effect on life span (Fig. 2). Of note, the life span extension effects may not be obvious when the ethical censoring rate for so-called end-of-life pathology was very high (unpublished data).

Increase in maximal life span is thought to be indicative of a protection of the underlying aging processes. Consistent with this, mCAT mice, which had robust mitochondrial catalase expression, especially in the heart and skeletal muscle, displayed reduced cardiac pathology in terms of attenuated age-dependent mineralization, arteriosclerosis, and myopathy phenotypes. Initial characterization of this reduced cardiac pathology was associated with reduction in H_2O_2 release from isolated cardiac mitochondria, decreased accumulation of oxidized DNA, and decreased susceptibility of mitochondrial aconitase to H_2O_2-induced damage.[13] All of these beneficial effects are consistent with elevated antioxidant defenses in mitochondria.[14]

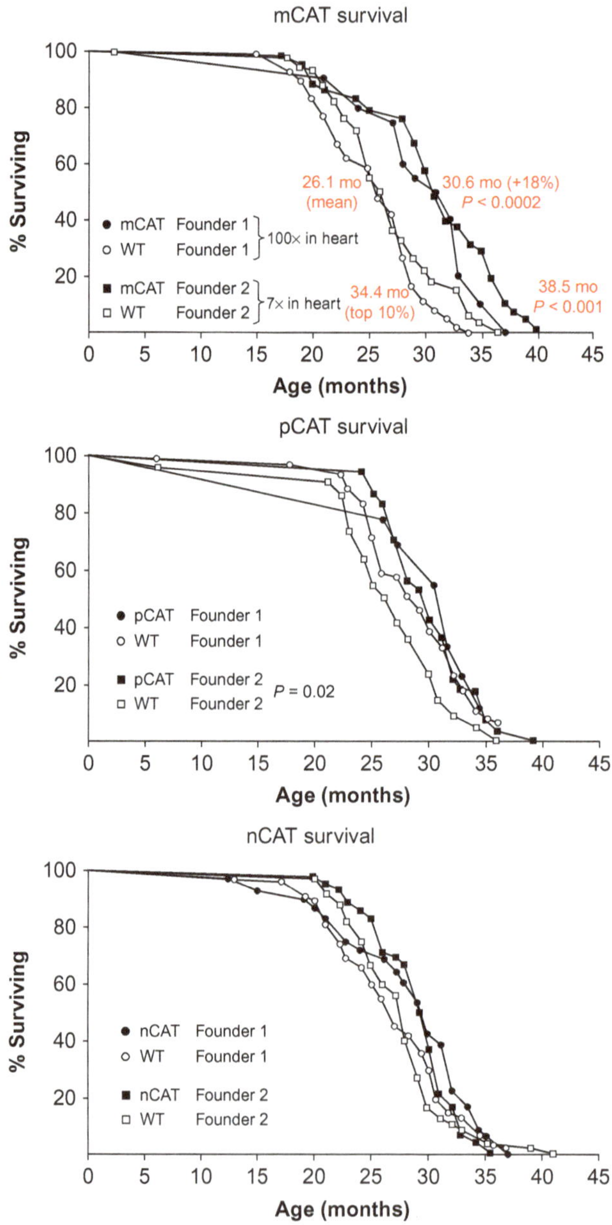

Fig. 2 Survival studies of catalase-overexpressing mice. *pCAT*, peroxisomal catalase; *mCAT*, mitochondrial-targeted catalase; *nCAT*, nuclear-targeted catalase. *Adapted from Schriner SE, Linford NJ, Martin GM, et al. Extension of murine life span by overexpression of catalase targeted to mitochondria.* Science. *2005;308(5730):1909–1911.*

3. THE CONTRIBUTION OF mCAT MOUSE MODELS TO THE STUDY OF DISEASES

In the following sections, we discuss the role of mitochondrial oxidative stress in several diseases which have been supported by the application mCAT mouse models.

3.1 Metabolic Syndrome and Atherosclerosis

Metabolic syndrome (MetS) is a cluster of disorders that increase the risk of type 2 diabetes mellitus (T2DM) and cardiovascular diseases (CVD).[15] There are various definitions of MetS,[15,16] which generally include some combination of hypertension, low HDL cholesterol levels, insulin resistance (IR), dysglycemia (e.g., impaired fasting glucose), hypertriglyceridemia, and obesity, particularly abdominal obesity.[17] The WHO estimates that in 2014, over 600 million adults were obese, representing about 11% of men and 15% of women worldwide. The prevalence of MetS is also rising worldwide.[18,19] In obese patients, the presence of MetS is associated with unfavorable factors and is referred as unhealthy obesity, when compared with obese patients without MetS. The serum of MetS patients has higher levels of advanced glycation end products (AGEs), associated with increased IR and dysglycemia, oxidative stress, and inflammation.[20] Interestingly, AGE levels increase with normal aging and are associated with IR in apparently healthy elderly individuals.[21] Thus, chronic increase in oxidative stress is present in MetS patients and is thought to be the consequence of metabolic dysregulation.[22] The connection between IR, MetS, and oxidative stress implies that the reduction of oxidative stress using mCAT may be beneficial in these settings.

3.1.1 mCAT Ameliorates IR

There is much indirect, and some direct, evidence that oxidative stress contributes to IR, though the mechanisms remain uncertain.[23] Obese skeletal muscle also shows reduced mitochondrial function and may not have the capacity to oxidize fat at an appropriate rate. The resulting lipid accumulation intracellularly in skeletal muscle leads to the generation and accumulation of fatty acyl-CoAs and other proinflammatory metabolites which in turn both impair β-cell function and dysregulate insulin signing.[23,24] This dysregulated insulin sensitivity in turn leads to increased hydrogen peroxide emission in mitochondria, shifting the redox state toward a more oxidized profile and reducing the redox buffering capacity. Excitingly, Lee and colleagues provided evidence that mCAT mice exhibit a preservation of mitochondrial function, energy metabolism, and insulin sensitivity in skeletal muscle with age. They also note a decrease compared to WT mice in mitochondrial oxidative damage concurrent with stable mitochondrial respiration, mitogenesis, and ATP synthesis.[25] Another study also noted that mCAT mice demonstrate a preservation of insulin sensitivity even when on a western diet by attenuating mtH_2O_2 emission.[23]

3.1.2 Atherosclerosis and mCAT

MetS patients have an increased risk of atherosclerotic CVD when compared to obese patients without MetS,[26] as MetS is well known to be associated with inflammation, prothrombotic state, and dyslipidemia,[18,27] all of which play crucial roles in atherogenesis.

The mCAT may prevent or curtail progression of atherosclerosis through multiple mechanisms. First, it has been shown in mice that expression of mCAT in macrophages directly leads to a reduced rate of lesion development in western diet-fed $LDLR^{-/-}$ mice.[28] Wang and colleagues concluded that mCAT was suppressing NF-κB-mediated entry of monocytes into the atherosclerotic lesions,[28] providing evidence that mitochondrial protection represents a viable strategy to attenuate progression of atherosclerosis. Second, since mCAT reduces IR in skeletal muscle, the improvement of insulin sensitivity may be beneficial in slowing atherosclerosis.

3.2 Cardiac Aging and Heart Failure

Increased oxidative stress is well known to play an important role in the pathogenesis of several CVD, including hypertension, atherosclerosis, cardiac hypertrophy related to aging or pressure overload, cardiac ischemia–reperfusion injury, as well as heart failure. Several sources of ROS have been reported, including mitochondria, NADPH oxidases (NOX), xanthine oxidase, monoamine oxidase, and nitric oxide synthase. Free radicals generated by these sources are maintained at physiological levels by several endogenous antioxidant systems, including superoxide dismutase (SOD), catalase, thioredoxin (TRX), glutaredoxin (GRX), GPxs, and glutathione reductase (GR). At physiological levels, ROS acts as a signaling mediator, but at pathological levels the substantial increase of ROS may activate autophagy as a cellular defense mechanism to prevent propagation of ROS from damaged mitochondria. Further bursts in ROS may incite opening of the mitochondrial permeability transition pore (mPTP) leading to cytochrome *c* release and activation of apoptosis.

In aged hearts, there is an age-dependent increase in mtROS production and impaired ROS detoxification (reviewed in Ref. 29–31). Impaired electron transport function may directly lead to electron leakage and subsequent generation of mtROS. Since the heart is a highly metabolic, active organ dependent on ATP generated by mitochondria, it is particularly susceptible to mitochondrial oxidative damage. Consistent with this, impairment of mitochondrial energetics has been documented in human and experimental animals with heart failure.[32] The molecular mechanisms of this may include mitochondrial biogenesis that does not keep up with the increasing demand,[33] mitochondrial uncoupling and decreased substrate availability,[34] and increased mitochondrial DNA deletions.[35]

Studies from our laboratory demonstrated an increase in mitochondrial protein carbonylation in aged or failing mouse hearts, indicative of increased oxidative damage to mitochondrial proteins.[36,37] We further showed that aged hearts had a three- to four-fold increase in mitochondrial DNA point mutation and deletion frequencies (minimum estimates, as newer methodologies have increased the dynamic range of assays). This mitochondrial damage, presumably related to oxidative stress, is expected to stimulate signaling for mitochondrial biogenesis, manifest in the aged heart by an increase in mtDNA copy number concomitant with significant upregulation of the master regulator PPAR-γ coactivator-1-α (PGC-1α), and its downstream transcription factors.[36] Using mCAT mice, we demonstrated that reduction of age-dependent mitochondrial oxidative damage in mitochondrial protein and mtDNA in mCAT mouse hearts was associated with significant amelioration of cardiac aging phenotypes. These beneficial effects included attenuation of age-dependent left ventricular hypertrophy and diastolic dysfunction, as well as improvement of overall myocardial performance.[36] These cardioprotective effects in aging mice were observed in C57BL/6,[36] as well as C3HF1 and BalbCF1 mice (unpublished observations).

The critical role of mitochondria in aging is further reinforced by mice with proofreading-deficient homozygous mutation of mitochondrial polymerase gamma ($Polg^{D257A/D257A}$ designated as $Polg^{m/m}$).[38,39] These mice had shortened life span and displayed substantial age-dependent increase in mtDNA point mutations and deletions, in parallel with "accelerated aging-like" phenotypes, including kyphosis, graying and loss of hair, anemia, osteoporosis, and age-dependent cardiomyopathy.[37,38] The observations that mitochondrial damage and cardiomyopathy in these mice can be partially rescued by concomitant mCAT overexpression in double-transgenic mice suggest that mtROS and mitochondrial DNA damage are part of a vicious cycle of ROS-induced ROS release (Fig. 1).[37] This mechanism of mtROS amplification may explain the observations that damaged mitochondria from aged or failing mouse hearts produce more ROS than healthy mitochondria in young hearts. In Polg$^{D257A/D257A}$ mutant mice, endurance exercise has been shown to enhance the performance of skeletal muscle and attenuate some phenotypes of cardiac dysfunction.[40] Possible mechanism includes exercise-induced augmentation of mitochondrial biogenesis, which may compensate the mitochondrial dysfunction in these mice. Relevant to this, accumulation of mtDNA deletions in old age has been documented in various tissues in man, including the heart.[41,42]

Genetic mutation of mitochondrial enzymes may manifest as idiopathic hypertrophic and dilated cardiomyopathies in human patients.[43] Consistently, mitochondrial DNA and protein oxidative damage have been reported in various experimental models of cardiac hypertrophy and heart failure,[44] including chronic infusion of angiotensin II and Gαq overexpression in mice.[35] By comparing the effect of mCAT and pCAT, we demonstrated that mCAT, but not pCAT, are protective against cardiac hypertrophy, fibrosis, and diastolic dysfunction induced by angiotensin II as well as heart failure phenotypes in Gαq-overexpressing mice.[35] These findings emphasize the central role of mtROS in cardiac hypertrophy and heart failure.

Our findings were consistent with the mechanism of ROS amplification within mitochondria, as illustrated in Fig. 1. Briefly, angiotensin II binds to ATR1, a Gαq-coupled receptor, and activates NOX through a PKC-dependent manner.[45] ROS generated from NOX2 at the cell membrane and/or from NOX4 at the mitochondrial membrane can stimulate electron leakage from respiratory complexes and induce further mtROS production.[35,46,47] Mechanisms of mtROS amplification may include ROS-induced ROS release as well as a ROS–mtDNA damage vicious cycle. The latter is supported by the observations that primary damage to mtDNA, either in Polg$^{m/m}$ or by administration of azidothymidine (AZT), is sufficient to elevate ROS, cause cardiac hypertrophy leading to heart failure.[35,36,48] AZT is a nucleoside analog inhibiting retroviral reverse transcriptase as the mechanism of anti-HIV. AZT-induced cardiomyopathy, like that in the Polg$^{m/m}$ mouse, is attenuated by mCAT.[48] Thus, breaking the ROS vicious cycle within mitochondria by mCAT or mitochondrial-targeted antioxidants (see Section 5) is effective in attenuating both cardiac hypertrophy and failure (Fig. 1).

The fact that NOX4 localized to the mitochondrial membrane reinforces the importance of mtROS in relation to nicotinamide adenine dinucleotide (NAD) metabolism in models of cardiac hypertrophy and failure. NOX4 activation consumes NADPH and directly generates superoxide anions leading to mitochondrial oxidative damage.[49,50] In mitochondria, superoxide anions are converted by SOD to become hydrogen peroxide, which is physiologically detoxified by peroxiredoxin-3 (PRx-3) and GPx. After their oxidation by hydrogen peroxide these enzymes are replenished using the reductive power of NADPH. However, the consumption of NADPH by NOX4 establishes another potential mitochondrial vicious cycle (Fig. 1). Enhancing mitochondrial antioxidants using mCAT or other small-molecule approach can break this vicious cycle by removing superoxide or hydrogen peroxide without consuming glutathione or NADPH. NADPH can itself be regenerated from $NADP^+$ by electron exchange with NADH, catalyzed by nicotinamide nucleotide transhydrogenase (Nnt). Thus, cardiomyocyte mitochondrial redox status is closely interrelated with NAD metabolism. This further implicates sirtuins (sensors of the ratio of NAD+/NADH), particularly mitochondrial SIRT3, in an epigenetic cardiac response to stress.

3.2.1 Ischemia–Reperfusion Injury

Oxidative stress is well known to mediate ischemia–reperfusion injury. ROS begins to accumulate during ischemia,[51] causing mitochondrial respiratory complex dysfunction, which produces a burst of ROS during reperfusion. In addition, the acidosis induced by ischemia may facilitate the conversion of the superoxide and hydrogen peroxide to the highly reactive peroxynitrite and hydroxyl-free radicals. Indeed, several conditions associated with ischemia reperfusion, including ROS accumulation, acidic pH, and a rise in $[Ca^{2+}]_i$, may open the mPTP, leading to more mtROS generation. This is consistent with a mtROS-induced ROS release mechanism discussed earlier.[52] Although mCAT has not been tested in ischemia–reperfusion injury, several mitochondrial-targeted molecules (e.g., SS-31) have shown beneficial effects in various experimental models and in a small clinical trial (cyclosporine)[53] (see Section 5).

3.2.2 mCAT in Pulmonary Hypertension

Pulmonary hypertension is characterized by increased pulmonary pressures due to vascular remodeling. Despite the advance of vasodilator and oxygen therapy, the mortality and morbidity of pulmonary hypertension remains high. ROS has been implicated in the pathogenesis of vascular injury and remodeling. In a hypoxia-induced pulmonary hypertension model in mice, mCAT expression was sufficient to attenuate NOX expression and it's downstream signaling, preventing the pathogenesis of pulmonary hypertension.[54]

3.3 Skeletal Muscle Pathology

Skeletal muscle, like the heart and brain, is a high metabolic demand tissue that depends on mitochondrial ATP production to meet the energetic costs of muscle contraction. However, in skeletal muscle, the energetic demand is very dynamic, differing over an order of magnitude between resting and intense muscle contraction.[55,56] The mitochondrial electron transport chain produces more ROS under low-flux conditions.[57] One consequence of this is that skeletal muscle physiology is characterized by periods of low metabolic flux and relatively higher superoxide production broken up by shorter periods of high metabolic flux during sustained muscle contraction where the majority of ROS production appears to come from nonmitochondrial sources such as NOX and xanthine oxidase.[58] These nonmitochondrial sources of ROS during muscle contraction play an important role in muscle adaptation to exercise,[58,59] while growing evidence implicates mitochondrial oxidative stress in skeletal muscle dysfunction, atrophy, and sarcopenia.[60–64]

Analyses of skeletal muscle dysfunction in mCAT mice support an important role for mitochondrial oxidative stress in skeletal muscle dysfunction with age and pathological stress. As might be expected, mitochondrial dysfunction is reduced or delayed in aging skeletal muscle expressing catalase. One characteristic of aging muscle is a decline in the quality of the mitochondria. In the EDL muscle from C57BL/6 mice aging was associated with a disruption in the stoichiometry of the electron transport system, particularly and increased expression of complex I proteins relative to other ETS subunits,[65] which led to a decline in respiratory efficiency (reduced flux per unit mitochondria). This shift with age was prevented in mCAT mice resulting in improved mitochondrial function, suggesting a causative link between mtH_2O_2 production and the development of mitochondrial deficits.

In addition to contributing to further declines in mitochondrial function in skeletal muscle, elevated mitochondrial oxidative stress disrupts diverse aspects of skeletal muscle physiology. The preservation of mitochondrial function with age has implications beyond muscle energetics. Lee *et al.* reported a significant decline in insulin sensitivity, glucose metabolism, and increased accumulation of intramyocellular lipid in aged skeletal muscle that was prevented in mCAT mice,[25] as well as attenuation of age-related decline in state 3 respiration (maximum ADP stimulated) in isolated mitochondria from C57BL/6 gastrocnemius muscles and elevated protein oxidative damage in both mitochondria and whole-muscle homogenates. Similar protection of skeletal muscle insulin sensitivity by mCAT was observed in mice fed a high-fat diet.[23] In this study WT mice fed a high-fat diet for 12 weeks demonstrated elevated mitochondrial hydrogen peroxide (mtH_2O_2) production, a more oxidized glutathione redox state, and reduced insulin sensitivity and glucose uptake. The presence of mCAT reduced the elevated mtH_2O_2, normalized redox state, and improved glucose uptake on the high-fat diet. These data clearly point to an important role for mtH_2O_2 in insulin action and glucose uptake in skeletal muscle.

Hydrogen peroxide provides an important target for redox regulation of myofiber physiology beyond oxidative damage due to its relative stability compared to other ROS.[66] This relative stability makes hydrogen peroxide a better candidate than other ROS for redox signaling, due the increased specificity resulting from the need to interact with a transition metal or thiol group.[66] A series of papers comparing mitochondrial superoxide vs mtH_2O_2 scavenging on skeletal muscle glucose uptake support the direct targeting of mtH_2O_2 in skeletal muscle as a more effective strategy to improve muscle physiology.[67–69] Heterozygous deletion of the mitochondrial isoform of superoxide dismutase (SOD2) resulted in elevated redox stress in islet cells and reduced insulin secretion without an effect on glucose uptake by the skeletal muscle.[67] Although reduction of SOD2 led to a more oxidized GSH redox status in skeletal muscle in the chow-fed mice, the effect on GSH redox was not significantly different in the high-fat fed mice. Additionally, there was no difference in maximal potential for muscle mtH_2O_2 production (succinate supported state 4) in the reduced SOD2 mice on either diet.[67] Further support for specific targeting of H_2O_2 scavenging comes from overexpression of SOD2. In this case elevating superoxide scavenging in the mitochondria does not protect against high-fat diet-induced IR in the skeletal muscle, nor does it provide any additive effect in combination with scavenging of mtH_2O_2 by mCAT[68,69] in sedentary mice. In contrast, during exercise elevated SOD2 activity does lead to an increase in muscle glucose uptake under both chow- and high-fat-fed conditions.[69]

Reducing mtH_2O_2 with mCAT also reduces muscle weakness and atrophy with chronic disease. Muscle atrophy, weakness, and reduced exercise tolerance are associated with a leaky sarcoplasmic reticulum (SR) calcium release channel (RyR1) in aged mouse muscles. Oxidation-dependent post-translational modifications destabilize the interaction between RyR1 and calstabin1 and lead to increased calcium leak.[70] This increased calcium leak reduces SR calcium loading and calcium release in response to muscle activation resulting in reduced force production. Increased calcium leak also leads to increased mitochondrial calcium uptake. Under acute conditions increased calcium uptake can stimulate TCA dehydrogenases and increase mitochondrial ATP production.[71,72] However, under chronic conditions of low metabolic demand, increased mitochondrial calcium increases mitochondrial superoxide production.[71,73] Thus, this elevated calcium leak can initiate a feedforward cycle that where the increased mtROS production induces an oxidative stress and further RyR1 calcium leak. Stabilizing the RyR1 and calstabin1 interaction improves muscle performance by reducing calcium leak.[70] In mCAT mice, this cycle was prevented.[74,75] Aged mCAT mice had higher specific force, increased calcium release amplitude, reduced calcium leak, and increased SR loading in flexor digitorum brevis muscle fibers compared to age-matched WT mice. These parameters were not different between WT and mCAT in young adult muscle fibers. This same group has identified RyR1 calcium leak as an important mechanism underlying skeletal muscle dysfunction in heart failure and some muscular dystrophies, although the efficacy of mCAT in ameliorating contractile dysfunction in these models has not been tested in this system.

Muscle weakness and wasting associated with cancer and chemotherapy are another area that is receiving increased attention as an important contributor to reduced quality of life and frailty.[76,77] The associated fatigue is rated as a significant factor impacting quality of life for cancer survivors that can persist several years postdiagnosis.[78] Cancer and chemotherapeutic agents, especially the anthracycline agents (doxorubicin),[61,79,80] can both contribute to skeletal muscle dysfunction independently. The combined effects on muscle function are relatively less studied, but the dual stressors are expected to have a synergistic effect on skeletal muscle function and fatigue. As in the examples earlier, mitochondrial oxidative stress has been implicated as a causative factor in muscle atrophy and weakness following exposure to anthracyclines.[61,80] Gilliam recently tested whether reducing mitochondrial oxidative stress with mCAT could prevent skeletal muscle dysfunction in mice inoculated with a breast cancer cell line, treated with a single dose of doxorubicin, and those receiving a combination of stressors. Tumors were allowed to grow for 17 days, followed by injection of either saline or doxorubicin on day 17. On day 20 mitochondrial respiration was elevated and mtH_2O_2 production was reduced in permeabilized muscle fibers from the mCAT mice compared to WT. This was associated with reduced protein oxidative damage and preservation of muscle mass and force production. These data are consistent with other studies demonstrating protection of skeletal muscle function following doxorubicin treatment using mitochondrial-targeted small molecules to reduce oxidative stress.[61]

The reports described earlier from multiple models of chronic stress, including aging, heart failure, and cancer, highlight an important role for mitochondrial quality in regulating muscle physiology beyond energy metabolism by controlling cellular redox homeostasis. The efficacy of mCAT in ameliorating skeletal muscle pathology in these diverse systems points to an important role of mtH_2O_2 production as a driver of muscle pathology and indicates that direct targeting mitochondrial oxidative stress, especially mtH_2O_2, is a promising intervention strategy for reducing muscle dysfunction and frailty.

3.4 Sensory Defects

3.4.1 Age-Related Sensorineural Hearing Loss

Age-dependent sensorineural hearing loss (presbycusis) is prevalent in the elderly, estimated to be 30%–35% in population aged 65–75 years and 40%–50% in those older than 75 years of age.[81] Presbycusis is characterized by a more severe hearing loss for high-pitched sound, which leads to difficulty in understanding speech. The pathology displays slowly progressive loss of sensory hair cells, spiral ganglion neurons, and stria vascularis cells in the inner ear cochlea. Mice with the deletion of the mitochondrial proapoptotic gene Bak have attenuated age-dependent apoptotic cell deaths and prevented presbycusis.[82] Oxidative stress was shown to induce Bak expression in primary cochlear cells; conversely, mCAT suppressed Bak expression, reduced cell death, and subsequently prevented presbycusis. These findings reinforce a central role of mtROS-induced apoptotic pathway in presbycusis.[82] Furthermore, CR prevents presbycusis via reduction of oxidative damage by upregulation of mitochondrial deacetylase SIRT3. In response to CR, SIRT3 directly deacetylates and activates mitochondrial isocitrate dehydrogenase 2, leading to increased NADPH levels and an increased ratio of reduced-to-oxidized glutathione in mitochondria and thereby enhancing the mitochondrial glutathione antioxidant defense system.[83]

3.4.2 Retinitis Pigmentosa

Retinitis pigmentosa (RP) is a group of inherited diseases causing various retinal degeneration results in death of rod and/or cone photoreceptor cells. Symptoms depend on whether rods or cones are initially involved. Night blindness is one of the earliest symptoms of RP because in most RP rod cells are affected first, followed by progressive involvement of cone cells, which are responsible for color vision, visual acuity, and sight in the central visual field. Mutation of several genes has been described in association with RP, including the most common mutation of rhodopsin gene, which accounts for 15%–25% of RP cases. Mutation of this gene leads to rhodopsin protein misfolding and may cause progressive loss of rod cells. As rod cells are the most numerous cell type in the retina and also the highly metabolic active using high oxygen, rod cell stress in the outer retina would increase the overall oxidative stress. This oxidative stress has been implicated in the progression of RP, especially in the later stage of cone cells death. As such, antioxidants treatment with various vitamins and superoxide dismutase mimetics have been shown to delay photoreceptor cell death in mouse models of RP,[84] both in early onset *rd1/rd1* and in late onset *rd10/rd10* mice. Using *rd10/rd10* mouse model of RP, Usui *et al.* reported that inducible expression of both SOD2 and mCAT, but not either alone, significantly reduce protein carbonylation (oxidative damage) and ameliorate cone cells death. Surprisingly, overexpression of SOD1 alone increased oxidative stress and accelerated cone cell death.[85] This study emphasizes the crucial roles of mitochondrial oxidative stress in the progression of cone cells death and suggests potential application of mitochondrial antioxidants as future treatment of RP, which has no effective treatment to date.

3.5 Neurodegenerative Disorders

Given the widespread interest in the potential roles of mitochondrial dysfunction in neurodegenerative disorders, especially in Parkinson's disease (PD),[86,87] it was no surprise to find a large number of citations to the Schriner *et al.* paper[13] by neurobiologists, second only to the number of citations related to cardiac disorders. The potential importance of this mCAT paper for neurodegenerative disorders was quickly noted by a leading expert on PD, Professor M.F. Beal.[88] Among the early references to the mCAT paper that was of direct relevance to Alzheimer's disease (AD) was an updated Journal of Biochemistry report showing evidence of a direct link between peroxisomal proliferation, increased catalase expression, and neuroprotection from Abeta-induced degenerative changes observed in primary cultures of rat hippocampal neurons.[89] This was also not surprising, given the fact that AD is widely reported to be the most common of all dementias among many geriatric populations, although autopsy studies have revealed that many cases appear to represent a mix of AD and other types of pathologies associated with dementia of the elderly—see, for example, Ref. 90. This and the research that followed have helped to broaden our understanding of the role of oxidative stress in the mitochondrial dysfunctions associated with age-related neurodegenerative disorders.

3.5.1 PD: Highlights of mCAT Citations

A group of Polish scientists were the first to comment on the relevance of mCAT for the pathogenesis of PD.[91] They also pointed to its relevance for AD and emphasized that features of PD and AD may be seen in the same individual, raising the question of some commonalities of pathogenesis, presumably involving mitochondrial dysfunction, as well as contributions from the nuclear genome. Two valuable reviews on the role of mitochondrial alterations in the pathogenesis of PD were also published in 2008.[92,93] The second of these concluded that "mitochondrial dysfunction remains at the forefront of PD research." In 2010, members of this Spanish group and their colleagues utilized mCAT mice to demonstrate a marked attenuation of mtROS production and dopaminergic cell death when challenged with a neurotoxin precursor capable of inducing PD phenotypes in both humans and experimental animals (1-methyl-4-phenyl-1,2,3,6-tetrahydropyridine, or MPTP).[94] In a 2012 review that focused upon the role of altered mitochondrial function upon neurophysiological parameters, the authors raised the question of the role of the genetic background upon the longevity-enhancing effects of the mCAT transgene.[95] Data addressing this question have yet to be published, but given the well-established differential impacts of genetic backgrounds upon many phenotypes, one might anticipate both enhancements and attenuations of mCAT upon life spans across a range of genetic backgrounds.

3.5.2 PD: Current Status of the Role of Mitochondrial Functions

There is by now widespread recognition that PD results from the gradual loss of dopaminergic neurons of the pars compacta of that structure and that treatment with L-DOPA is effective.[96] The canonical diagnostic lesions in this structure are intranuclear inclusions consisting of aggregates of an altered structure of alpha synuclein, a molecule that is localized to the mitochondria.[97]

It is perhaps less well appreciated, however, that nonmotor symptoms occur early in the course of the disease and appear to involve the olfactory bulb, the brain stem, and the intestinal tract (prodromal PD).[98] A theory that invokes the gradual trans-synaptic spreading of the disease to connected structures—i.e., via an infectious protein or prion mechanism,[99] has recently received robust experimental support.[100] This scenario fits within the general concept of altered proteostasis as one of the fundamental mechanisms of aging.[101] How then can this be reconciled with another fundamental mechanism of aging, mitochondrial dysfunction[101]? That these two phenomena are not mutually exclusive is supported by several studies on the original prion disease, Scrapie.[102–104] While the Park *et al.* paper is consistent with a relatively late contribution of mitochondrial dysfunction, this does not rule out the possibility that some forms of PD may in fact be driven by mitochondrial abnormalities as a primary mechanism. Evidence to that effect comes from several of the growing number of gene mutations and chromosomal regions associated with PD (currently $N=22$; http://www.ncbi.nlm.nih.gov/omim). A number of papers have focused upon a group of recessive mutations: Parkin (PARK2), a E3 ubiquitin ligase complex whose functions are thought to include participation in the autophagic degradation of dysfunctional depolarized mitochondria (i.e., mitophagy) (http://www.uniprot.org/uniprot/O60260#section_comments); PINK1, a mitochondrial serine/threonine-protein kinase activator of Parkin (http://www.uniprot.org/uniprot/Q9BXM7); DJ-1 (PARK7), a protein deglycase thought to act as an oxidative stress sensor and redox-sensitive chaperone and protease.[105–112] Much more remains to be learned about the roles of oxidative stress and mitochondrial dysfunction in the pathogenesis of PD; it is surely "much more than mitophagy"[113]; for example, there is evidence of a role for the mitochondrial modulation of neuroinflammation.[114]

3.5.3 AD: Highlights of mCAT Citations

Papers citing the 2005 Schriner *et al.* mCAT publication that emphasize the pathogenesis of AD greatly outnumbered those with a relatively stronger focus on PD. Some may regard this as counterintuitive, as many will agree that the evidence for a primary role for mitochondrial dysfunction in PD is much stronger. Many of the papers being reviewed in this section of the manuscript consider both PD and AD.

Several mCAT citations have followed from the implications of the evidence that Abeta can be found in mitochondria.[115–118] The most important contribution was the 2007 paper of Reddy *et al.* These authors crossed mCAT mice with mice-overexpressing pathogenic Abeta peptides. These mice had reductions in the levels of the full-length amyloid precursor protein, its C-terminal fragments, the levels of an enzyme (BACE1) that abrogates the synthesis of toxic Abeta peptides by proteolytic cleavage of that portion of the precursor protein, and significantly, the levels of the Abeta peptides that are found in the neuritic plaques of patients with AD, Abeta 40 and 42. Two 2014 mCAT citations reviewed the status of genetic aspects of AD and mitochondria.[119,120] The Dhillon *et al.* review is particularly valuable for its discussion of the complex interactions of nuclear and mitochondrial mutations and micronutrient deficiencies upon the mtDNA genome. Of special note is their reference to the impact of TOMM40 mutations; that locus is closely linked to the APOE locus, polymorphic variants of which are the most significant modulators of the risk of the common late onset sporadic forms of AD. This paper (as is the case for many citations to the 2005 Schriner *et al.* paper) also discusses a wide range of non-AD neurodegenerative disorders. The Bilkei-Gorzo *et al.* paper is of special interest because of its careful evaluation of the relative strengths of a wide range of putative mouse models of AD.

The great majority of all mCAT citations refers to aspects of the role of oxidative stress in the pathogenesis of AD, but four have that subject as their major themes.[121–124] Reddy's 2006 approach was to consider therapeutic options for interventions in AD, given the evidence of an important role of oxidative stress and mitochondrial dysfunction, including a particularly promising member of this group, the Szeto–Schiller (SS) peptides developed by Szeto and her colleagues (see Section 5). Ohta and Ohsawa outline an interesting model of pathogenesis that emphasizes the role of suppressed levels of mitochondrial aldehyde dehydrogenase (ALDH2), a mitochondrial matrix protein, on the generation of oxidizing toxic aldehydes, notably 4-hydroxy-2-nonenal.

As noted in Section 1, the first paper to cite the 2005 mCAT Schriner *et al.* paper from the point of view of neurodegeneration dealt with the pathobiology of peroxisomes.[89] The authors provided evidence of a "direct link" between the extent of the proliferation of peroxisomes and the degree neuroprotection from neurodegenerative alterations associated with Abeta peptides. This interest in peroxisomes was reencountered in a

2011 citation to the Schriner *et al.* paper.[125] Using human postmortem material from AD patients, these authors concluded that there was evidence of "substantial peroxisome-related alterations in AD."

Two citations considered the role of metals in AD pathophysiology.[126,127] The Adlard and Bush paper argue for an important role for the dysregulation of metal ion homeostasis in both aging and AD, specifically for the case of iron and zinc ions, which interact with both the beta-amyloid precursor protein and Abeta peptides. In a chemically sophisticated "tour-de-force" publication with almost 1000 references, Kepp attempts to reconcile the roles of the beta-amyloid cascade hypothesis, the oxidative stress hypothesis, and the metal ion dysregulation hypothesis in the initiation and progression of AD.

A 2007 paper emphasizing the importance of diminished autophagy in aging and AD (the mitochondrial–lysosomal axis theory of aging) cites the Schriner *et al.* paper and states that "...a big advance occurred with the development of mitochondrial-targeted antioxidants, which preferentially enter mitochondria at several hundred-fold more than natural antioxidants."[128]

In keeping with the growing interest in the role of microglia in the pathogenesis of AD, Nam *et al.*[129] observed, using a model of mitochondrial membrane potential loss in neuronal cultures, a marked increase in the degree of neuronal death when these neurons were cocultured with microglia. They cited the Schriner *et al.* paper in their discussion of age-related losses of mitochondrial membrane potential and their association with neuronal depolarization and neuronal death. They emphasized that such alterations are also observed in age-related neurodegenerative disorders, including AD.

Two interesting papers on sirtuins cited the Schriner *et al.* paper.[130,131] The 2006 paper reviews the state of the field for the case of all seven sirtuin genes and proposes roles for SIRT1, 2, and 3 as playing important roles in neuroprotection against AD (their Fig. 1). The 2013 paper narrows that focus to SIRT3, pointing out the extensive evidence for its mitochondrial localization, but pointing to references that it may also have nuclear and cytoplasmic functions by promoting activation of antioxidant systems, fatty acid oxidation, and associated neuroprotection.

The loss of proteostasis has become a major theme of current aging research, so it is not surprising that there is at least one citation to the Schriner *et al.* paper that links oxidative stress to the loss of proteostasis in AD.[132]

The authors emphasize the endoplasmic reticulum–mitochondrial unfolded protein response to abnormal configurations of Abeta and tau.

Using mCAT mice, Olsen *et al.*[133] provided evidence that the overexpression of catalase did not alter markers of oxidative stress, yet did result in improved cognition and decreased anxiety. They therefore suggested that the improved cognitive health may be the result of alterations in redox signaling.

A group of investigators concerned with radiation oncology and the effects of ionizing irradiation on health published two papers demonstrating protection, by mCAT, from proton irradiation, including effects associated with clinically relevant doses.[134,135] The authors pointed to the relevance of their research for those involved in space travel—evidence of the universal impact of the 2005 paper by Schriner *et al.*!

3.5.4 AD: Current Status of the Role of Mitochondrial Functions

Given the fact that beta-amyloid peptides can be found within mitochondria, there is now a surge of interest on the role of mitochondrial dysfunction in AD, as exemplified by a recent comprehensive review of that subject.[136] These authors emphasize the complex set of interactions between beta-amyloid, altered tau, and mitochondria with activated microglia and astrocytes (and embrace beta-amyloid as the initial pathogenic molecule), thus integrating the mitochondrial hypothesis with the beta-amyloid cascade hypothesis of pathogenesis.

3.5.5 Highlights of mCAT Citations to Papers on Other Neurodegenerative Disorders

It is gratifying to note that, in addition to the special attention given to the 2005 Schriner *et al.* paper on the subjects of PD and AD, neurobiologists concerned with a range of other neurodegenerative disorders have cited that paper. Examples can be found in the following publications.[88,137,138]

3.6 Cancer

Oxidative damage to nucleic acids and proteins is widely documented in carcinogenesis. Mitochondria have also been implicated in carcinogenesis. For example, in human patients with ulcerative colitis, loss of mitochondrial cytochrome oxidase has been shown to associate with the development of colonic dysplasia (precancerous state), linking mitochondrial damage to carcinogenesis in human.[139]

The role of mitochondrial oxidative stress is supported by the use of mCAT mouse models to investigate carcinogenesis and cancer progression. The mCAT mice were shown to have reduced nonhematopoietic tumor burden in a mouse end-of-life pathology study.[140] The mCAT expression has also been shown to ameliorate metastatic breast cancer in the PyMT mouse model of breast cancer. The mCAT mice displayed reduction of invasive primary breast tumor and had a 30% lower incidence of pulmonary metastasis. Both tumor cells and lung fibroblasts in mCAT/PyMT double-transgenic mice demonstrated reduced ROS and enhanced resistance to H_2O_2-induced cell death, which may confer the protective effects in mCAT mice.[141]

Additional evidence of mtROS in carcinogenesis was shown in ataxia telangiectasia mutated null mice ($ATM^{-/-}$). ATM kinase plays a central role in the DNA damage response and redox sensing by the phosphorylation of many key proteins that initiate activation of the DNA damage checkpoint, leading to cell cycle arrest, DNA repair or apoptosis. Patients with ATM mutation may present with severe ataxia due to cerebellar degeneration, immune defect, as well as increased risk of lymphomas and leukemias.[137] Consistent with this, $ATM^{-/-}$ mice displayed thymic lymphomas and mild neurodegenerative phenotypes. Reducing mtROS by mCAT in $ATM^{-/-}$ mice reduced propensity to develop thymic lymphoma, improved bone marrow hematopoiesis and macrophage differentiation in vitro, and partially rescued memory T-cell development.[137]

4. PLEOTROPIC OR ADVERSE EFFECTS OF mCAT EXPRESSION

While most of the literature associated with mitochondria-targeted catalase highlights its positive effects and potential therapeutic uses, this abundance of benefits also begs an important question: If the removal of mitochondrial oxidants is good for organismal health, then why have not mitochondrial antioxidants like mCAT evolved in nature? Presumably, the potential benefits of mCAT would only serve to increase the fitness of organisms, yet no organisms described to date naturally express high levels of catalase in mitochondria. Additionally, a number of studies have suggested that removal of ROS could have negative outcomes, and some have suggested that ROS and mitochondrial dysfunction can be beneficial to longevity and health. In this section, we highlight these studies and how they can be reconciled with other findings.

4.1 General Antioxidants

Numerous studies in model organisms have led to skepticism about the importance of antioxidants in longevity and health. In *Caenorhabditis elegans*, an invertebrate model of aging, several strains deficient in respiratory proteins that produce excess ROS are in fact longer lived than WT controls,[142] and ROS has been shown to be indispensable for the increased life span of glycolysis-deficient worms or worms on glucose restriction.[143] In mice, deletion of many antioxidant enzymes has little effect on life span and, importantly, overexpression of several antioxidants including superoxide dismutase and peroxisomal catalase has failed to extend life span.[2] In fact, mice lacking the cellular antioxidant GPx1 are protected from high-fat-induced IR, and this benefit is lost upon the administration of an antioxidant.[144] Compared to mice with a median life span just over 2 years, the naked mole rat shows remarkable longevity, living 10–30 years, in spite of similar rates of ROS production and more extensive oxidative damage to its tissues over its lifetime,[145] suggesting that ROS may not necessarily play a causative role in aging. Additionally, while mtDNA mutations increase with age, the characteristic mutations created by ROS are not among those most seen by state-of-the-art duplex sequencing in *Drosophila*, suggesting that ROS may not be a driver of somatic mutations in aging.[146] Human clinical trials testing the therapeutic potential of dietary antioxidants in a wide range of diseases including cancer,[147] gastrointestinal,[148] neurological,[149] rheumatoid,[150] endocrine,[151] and CVD[152] have thus far shown little to no efficacy. Some have shown adverse outcomes.[153]

4.2 Mitochondrial Antioxidants

Unlike the transgenic antioxidant overexpressors that came before it, mCAT mice benefit from increased mitochondrial antioxidant expression by many metrics, including increased median and maximal life span.[13] The potential benefits of pharmacological mitochondrial antioxidants have not been tested to the same extent as general antioxidants. However, a few studies have shown that some considerations should be made for potential negative effects when targeting mtROS. Song *et al.* reported that while low levels of transgenic mCAT expression were beneficial in a cardiac Mfn2-knockout background, "supersuppression" of ROS by higher levels of mCAT exacerbated the cardiac phenotype and suppressed compensatory autophagy.[154] A separate study found that bactericidal activity is impaired in young mCAT mice.[155] In a comprehensive proteomic analysis, we have recently shown that while the proteome turnover and composition in old mCAT mice resemble that of young controls, the young mCAT mouse proteome recapitulates features of an old wild-type mouse.[156]

4.3 ROS and Antagonistic Pleiotropy

Antagonistic pleiotropy is an effect that is beneficial to an organism's fitness early in life, but which causes functional decline and aging phenotypes later in life.[157] The seemingly inconsistent results emerging in studies of antioxidants and mitochondrial antioxidants are consistent with a pattern of age-dependent pleiotropy. A model of a continuum of ROS, at the center of which is a physiologically necessary level of mtROS,[158] may underlie this pattern. At moderate or low levels, ROS are increasingly being found to serve important physiological signaling roles that may be important for metabolism, protein turnover, cellular differentiation, stress response, and apoptosis.[159,160] Low-level ROS is also necessary for the renewal of stem cells, and it has been shown that oversuppression of ROS significantly limits renewal of stem cells and reduces neurogenesis in mice.[161] On the other end of the spectrum, "pathological," or high levels of ROS, may react with proteins, DNA, and lipids to damage important components of the cell and has been linked to aging and numerous diseases. In aging, increasing levels of ROS and oxidative damage are widely documented and were part of the support for the "free-radical theory of aging."[162] However, the mCAT effect on the proteome of young mice is consistent with an adverse impact on the more beneficial effects of ROS, while mCAT suppression of pathological levels of ROS in old animals is protective. This also suggests that ROS itself exhibits conventional antagonistic pleiotropy and would explain why stronger antioxidant mechanisms, such as mCAT, have not evolved under natural selection of young animals in nature.

It would be of interest to pursue assays of reproductive fitness in mCAT vs WT mice as a further test of the hypothesis of reverse antagonistic pleiotropy. We have not detected such an effect in the laboratory; however, such trade-offs in fitness may only be present in natural environments. Alternatively, one could pursue surrogate functional assays, such as tests of fighting behavior or endurance on treadmills.[163]

In related studies, loss of the mitochondrial antioxidant SOD2 has been shown to exhibit antagonistic pleiotropy.[164] Using an inducible mouse model of SOD2 loss in epidermal stem/progenitor cells, which induces senescence and slows proliferation in a fraction of keratinocytes,

Velarde *et al.* measured rates of wound closure in young and old WT and SOD2-deficient mice. Old SOD2-deficient mice showed delayed wound closure, reduced epidermal thickness, and stem cell exhaustion. In young mice, however, SOD2 deficiency accelerated wound closure and increased epidermal differentiation and epithelialization, in spite of slower proliferation rates. Interestingly, the proliferation-promoting agent 12-*O*-tetradecanoylphorbol-13-acetate, which normally increases epidermal thickening in young mice, caused accelerated epidermal thinning in young SOD2-deficient mice and phenocopied the old SOD2-deficient mouse phenotype. These findings demonstrate that mtROS can serve a beneficial role and increase fitness at a younger age while later resulting in age-related phenotypes.

Taken together, the studies discussed here suggest that mitochondrial antioxidant may not be universally beneficial, and the beneficial effects are observed in a setting when "pathological" oxidative stress or a high burst of ROS is anticipated. Thus, as with many drugs, mitochondrial antioxidants likely have a therapeutic windows and this may be age dependent. It is also likely that such therapeutic windows vary by genetic background, cell type, and organism.

5. PHARMACOLOGIC ANALOGS OF mCAT EXPRESSION

The beneficial effects of mCAT expression in protection of multiple disease models implicate a critical role of mitochondrial oxidative stress and damage in pathogenesis of multiple diseases. While gene therapy of mCAT expression is a potential strategy to translate the protective effects of mCAT, pharmacologic analogs of mCAT expression, i.e., mitochondrial-targeted antioxidants, provide an alternative therapeutic strategy.

Two main approaches have been used to deliver pharmacologic compounds to mitochondria. The first approach is by conjugating redox agents to triphenylphosphonium ion (TPP^+) and taking advantage of the potential gradient across the inner mitochondrial membrane (IMM). MitoQ and SkQ1 are TPP^+ conjugated to ubiquinone and plastoquinone, respectively, and they utilize the mitochondrial membrane potential across the IMM to deliver these redox-active compounds into the mitochondrial matrix.[165,166] The second approach is utilizing by the affinity to a mitochondrial component to target the mitochondria without relying on mitochondrial potential. The SS compounds are tetrapeptides with an alternating aromatic–cationic amino acids motif that selectively bind to cardiolipin (CL) on the IMM[167–169] to target delivery to the mitochondria. The effects of these pharmacologic interventions on longevity and healthspan have been studied by different researchers, as summarized below.

5.1 TPP$^+$-Conjugated Antioxidants

MitoQ, 10-(6′-ubiquinonyl)decyltriphenylphosphonium bromide, selectively concentrates in the mitochondria and prevents mitochondrial oxidative damage.[166] Magwere and colleagues showed that MitoQ prolongs life span of SOD-deficient flies but not normal WT flies and improves pathology associated with antioxidant deficiency.[170]

In a *C. elegans* model of AD, transgenic *C. elegans* with muscle-specific expression of human Aβ, MitoQ extends life span and improves healthspan without reducing ROS production, protein carbonyl content, and mtDNA damage burden.[171] The protective effect of MitoQ against Aβ toxicity had previously been demonstrated in primary neurons from amyloid-beta precursor protein transgenic mice and neuroblastoma cells incubated with Aβ.[172] In a triple-transgenic mouse model of AD (3xTg-AD), MitoQ treatment for 5 months prevents cognitive decline and AD-like neuropathology, supporting the therapeutic potential of MitoQ in AD.[173] MitoQ treatment has also been shown to confer neuroprotection in cell culture and mouse models of PD.[174]

In addition to its neuroprotective effects, MitoQ treatment has been shown to confer cardioprotection in multiple models.[175–178] Adlam *et al.* showed that MitoQ decreased IR-induced mitochondrial damage and cardiac dysfunction,[175] and Graham *et al.* showed that MitoQ treatment for 8 weeks reduced systolic blood pressure and attenuated cardiac hypertrophy in spontaneous hypertensive rats.[177] A recent study showed that administration of MitoQ to the storage solution of donor hearts prevents IR-related injury after heart transplantation.[176] MitoQ also attenuates oxidative damage and liver dysfunction in a murine hepatic IR model.[179]

SkQ1, another TPP$^+$-conjugated mitochondrial-targeted antioxidant, has been shown to prolong life span of *Podospora*, *Ceriodaphnia*, *Drosophila*, and female outbred SHR mice.[180] A recent study reported that SkQ1 can also extend life span of male BALB/c and C57BL/6 mice.[181] Like MitoQ, SkQ1 has been shown to have beneficial effects in models of cardiac ischemia–reperfusion.[182] Moreover, it is also protective against renal and brain ischemic injuries.[182] Interestingly, a study showed that administration of

SkQ1 via diet can prevent the age-induced cataract and retinopathies in senescence-accelerated OXYS rats, and SkQ1 eye drops can reverse cataract in middle-aged OXYS rats and Wistar rats.[183]

One limitation of TPP^+-conjugated antioxidants is their dependence on mitochondrial membrane potential to penetrate the mitochondria, given that mitochondrial membrane potential is often compromised in pathological conditions. Moreover, MitoQ and SkQ have also been shown to inhibit respiration and disrupt mitochondrial membrane potential at concentrations above 5–25 μM.[165,166] Because MitoQ and SkQ are both quinone derivatives that process prooxidant properties, optimal dosages that exert antioxidant effect but not prooxidant activities must be carefully evaluated before using these interventions.

5.2 SS Peptides

The SS tetrapeptides have an alternating aromatic–cationic amino acids motif, and they have been shown to preferentially concentrate in the IMM over 1000-fold compared with the cytosolic concentration.[182,184,185] Unlike MitoQ and SkQ1, the mitochondrial uptake of SS peptides is not dependent on mitochondrial membrane potential, and they can penetrate depolarized mitochondria.[184,185] The SS-31 peptide (H-D-Arg-Dmt-Lys-Phe-NH_2), also called Bendavia, Elamipretide, or MTP-131, is the best characterized of these peptides. SS-31 was initially thought to exert its protective effect by the ROS-scavenging activity of the dimethyltyrosine residue.[177] However, more recent studies revealed a novel mechanism of SS-31 action.[167–169] Birk *et al.* showed that SS-31 selectively interacts with CL in liposomes, bicelles, and mitoplasts in vitro.[169] They also showed that SS-31 can abolish inhibitory effects of CL on cytochrome *c* reduction and electron transport in mitoplasts and that SS-31 can increase oxygen consumption and ATP production in isolated mitochondria.[169] They proposed that the binding of SS-31 to CL on the IMM alters the interaction of CL with cytochrome *c*.[169] This altered interaction preserves Met80-heme ligation of cytochrome *c* and favors cytochrome *c* electron carrier activity while inhibiting its peroxidase activity.[168,169] In a renal IR model, they showed that SS-31 treatment can increase ATP production and reduce ROS generation post-IR, preventing CL peroxidation and preserving cristae membrane integrity.[169] These findings suggest that ROS-independent mechanisms may contribute to the protective effects of SS-31, with reduced ROS production as a secondary benefit. This mechanism may explain how

SS-31 protects mitochondrial cristae architecture, prevents mitochondrial swelling, and attenuates renal dysfunction after ischemia or IR.[167,186]

The SS-31 peptide has also been shown to be protective in many models of age-related diseases, including AD.[172,187] Similar to MitoQ, SS-31 prevents Aβ toxicity in primary neurons from Aβ precursor protein transgenic mice and neuroblastoma cells incubated with Aβ.[172] SS-31 treatment rescues the impairment of mitochondrial dynamics and antegrade transport and prevents synaptic degeneration caused by Aβ toxicity.[187] The neuroprotective effect of SS-31 was also demonstrated in a mouse model of PD. Yang *et al.* showed that SS-31 can dose-dependently protect dopaminergic neurons and preserve striatal dopamine levels in mice treated with MPTP.[188] SS-20, a version of SS peptide without dimethyltyrosine residue, also showed neuroprotection on dopaminergic neurons of MPTP-treated mice. The same study also showed that SS-31 and SS-20 prevented the inhibition of oxygen consumption, ATP production, and mitochondrial swelling by MPTP treatment in isolated mitochondria. These findings suggest that the ROS-scavenging activity of SS-31 is not necessary for the neuroprotection and that preservation of mitochondrial ATP production and inhibition of mitochondrial permeability transition may mediate the neuroprotective effects of SS peptides.[188]

Beside neuroprotection, SS peptides also exert cardioprotection in multiple disease models. SS-31 has been demonstrated to offer cardioprotective effects in cardiac ischemia–reperfusion injury, reperfusion arrhythmia, and myocardial infarction models.[189–194] SS-31 also protects against cardiac hypertrophy and heart failure. In a similar manner to mCAT expression, SS-31 prevents Ang II-induced cardiac hypertrophy and preserves cardiac function in TAC-induced heart failure model.[195–197] Electron microscopy analysis showed that SS-31 preserves cardiac mitochondria from the TAC-induced abnormalities, and proteomic analysis showed that SS-31 attenuates TAC-induced changes in mitochondrial and nonmitochondrial proteins.[196] Using a canine microembolization-induced heart failure model, Sabbah *et al.* demonstrated that administration of SS-31 for 3 months significantly improved systolic function. They further demonstrated that the treatment normalized levels of plasma biomarkers and preserved mitochondrial function and bioenergetics in the myocardium of dogs with advanced HF.[198] Although the effect of SS-31 on cardiac aging has not been previously reported, unpublished data from our laboratory showed that SS-31 treatment for 8 weeks can reverse age-related diastolic dysfunction in old mice (Y.A. Chiao, unpublished data), highlighting the therapeutic potential of SS-31 in cardiac aging.

Similar to the heart, skeletal muscle has a high energy demand and is highly dependent on mitochondrial energy production to function. SS-31 has been shown to acutely reverse age-related impairment in mitochondrial energetics and led to improved muscle performance.[60] The effect of longer term SS-31 treatment on skeletal muscle aging is under further investigation. In a hindlimb immobilization model and a casting model, SS-31 has been shown to prevent disuse skeletal muscle atrophy by attenuating the ROS production and protease activation.[62,199] Similar protective effects of SS-31 have also been demonstrated in an inactivity-induced diaphragm dysfunction.[200] Doxorubicin is anthracycline cancer chemotherapy drug that has been shown to cause cardiac and skeletal muscle myopathy. SS-31 protects cardiac and skeletal muscles from doxorubicin-induced mtROS production and prevents doxorubicin-induced atrophy and dysfunction.[61] These findings implicate the role of mtROS in muscle weakness and dysfunction and support the potential of mitochondrial targeting antioxidants as therapies.

The promising results of preclinical studies of mitochondrial-targeted antioxidants such as the above have led to clinical trials on neurodegenerative, cardiorenal, skeletal muscle, and ocular diseases.[122,201] The roles of these interventions on aging and aging-associated diseases will be further evaluated; however, the findings to date have demonstrated the potentially high therapeutic potential of mitochondrial-targeted antioxidants.

REFERENCES

1. Harman D. Aging: a theory based on free radical and radiation chemistry. *J Gerontol.* 1956;11(3):298–300.
2. Perez VI, Van Remmen H, Bokov A, Epstein CJ, Vijg J, Richardson A. The overexpression of major antioxidant enzymes does not extend the lifespan of mice. *Aging Cell.* 2009;8(1):73–75.
3. Perez VI, Bokov A, Van Remmen H, et al. Is the oxidative stress theory of aging dead? *Biochim Biophys Acta.* 2009;1790(10):1005–1014.
4. Bjelakovic G, Nikolova D, Gluud LL, Simonetti RG, Gluud C. Mortality in randomized trials of antioxidant supplements for primary and secondary prevention: systematic review and meta-analysis. *JAMA.* 2007;297(8):842–857.
5. Myung SK, Ju W, Cho B, et al. Efficacy of vitamin and antioxidant supplements in prevention of cardiovascular disease: systematic review and meta-analysis of randomised controlled trials. *BMJ.* 2013;346:f10.
6. Harman D. The biologic clock: the mitochondria? *J Am Geriatr Soc.* 1972;20(4):145–147.
7. Bause AS, Haigis MC. SIRT3 regulation of mitochondrial oxidative stress. *Exp Gerontol.* 2013;48(7):634–639.

8. Ristow M, Zarse K. How increased oxidative stress promotes longevity and metabolic health: the concept of mitochondrial hormesis (mitohormesis). *Exp Gerontol.* 2010;45(6):410–418.
9. Abe K, Makino N, Anan FK. pH dependency of kinetic parameters and reaction mechanism of beef liver catalase. *J Biochem.* 1979;85(2):473–479.
10. Agar NS, Sadrzadeh SM, Hallaway PE, Eaton JW. Erythrocyte catalase. A somatic oxidant defense? *J Clin Invest.* 1986;77(1):319–321.
11. Percy ME. Catalase: an old enzyme with a new role? *Can J Biochem Cell Biol.* 1984;62(10):1006–1014.
12. Schriner SE, Linford NJ. Extension of mouse lifespan by overexpression of catalase. *Age (Dordr).* 2006;28(2):209–218.
13. Schriner SE, Linford NJ, Martin GM, et al. Extension of murine life span by overexpression of catalase targeted to mitochondria. *Science.* 2005;308(5730):1909–1911.
14. Linford NJ, Schriner SE, Rabinovitch PS. Oxidative damage and aging: spotlight on mitochondria. *Cancer Res.* 2006;66(5):2497–2499.
15. Alberti KG, Eckel RH, Grundy SM, et al. Harmonizing the metabolic syndrome: a joint interim statement of the International Diabetes Federation Task Force on Epidemiology and Prevention; National Heart, Lung, and Blood Institute; American Heart Association; World Heart Federation; International Atherosclerosis Society; and International Association for the Study of Obesity. *Circulation.* 2009;120(16):1640–1645.
16. Alberti KG, Zimmet P. The metabolic syndrome: time to reflect. *Curr Diab Rep.* 2006;6(4):259–261.
17. O'Neill S, O'Driscoll L. Metabolic syndrome: a closer look at the growing epidemic and its associated pathologies. *Obes Rev.* 2015;16(1):1–12.
18. Kaur J. A comprehensive review on metabolic syndrome. *Cardiol Res Pract.* 2014;2014:943162.
19. Ervin RB. Prevalence of metabolic syndrome among adults 20 years of age and over, by sex, age, race and ethnicity, and body mass index: United States, 2003–2006. *Natl Health Stat Report.* 2009;(13):1–7.
20. Vlassara H, Cai W, Tripp E, et al. Oral AGE restriction ameliorates insulin resistance in obese individuals with the metabolic syndrome: a randomised controlled trial. *Diabetologia.* 2016;59:2181–2192.
21. Tan KC, Shiu SW, Wong Y, Tam X. Serum advanced glycation end products (AGEs) are associated with insulin resistance. *Diabetes Metab Res Rev.* 2011;27(5):488–492.
22. Hohn A, Konig J, Jung T. Metabolic syndrome, redox state, and the proteasomal system. *Antioxid Redox Signal.* 2016;25:902–917.
23. Anderson EJ, Lustig ME, Boyle KE, et al. Mitochondrial H2O2 emission and cellular redox state link excess fat intake to insulin resistance in both rodents and humans. *J Clin Invest.* 2009;119(3):573–581.
24. Kahn SE, Hull RL, Utzschneider KM. Mechanisms linking obesity to insulin resistance and type 2 diabetes. *Nature.* 2006;444(7121):840–846.
25. Lee HY, Choi CS, Birkenfeld AL, et al. Targeted expression of catalase to mitochondria prevents age-associated reductions in mitochondrial function and insulin resistance. *Cell Metab.* 2010;12(6):668–674.
26. Vassallo P, Driver SL, Stone NJ. Metabolic syndrome: an evolving clinical construct. *Prog Cardiovasc Dis.* 2016;59:172–177.
27. Expert Panel on Detection, Evaluation, and Treatment of High Blood Cholesterol in Adults. Executive Summary of the Third Report of The National Cholesterol Education Program (NCEP) Expert Panel on Detection, Evaluation, and Treatment of High Blood Cholesterol in Adults (Adult Treatment Panel III). *JAMA.* 2001;285(19):2486–2497.

28. Wang Y, Wang GZ, Rabinovitch PS, Tabas I. Macrophage mitochondrial oxidative stress promotes atherosclerosis and nuclear factor-kappaB-mediated inflammation in macrophages. *Circ Res*. 2014;114(3):421–433.
29. Mammucari C, Rizzuto R. Signaling pathways in mitochondrial dysfunction and aging. *Mech Ageing Dev*. 2010;131(7–8):536–543.
30. Trifunovic A, Larsson NG. Mitochondrial dysfunction as a cause of ageing. *J Intern Med*. 2008;263(2):167–178.
31. Terzioglu M, Larsson NG. Mitochondrial dysfunction in mammalian ageing. *Novartis Found Symp*. 2007;287:197–208. discussion 208–113.
32. Ventura-Clapier R, Garnier A, Veksler V. Transcriptional control of mitochondrial biogenesis: the central role of PGC-1alpha. *Cardiovasc Res*. 2008;79(2):208–217.
33. Goffart S, von Kleist-Retzow J-C, Wiesner RJ. Regulation of mitochondrial proliferation in the heart: power-plant failure contributes to cardiac failure in hypertrophy. *Cardiovasc Res*. 2004;64(2):198–207.
34. Murray AJ, Anderson RE, Watson GC, Radda GK, Clarke K. Uncoupling proteins in human heart. *Lancet*. 2004;364(9447):1786–1788.
35. Dai DF, Johnson SC, Villarin JJ, et al. Mitochondrial oxidative stress mediates angiotensin II-induced cardiac hypertrophy and G{alpha}q overexpression-induced heart failure. *Circ Res*. 2011;108(7):837–846.
36. Dai DF, Santana LF, Vermulst M, et al. Overexpression of catalase targeted to mitochondria attenuates murine cardiac aging. *Circulation*. 2009;119(21):2789–2797.
37. Dai DF, Chen T, Wanagat J, et al. Age-dependent cardiomyopathy in mitochondrial mutator mice is attenuated by overexpression of catalase targeted to mitochondria. *Aging Cell*. 2010;9(4):536–544.
38. Trifunovic A, Wredenberg A, Falkenberg M, et al. Premature ageing in mice expressing defective mitochondrial DNA polymerase. *Nature*. 2004;429(6990):417–423.
39. Kujoth GC, Hiona A, Pugh TD, et al. Mitochondrial DNA mutations, oxidative stress, and apoptosis in mammalian aging. *Science*. 2005;309(5733):481–484.
40. Safdar A, Bourgeois JM, Ogborn DI, et al. Endurance exercise rescues progeroid aging and induces systemic mitochondrial rejuvenation in mtDNA mutator mice. *Proc Natl Acad Sci USA*. 2011;108:4135–4140.
41. Corral-Debrinski M, Stepien G, Shoffner JM, Lott MT, Kanter K, Wallace DC. Hypoxemia is associated with mitochondrial DNA damage and gene induction. Implications for cardiac disease. *JAMA*. 1991;266(13):1812–1816.
42. Zhang C, Bills M, Quigley A, Maxwell RJ, Linnane AW, Nagley P. Varied prevalence of age-associated mitochondrial DNA deletions in different species and tissues: a comparison between human and rat. *Biochem Biophys Res Commun*. 1997;230(3): 630–635.
43. DiMauro S, Schon EA. Mitochondrial respiratory-chain diseases. *N Engl J Med*. 2003;348(26):2656–2668.
44. Marin-Garcia J, Goldenthal MJ, Moe GW. Abnormal cardiac and skeletal muscle mitochondrial function in pacing-induced cardiac failure. *Cardiovasc Res*. 2001;52(1): 103–110.
45. Mollnau H, Wendt M, Szocs K, et al. Effects of angiotensin II infusion on the expression and function of NAD(P)H oxidase and components of nitric oxide/cGMP signaling. *Circ Res*. 2002;90(4):E58–E65.
46. Doughan AK, Harrison DG, Dikalov SI. Molecular mechanisms of angiotensin II-mediated mitochondrial dysfunction: linking mitochondrial oxidative damage and vascular endothelial dysfunction. *Circ Res*. 2008;102(4):488–496.
47. Kimura S, Zhang GX, Nishiyama A, et al. Mitochondria-derived reactive oxygen species and vascular MAP kinases: comparison of angiotensin II and diazoxide. *Hypertension*. 2005;45(3):438–444.

48. Kohler JJ, Cucoranu I, Fields E, et al. Transgenic mitochondrial superoxide dismutase and mitochondrially targeted catalase prevent antiretroviral-induced oxidative stress and cardiomyopathy. *Lab Invest*. 2009;89(7):782–790.
49. Ago T, Kuroda J, Pain J, Fu C, Li H, Sadoshima J. Upregulation of Nox4 by hypertrophic stimuli promotes apoptosis and mitochondrial dysfunction in cardiac myocytes. *Circ Res*. 2010;106(7):1253–1264.
50. Kuroda J, Ago T, Matsushima S, Zhai P, Schneider MD, Sadoshima J. NADPH oxidase 4 (Nox4) is a major source of oxidative stress in the failing heart. *Proc Natl Acad Sci USA*. 2010;107(35):15565–15570.
51. Becker LB, vanden Hoek TL, Shao ZH, Li CQ, Schumacker PT. Generation of superoxide in cardiomyocytes during ischemia before reperfusion. *Am J Physiol*. 1999;277(6 pt 2):H2240–H2246.
52. Zorov DB, Juhaszova M, Sollott SJ. Mitochondrial ROS-induced ROS release: an update and review. *Biochim Biophys Acta*. 2006;1757(5–6):509–517.
53. Piot C, Croisille P, Staat P, et al. Effect of cyclosporine on reperfusion injury in acute myocardial infarction. *N Engl J Med*. 2008;359(5):473–481.
54. Adesina SE, Kang BY, Bijli KM, et al. Targeting mitochondrial reactive oxygen species to modulate hypoxia-induced pulmonary hypertension. *Free Radic Biol Med*. 2015;87:36–47.
55. Conley KE. Cellular energetics during exercise. In: Jones JH, ed. San Diego, CA: Academic Press, Inc.; 1994:1–39. Comparative Vertebrate Exercise Physiology: Unifying Physiological Principles; vol 38A.
56. Conley KE, Jubrias SA, Esselman PC. Oxidative capacity and ageing in human muscle. *J Physiol*. 2000;526(pt 1):203–210.
57. Goncalves RL, Quinlan CL, Perevoshchikova IV, Hey-Mogensen M, Brand MD. Sites of superoxide and hydrogen peroxide production by muscle mitochondria assessed ex vivo under conditions mimicking rest and exercise. *J Biol Chem*. 2015;290(1):209–227.
58. Powers SK, Jackson MJ. Exercise-induced oxidative stress: cellular mechanisms and impact on muscle force production. *Physiol Rev*. 2008;88(4):1243–1276.
59. Ristow M, Zarse K, Oberbach A, et al. Antioxidants prevent health-promoting effects of physical exercise in humans. *Proc Natl Acad Sci USA*. 2009;106:8665–8670.
60. Siegel MP, Kruse SE, Percival JM, et al. Mitochondrial-targeted peptide rapidly improves mitochondrial energetics and skeletal muscle performance in aged mice. *Aging Cell*. 2013;12(5):763–771.
61. Min K, Kwon OS, Smuder AJ, et al. Increased mitochondrial emission of reactive oxygen species and calpain activation are required for doxorubicin-induced cardiac and skeletal muscle myopathy. *J Physiol*. 2015;593(8):2017–2036.
62. Min K, Smuder AJ, Kwon OS, Kavazis AN, Szeto HH, Powers SK. Mitochondrial-targeted antioxidants protect skeletal muscle against immobilization-induced muscle atrophy. *J Appl Physiol (1985)*. 2011;111(5):1459–1466.
63. McClung JM, Kavazis AN, Whidden MA, et al. Antioxidant administration attenuates mechanical ventilation-induced rat diaphragm muscle atrophy independent of protein kinase B (PKB Akt) signalling. *J Physiol*. 2007;585(pt 1):203–215.
64. Laitano O, Ahn B, Patel N, et al. Pharmacological targeting of mitochondrial reactive oxygen species counteracts diaphragm weakness in chronic heart failure. *J Appl Physiol (1985)*. 2016;120(7):733–742.
65. Kruse SE, Karunadharma PP, Basisty N, et al. Age modifies respiratory complex I and protein homeostasis in a muscle type-specific manner. *Aging Cell*. 2015;15:89–99.
66. Forman HJ, Ursini F, Maiorino M. An overview of mechanisms of redox signaling. *J Mol Cell Cardiol*. 2014;73:2–9.

67. Kang L, Dai C, Lustig ME, et al. Heterozygous SOD2 deletion impairs glucose-stimulated insulin secretion, but not insulin action, in high-fat-fed mice. *Diabetes*. 2014;63(11):3699–3710.
68. Lark DS, Kang L, Lustig ME, et al. Enhanced mitochondrial superoxide scavenging does not improve muscle insulin action in the high fat-fed mouse. *PLoS One*. 2015;10(5), e0126732.
69. Kang L, Lustig ME, Bonner JS, et al. Mitochondrial antioxidative capacity regulates muscle glucose uptake in the conscious mouse: effect of exercise and diet. *J Appl Physiol (1985)*. 2012;113(8):1173–1183.
70. Andersson DC, Betzenhauser MJ, Reiken S, et al. Ryanodine receptor oxidation causes intracellular calcium leak and muscle weakness in aging. *Cell Metab*. 2011;14(2): 196–207.
71. Brookes PS, Darley-Usmar VM. Role of calcium and superoxide dismutase in sensitizing mitochondria to peroxynitrite-induced permeability transition. *Am J Physiol Heart Circ Physiol*. 2004;286(1):H39–H46.
72. Kavanaugh NI, Ainscow EK, Brand MD. Calcium regulation of oxidative phosphorylation in rat skeletal muscle mitochondria. *Biochim Biophys Acta*. 2000;1457:57–70.
73. Gutierrez-Perez A, Cortes-Rojo C, Noriega-Cisneros R, et al. Protective effects of resveratrol on calcium-induced oxidative stress in rat heart mitochondria. *J Bioenerg Biomembr*. 2011;43(2):101–107.
74. Andersson DC, Marks AR. Fixing ryanodine receptor Ca leak—a novel therapeutic strategy for contractile failure in heart and skeletal muscle. *Drug Discov Today Dis Mech*. 2010;7(2):e151–e157.
75. Andersson DC, Meli AC, Reiken S, et al. Leaky ryanodine receptors in beta-sarcoglycan deficient mice: a potential common defect in muscular dystrophy. *Skelet Muscle*. 2012;2(1):9.
76. Butt Z, Rosenbloom SK, Abernethy AP, et al. Fatigue is the most important symptom for advanced cancer patients who have had chemotherapy. *J Natl Compr Canc Netw*. 2008;6(5):448–455.
77. Bower JE, Ganz PA, Desmond KA, Rowland JH, Meyerowitz BE, Belin TR. Fatigue in breast cancer survivors: occurrence, correlates, and impact on quality of life. *J Clin Oncol*. 2000;18(4):743–753.
78. Bower JE, Ganz PA, Desmond KA, et al. Fatigue in long-term breast carcinoma survivors: a longitudinal investigation. *Cancer*. 2006;106(4):751–758.
79. Gilliam LA, Lark DS, Reese LR, et al. Targeted overexpression of mitochondrial catalase protects against cancer chemotherapy-induced skeletal muscle dysfunction. *Am J Physiol Endocrinol Metab*. 2016;311(2):E293–E301.
80. Gilliam LA, Fisher-Wellman KH, Lin CT, Maples JM, Cathey BL, Neufer PD. The anticancer agent doxorubicin disrupts mitochondrial energy metabolism and redox balance in skeletal muscle. *Free Radic Biol Med*. 2013;65:988–996.
81. Yueh B, Shapiro N, MacLean CH, Shekelle PG. Screening and management of adult hearing loss in primary care: scientific review. *JAMA*. 2003;289(15):1976–1985.
82. Someya S, Xu J, Kondo K, et al. Age-related hearing loss in C57BL/6J mice is mediated by Bak-dependent mitochondrial apoptosis. *Proc Natl Acad Sci USA*. 2009;106(46): 19432–19437.
83. Someya S, Yu W, Hallows WC, et al. Sirt3 mediates reduction of oxidative damage and prevention of age-related hearing loss under caloric restriction. *Cell*. 2010;143(5): 802–812.
84. Komeima K, Rogers BS, Campochiaro PA. Antioxidants slow photoreceptor cell death in mouse models of retinitis pigmentosa. *J Cell Physiol*. 2007;213(3):809–815.
85. Usui S, Komeima K, Lee SY, et al. Increased expression of catalase and superoxide dismutase 2 reduces cone cell death in retinitis pigmentosa. *Mol Ther*. 2009;17(5):778–786.

86. Betarbet R, Sherer TB, MacKenzie G, Garcia-Osuna M, Panov AV, Greenamyre JT. Chronic systemic pesticide exposure reproduces features of Parkinson's disease. *Nat Neurosci*. 2000;3(12):1301–1306.
87. Valente EM, Abou-Sleiman PM, Caputo V, et al. Hereditary early-onset Parkinson's disease caused by mutations in PINK1. *Science*. 2004;304(5674):1158–1160.
88. Beal MF. Mitochondria take center stage in aging and neurodegeneration. *Ann Neurol*. 2005;58(4):495–505.
89. Santos MJ, Quintanilla RA, Toro A, et al. Peroxisomal proliferation protects from beta-amyloid neurodegeneration. *J Biol Chem*. 2005;280(49):41057–41068.
90. White L, Small BJ, Petrovitch H, et al. Recent clinical-pathologic research on the causes of dementia in late life: update from the Honolulu-Asia Aging Study. *J Geriatr Psychiatry Neurol*. 2005;18(4):224–227.
91. Maruszak A, Gaweda-Walerych K, Soltyszewski I, Zekanowski C. Mitochondrial DNA in pathogenesis of Alzheimer's and Parkinson's diseases. *Acta Neurobiol Exp (Wars)*. 2006;66(2):153–176.
92. Khusnutdinova E, Gilyazova I, Ruiz-Pesini E, et al. A mitochondrial etiology of neurodegenerative diseases: evidence from Parkinson's disease. *Ann N Y Acad Sci*. 2008;1147:1–20.
93. Vila M, Ramonet D, Perier C. Mitochondrial alterations in Parkinson's disease: new clues. *J Neurochem*. 2008;107(2):317–328.
94. Perier C, Bove J, Dehay B, et al. Apoptosis-inducing factor deficiency sensitizes dopaminergic neurons to parkinsonian neurotoxins. *Ann Neurol*. 2010;68(2):184–192.
95. Surmeier DJ, Guzman JN, Sanchez J, Schumacker PT. Physiological phenotype and vulnerability in Parkinson's disease. *Cold Spring Harb Perspect Med*. 2012;2(7):a009290.
96. Bizzarri BM, Tortolini S, Rotelli L, Botta G, Saladino R. Current advances in L-DOPA and DOPA-peptidomimetics: chemistry, applications and biological activity. *Curr Med Chem*. 2015;22(36):4138–4165.
97. Guardia-Laguarta C, Area-Gomez E, Schon EA, Przedborski S. Novel subcellular localization for alpha-synuclein: possible functional consequences. *Front Neuroanat*. 2015;9:17.
98. Sauerbier A, Qamar MA, Rajah T, Chaudhuri KR. New concepts in the pathogenesis and presentation of Parkinson's disease. *Clin Med (Lond)*. 2016;16(4):365–370.
99. Watts JC, Prusiner SB. Mouse models for studying the formation and propagation of prions. *J Biol Chem*. 2014;289(29):19841–19849.
100. Rey NL, Steiner JA, Maroof N, et al. Widespread transneuronal propagation of alpha-synucleinopathy triggered in olfactory bulb mimics prodromal Parkinson's disease. *J Exp Med*. 2016;213:1759–1778.
101. Lopez-Otin C, Blasco MA, Partridge L, Serrano M, Kroemer G. The hallmarks of aging. *Cell*. 2013;153(6):1194–1217.
102. Choi HS, Choi YG, Shin HY, et al. Dysfunction of mitochondrial dynamics in the brains of scrapie-infected mice. *Biochem Biophys Res Commun*. 2014;448(2):157–162.
103. Choi SI, Ju WK, Choi EK, et al. Mitochondrial dysfunction induced by oxidative stress in the brains of hamsters infected with the 263K scrapie agent. *Acta Neuropathol*. 1998;96(3):279–286.
104. Park JH, Kim BH, Park SJ, et al. Association of endothelial nitric oxide synthase and mitochondrial dysfunction in the hippocampus of scrapie-infected mice. *Hippocampus*. 2011;21(3):319–333.
105. Ashrafi G, Schwarz TL. PINK1- and PARK2-mediated local mitophagy in distal neuronal axons. *Autophagy*. 2015;11(1):187–189.
106. Corti O, Brice A. Mitochondrial quality control turns out to be the principal suspect in parkin and PINK1-related autosomal recessive Parkinson's disease. *Curr Opin Neurobiol*. 2013;23(1):100–108.

107. Durcan TM, Fon EA. The three 'P's of mitophagy: PARKIN, PINK1, and post-translational modifications. *Genes Dev.* 2015;29(10):989–999.
108. Erpapazoglou Z, Corti O. The endoplasmic reticulum/mitochondria interface: a sub-cellular platform for the orchestration of the functions of the PINK1-Parkin pathway? *Biochem Soc Trans.* 2015;43(2):297–301.
109. Haelterman NA, Yoon WH, Sandoval H, Jaiswal M, Shulman JM, Bellen HJ. A mitocentric view of Parkinson's disease. *Annu Rev Neurosci.* 2014;37:137–159.
110. Han JY, Kim JS, Son JH. Mitochondrial homeostasis molecules: regulation by a trio of recessive Parkinson's disease genes. *Exp Neurobiol.* 2014;23(4):345–351.
111. Pickrell AM, Youle RJ. The roles of PINK1, parkin, and mitochondrial fidelity in Parkinson's disease. *Neuron.* 2015;85(2):257–273.
112. Trempe JF, Fon EA. Structure and function of Parkin, PINK1, and DJ-1, the three musketeers of neuroprotection. *Front Neurol.* 2013;4:38.
113. Scarffe LA, Stevens DA, Dawson VL, Dawson TM. Parkin and PINK1: much more than mitophagy. *Trends Neurosci.* 2014;37(6):315–324.
114. Trudler D, Nash Y, Frenkel D. New insights on Parkinson's disease genes: the link between mitochondria impairment and neuroinflammation. *J Neural Transm (Vienna).* 2015;122(10):1409–1419.
115. Mao P, Manczak M, Calkins MJ, et al. Mitochondria-targeted catalase reduces abnormal APP processing, amyloid beta production and BACE1 in a mouse model of Alzheimer's disease: implications for neuroprotection and lifespan extension. *Hum Mol Genet.* 2012;21(13):2973–2990.
116. Pavlov PF, Hansson Petersen C, Glaser E, Ankarcrona M. Mitochondrial accumulation of APP and Abeta: significance for Alzheimer disease pathogenesis. *J Cell Mol Med.* 2009;13(10):4137–4145.
117. Reddy PH. Mitochondrial dysfunction in aging and Alzheimer's disease: strategies to protect neurons. *Antioxid Redox Signal.* 2007;9(10):1647–1658.
118. Skaper SD. Alzheimer's disease and amyloid: culprit or coincidence? *Int Rev Neurobiol.* 2012;102:277–316.
119. Bilkei-Gorzo A. Genetic mouse models of brain ageing and Alzheimer's disease. *Pharmacol Ther.* 2014;142(2):244–257.
120. Dhillon VS, Fenech M. Mutations that affect mitochondrial functions and their association with neurodegenerative diseases. *Mutat Res Rev Mutat Res.* 2014;759:1–13.
121. Chen L, Yoo SE, Na R, Liu Y, Ran Q. Cognitive impairment and increased Abeta levels induced by paraquat exposure are attenuated by enhanced removal of mitochondrial H(2)O(2). *Neurobiol Aging.* 2012;33(2)432. e415-426.
122. Dai DF, Chiao YA, Marcinek DJ, Szeto HH, Rabinovitch PS. Mitochondrial oxidative stress in aging and healthspan. *Longev Healthspan.* 2014;3:6.
123. Ohta S, Ohsawa I. Dysfunction of mitochondria and oxidative stress in the pathogenesis of Alzheimer's disease: on defects in the cytochrome c oxidase complex and aldehyde detoxification. *J Alzheimers Dis.* 2006;9(2):155–166.
124. Reddy PH. Mitochondrial oxidative damage in aging and Alzheimer's disease: implications for mitochondrially targeted antioxidant therapeutics. *J Biomed Biotechnol.* 2006;2006(3):31372.
125. Kou J, Kovacs GG, Hoftberger R, et al. Peroxisomal alterations in Alzheimer's disease. *Acta Neuropathol.* 2011;122(3):271–283.
126. Adlard PA, Bush AI. Metals and Alzheimer's disease. *J Alzheimers Dis.* 2006;10(2–3): 145–163.
127. Kepp KP. Bioinorganic chemistry of Alzheimer's disease. *Chem Rev.* 2012;112(10): 5193–5239.
128. Moreira PI, Santos MS, Oliveira CR. Alzheimer's disease: a lesson from mitochondrial dysfunction. *Antioxid Redox Signal.* 2007;9(10):1621–1630.

129. Nam MK, Shin HA, Han JH, Park DW, Rhim H. Essential roles of mitochondrial depolarization in neuron loss through microglial activation and attraction toward neurons. *Brain Res*. 2013;1505:75–85.
130. Anekonda TS, Reddy PH. Neuronal protection by sirtuins in Alzheimer's disease. *J Neurochem*. 2006;96(2):305–313.
131. Kincaid B, Bossy-Wetzel E. Forever young: SIRT3 a shield against mitochondrial meltdown, aging, and neurodegeneration. *Front Aging Neurosci*. 2013;5:48.
132. Pajak B, Kania E, Orzechowski A. Killing me softly: connotations to unfolded protein response and oxidative stress in Alzheimer's disease. *Oxid Med Cell Longev*. 2016;2016: 1805304.
133. Olsen RH, Johnson LA, Zuloaga DG, Limoli CL, Raber J. Enhanced hippocampus-dependent memory and reduced anxiety in mice over-expressing human catalase in mitochondria. *J Neurochem*. 2013;125(2):303–313.
134. Liao AC, Craver BM, Tseng BP, et al. Mitochondrial-targeted human catalase affords neuroprotection from proton irradiation. *Radiat Res*. 2013;180(1):1–6.
135. Parihar VK, Allen BD, Tran KK, et al. Targeted overexpression of mitochondrial catalase prevents radiation-induced cognitive dysfunction. *Antioxid Redox Signal*. 2015;22(1):78–91.
136. Adiele RC, Adiele CA. Mitochondrial regulatory pathways in the pathogenesis of Alzheimer's disease. *J Alzheimers Dis*. 2016;53(4):1257–1270.
137. D'Souza AD, Parish IA, Krause DS, Kaech SM, Shadel GS. Reducing mitochondrial ROS improves disease-related pathology in a mouse model of ataxia-telangiectasia. *Mol Ther*. 2013;21(1):42–48.
138. Pehar M, Beeson G, Beeson CC, Johnson JA, Vargas MR. Mitochondria-targeted catalase reverts the neurotoxicity of hSOD1G(9)(3)A astrocytes without extending the survival of ALS-linked mutant hSOD1 mice. *PLoS One*. 2014;9(7)e103438
139. Ussakli CH, Ebaee A, Binkley J, et al. Mitochondria and tumor progression in ulcerative colitis. *J Natl Cancer Inst*. 2013;105(16):1239–1248.
140. Treuting PM, Linford NJ, Knoblaugh SE, et al. Reduction of age-associated pathology in old mice by overexpression of catalase in mitochondria. *J Gerontol A Biol Sci Med Sci*. 2008;63(8):813–822.
141. Goh J, Enns L, Fatemie S, et al. Mitochondrial targeted catalase suppresses invasive breast cancer in mice. *BMC Cancer*. 2011;11:191.
142. Lee SJ, Hwang AB, Kenyon C. Inhibition of respiration extends C. elegans life span via reactive oxygen species that increase HIF-1 activity. *Curr Biol*. 2010;20(23): 2131–2136.
143. Schulz TJ, Zarse K, Voigt A, Urban N, Birringer M, Ristow M. Glucose restriction extends Caenorhabditis elegans life span by inducing mitochondrial respiration and increasing oxidative stress. *Cell Metab*. 2007;6(4):280–293.
144. Loh K, Deng H, Fukushima A, et al. Reactive oxygen species enhance insulin sensitivity. *Cell Metab*. 2009;10(4):260–272.
145. Lewis KN, Andziak B, Yang T, Buffenstein R. The naked mole-rat response to oxidative stress: just deal with it. *Antioxid Redox Signal*. 2013;19(12):1388–1399.
146. Itsara LS, Kennedy SR, Fox EJ, et al. Oxidative stress is not a major contributor to somatic mitochondrial DNA mutations. *PLoS Genet*. 2014;10(2), e1003974.
147. Goodman M, Bostick RM, Kucuk O, Jones DP. Clinical trials of antioxidants as cancer prevention agents: past, present, and future. *Free Radic Biol Med*. 2011;51(5): 1068–1084.
148. Bhardwaj P. Oxidative stress and antioxidants in gastrointestinal diseases. *Trop Gastroenterol*. 2008;29(3):129–135.
149. Mecocci P, Polidori MC. Antioxidant clinical trials in mild cognitive impairment and Alzheimer's disease. *Biochim Biophys Acta*. 2012;1822(5):631–638.

150. Jalili M, Kolahi S, Aref-Hosseini SR, Mamegani ME, Hekmatdoost A. Beneficial role of antioxidants on clinical outcomes and erythrocyte antioxidant parameters in rheumatoid arthritis patients. *Int J Prev Med.* 2014;5(7):835–840.
151. Golbidi S, Laher I. Antioxidant therapy in human endocrine disorders. *Med Sci Monit.* 2010;16(1):RA9–RA24.
152. Kris-Etherton PM, Lichtenstein AH, Howard BV, Steinberg D, Witztum JL. Nutrition Committee of the American Heart Association Council on Nutrition PyA, and Metabolism. Antioxidant vitamin supplements and cardiovascular disease. *Circulation.* 2004;110(5):637–641.
153. Bjelakovic G, Nikolova D, Gluud LL, Simonetti RG, Gluud C. Antioxidant supplements for prevention of mortality in healthy participants and patients with various diseases. *Cochrane Database Syst Rev.* 2012;(3):CD007176.
154. Song M, Chen Y, Gong G, Murphy E, Rabinovitch PS, Dorn GW. Super-suppression of mitochondrial reactive oxygen species signaling impairs compensatory autophagy in primary mitophagic cardiomyopathy. *Circ Res.* 2014;115(3): 348–353.
155. West AP, Brodsky IE, Rahner C, et al. TLR signalling augments macrophage bactericidal activity through mitochondrial ROS. *Nature.* 2011;472(7344):476–480.
156. Basisty N, Dai DF, Gagnidze A, et al. Mitochondrial-targeted catalase is good for the old mouse proteome, but not for the young: 'reverse' antagonistic pleiotropy? *Aging Cell.* 2016;15(4):634–645.
157. Williams GC. Pleiotropy, natural selection, and the evolution of senescence. *Evolution.* 1957;11(4):398–411.
158. Marcinek DJ, Siegel MP. Targeting redox biology to reverse mitochondrial dysfunction. *Aging (Albany NY).* 2013;5(8):588–589.
159. Starkov AA. The role of mitochondria in reactive oxygen species metabolism and signaling. *Ann N Y Acad Sci.* 2008;1147:37–52.
160. Reczek CR, Chandel NS. ROS-dependent signal transduction. *Curr Opin Cell Biol.* 2015;33:8–13.
161. Le Belle JE, Orozco NM, Paucar AA, et al. Proliferative neural stem cells have high endogenous ROS levels that regulate self-renewal and neurogenesis in a PI3K/Akt-dependant manner. *Cell Stem Cell.* 2011;8(1):59–71.
162. Shi Y, Buffenstein R, Pulliam DA, Van Remmen H. Comparative studies of oxidative stress and mitochondrial function in aging. *Integr Comp Biol.* 2010;50(5): 869–879.
163. Torma F, Koltai E, Nagy E, et al. Exercise increases markers of spermatogenesis in rats selectively bred for low running capacity. *PLoS One.* 2014;9(12), e114075.
164. Velarde MC, Demaria M, Melov S, Campisi J. Pleiotropic age-dependent effects of mitochondrial dysfunction on epidermal stem cells. *Proc Natl Acad Sci USA.* 2015;112(33):10407–10412.
165. Antonenko YN, Avetisyan AV, Bakeeva LE, et al. Mitochondria-targeted plastoquinone derivatives as tools to interrupt execution of the aging program. 1. Cationic plastoquinone derivatives: synthesis and in vitro studies. *Biochemistry (Mosc).* 2008;73(12):1273–1287.
166. Kelso GF, Porteous CM, Coulter CV, et al. Selective targeting of a redox-active ubiquinone to mitochondria within cells: antioxidant and antiapoptotic properties. *J Biol Chem.* 2001;276(7):4588–4596.
167. Birk AV, Liu S, Soong Y, et al. The mitochondrial-targeted compound SS-31 re-energizes ischemic mitochondria by interacting with cardiolipin. *J Am Soc Nephrol.* 2013;24(8):1250–1261.
168. Szeto HH. First-in-class cardiolipin therapeutic to restore mitochondrial bioenergetics. *Br J Pharmacol.* 2014;171:2029–2050.

169. Birk AV, Chao WM, Bracken WC, Warren JD, Szeto HH. Targeting mitochondrial cardiolipin and the cytochrome c/cardiolipin complex to promote electron transport and optimize mitochondrial ATP synthesis. *Br J Pharmacol.* 2014;171:2017–2028.
170. Magwere T, West M, Riyahi K, Murphy MP, Smith RA, Partridge L. The effects of exogenous antioxidants on lifespan and oxidative stress resistance in Drosophila melanogaster. *Mech Ageing Dev.* 2006;127(4):356–370.
171. Ng LF, Gruber J, Cheah IK, et al. The mitochondria-targeted antioxidant MitoQ extends lifespan and improves healthspan of a transgenic Caenorhabditis elegans model of Alzheimer disease. *Free Radic Biol Med.* 2014;71:390–401.
172. Manczak M, Mao P, Calkins MJ, et al. Mitochondria-targeted antioxidants protect against amyloid-beta toxicity in Alzheimer's disease neurons. *J Alzheimers Dis.* 2010;20(suppl 2):S609–S631.
173. McManus MJ, Murphy MP, Franklin JL. The mitochondria-targeted antioxidant MitoQ prevents loss of spatial memory retention and early neuropathology in a transgenic mouse model of Alzheimer's disease. *J Neurosci.* 2011;31(44):15703–15715.
174. Ghosh A, Chandran K, Kalivendi SV, et al. Neuroprotection by a mitochondria-targeted drug in a Parkinson's disease model. *Free Radic Biol Med.* 2010;49(11):1674–1684.
175. Adlam VJ, Harrison JC, Porteous CM, et al. Targeting an antioxidant to mitochondria decreases cardiac ischemia-reperfusion injury. *FASEB J.* 2005;19(9):1088–1095.
176. Dare AJ, Logan A, Prime TA, et al. The mitochondria-targeted anti-oxidant MitoQ decreases ischemia-reperfusion injury in a murine syngeneic heart transplant model. *J Heart Lung Transplant.* 2015;34(11):1471–1480.
177. Graham D, Huynh NN, Hamilton CA, et al. Mitochondria-targeted antioxidant MitoQ10 improves endothelial function and attenuates cardiac hypertrophy. *Hypertension.* 2009;54(2):322–328.
178. Supinski GS, Murphy MP, Callahan LA. MitoQ administration prevents endotoxin-induced cardiac dysfunction. *Am J Physiol Regul Integr Comp Physiol.* 2009;297(4):R1095–R1102.
179. Mukhopadhyay P, Horvath B, Zsengeller Z, et al. Mitochondrial reactive oxygen species generation triggers inflammatory response and tissue injury associated with hepatic ischemia-reperfusion: therapeutic potential of mitochondrially targeted antioxidants. *Free Radic Biol Med.* 2012;53(5):1123–1138.
180. Anisimov VN, Bakeeva LE, Egormin PA, et al. Mitochondria-targeted plastoquinone derivatives as tools to interrupt execution of the aging program. 5. SkQ1 prolongs lifespan and prevents development of traits of senescence. *Biochemistry (Mosc).* 2008;73(12):1329–1342.
181. Anisimov VN, Egorov MV, Krasilshchikova MS, et al. Effects of the mitochondria-targeted antioxidant SkQ1 on lifespan of rodents. *Aging (Albany NY).* 2011;3(11):1110–1119.
182. Bakeeva LE, Barskov IV, Egorov MV, et al. Mitochondria-targeted plastoquinone derivatives as tools to interrupt execution of the aging program. 2. Treatment of some ROS- and age-related diseases (heart arrhythmia, heart infarctions, kidney ischemia, and stroke). *Biochemistry (Mosc).* 2008;73(12):1288–1299.
183. Neroev VV, Archipova MM, Bakeeva LE, et al. Mitochondria-targeted plastoquinone derivatives as tools to interrupt execution of the aging program. 4. Age-related eye disease. SkQ1 returns vision to blind animals. *Biochemistry (Mosc).* 2008;73(12):1317–1328.
184. Zhao K, Zhao GM, Wu D, et al. Cell-permeable peptide antioxidants targeted to inner mitochondrial membrane inhibit mitochondrial swelling, oxidative cell death, and reperfusion injury. *J Biol Chem.* 2004;279(33):34682–34690.

185. Doughan AK, Dikalov SI. Mitochondrial redox cycling of mitoquinone leads to superoxide production and cellular apoptosis. *Antioxid Redox Signal.* 2007;9(11):1825–1836.
186. Szeto HH, Liu S, Soong Y, et al. Mitochondria-targeted peptide accelerates ATP recovery and reduces ischemic kidney injury. *J Am Soc Nephrol.* 2011;22(6):1041–1052.
187. Calkins MJ, Manczak M, Mao P, Shirendeb U, Reddy PH. Impaired mitochondrial biogenesis, defective axonal transport of mitochondria, abnormal mitochondrial dynamics and synaptic degeneration in a mouse model of Alzheimer's disease. *Hum Mol Genet.* 2011;20(23):4515–4529.
188. Yang L, Zhao K, Calingasan NY, Luo G, Szeto HH, Beal MF. Mitochondria targeted peptides protect against 1-methyl-4-phenyl-1,2,3,6-tetrahydropyridine neurotoxicity. *Antioxid Redox Signal.* 2009;11(9):2095–2104.
189. Brown DA, Hale SL, Baines CP, et al. Reduction of early reperfusion injury with the mitochondria-targeting peptide bendavia. *J Cardiovasc Pharmacol Ther.* 2014;19(1): 121–132.
190. Cho J, Won K, Wu D, et al. Potent mitochondria-targeted peptides reduce myocardial infarction in rats. *Coron Artery Dis.* 2007;18(3):215–220.
191. Dai W, Shi J, Gupta RC, Sabbah HN, Hale SL, Kloner RA. Bendavia, a mitochondria-targeting peptide, improves postinfarction cardiac function, prevents adverse left ventricular remodeling, and restores mitochondria-related gene expression in rats. *J Cardiovasc Pharmacol.* 2014;64(6):543–553.
192. Kloner RA, Hale SL, Dai W, et al. Reduction of ischemia/reperfusion injury with bendavia, a mitochondria-targeting cytoprotective Peptide. *J Am Heart Assoc.* 2012; 1(3), e001644.
193. Shi J, Dai W, Hale SL, et al. Bendavia restores mitochondrial energy metabolism gene expression and suppresses cardiac fibrosis in the border zone of the infracted heart. *Life Sci.* 2015;141:170–178.
194. Szeto HH. Mitochondria-targeted cytoprotective peptides for ischemia-reperfusion injury. *Antioxid Redox Signal.* 2008;10(3):601–619.
195. Dai DF, Chen T, Szeto H, et al. Mitochondrial targeted antioxidant peptide ameliorates hypertensive cardiomyopathy. *J Am Coll Cardiol.* 2011;58(1):73–82.
196. Dai DF, Hsieh EJ, Chen T, et al. Global proteomics and pathway analysis of pressure-overload-induced heart failure and its attenuation by mitochondrial-targeted peptides. *Circ Heart Fail.* 2013;6(5):1067–1076.
197. Dai DF, Hsieh EJ, Liu Y, et al. Mitochondrial proteome remodelling in pressure overload-induced heart failure: the role of mitochondrial oxidative stress. *Cardiovasc Res.* 2012;93(1):79–88.
198. Sabbah HN, Gupta RC, Kohli S, Wang M, Hachem S, Zhang K. Chronic therapy with Elamipretide (MTP-131), a novel mitochondria-targeting peptide, improves left ventricular and mitochondrial function in dogs with advanced heart failure. *Circ Heart Fail.* 2016;9(2)e002206
199. Talbert EE, Smuder AJ, Min K, Kwon OS, Szeto HH, Powers SK. Immobilization-induced activation of key proteolytic systems in skeletal muscles is prevented by a mitochondria-targeted antioxidant. *J Appl Physiol (1985).* 2013;115(4):529–538.
200. Powers SK, Hudson MB, Nelson WB, et al. Mitochondria-targeted antioxidants protect against mechanical ventilation-induced diaphragm weakness. *Crit Care Med.* 2011;39(7):1749–1759.
201. Szeto HH. First-in-class cardiolipin-protective compound as a therapeutic agent to restore mitochondrial bioenergetics. *Br J Pharmacol.* 2014;171(8):2029–2050.

CHAPTER EIGHT

Metabolic Syndrome and the Cellular Phase of Alzheimer's Disease

S. Pugazhenthi[*,†,1]
[*]University of Colorado, Aurora, CO, United States
[†]Eastern Colorado Health Care System, Denver, CO, United States
[1]Corresponding author: e-mail address: subbiah.pugazhenthi@ucdenver.edu

Contents

Abstract

Alzheimer's disease (AD) is characterized by cognitive dysfunction and progressive neurodegeneration. The major hallmarks of AD pathology are amyloid plaques and neurofibrillary tangles. However, AD often coexists with other brain microvascular lesions caused by comorbidities, including obesity, diabetes, hypertension, and cardiovascular diseases. The risk factors for these comorbidities are collectively referred to as metabolic syndrome (MetS). Clinical AD is preceded by decades of prodromal cellular phase. During this asymptomatic phase, systemic changes caused by MetS can play critical roles in driving neuroinflammation, an important cause of AD pathogenesis. Studies of MetS and AD have traditionally remained in distinct domains. The cross talk between MetS and the cellular phase of AD is an important area to be investigated. AD risk factors identified by genome-wide association studies (GWAS) have strongly suggested the role of microglia, the resident immune cells of the brain, in AD pathogenesis. Microglial dysregulation is caused not only by CNS-intrinsic factors but also by systemic changes. MetS appears to cause brain mitochondrial dysfunction through a defective NAD^+-sirtuin pathway. Sirtuins are a family of seven proteins that are involved in longevity and

ISSN 1877-1173
http://dx.doi.org/10.1016/bs.pmbts.2016.12.016

inflammation. Among them, SIRT3 is exclusively present in mitochondria, playing a significant role in metabolic adaptation. SIRT3 deacetylates and activates key metabolic enzymes and transcriptional regulators, utilizing NAD^+ in the process. MetS could prime microglia through the interface of blood–brain barrier (BBB). Age-dependent breakdown of the BBB has been reported in human subjects. The neurovascular unit at BBB consists of brain microvascular endothelial cells, end feet of astrocytes, and pericytes. Therapeutic targeting of the sirtuin pathway in AD with coexisting pathologies has the potential to produce profoundly beneficial effects in improving mitochondrial function and decreasing neuroinflammation.

1. ALZHEIMER'S DISEASE IS A CONVERGENT SYNDROME WITH MIXED PATHOLOGIES

Deposition of amyloid plaques and formation of neurofibrillary tangles are important causes of Alzheimer's disease (AD).[1] However, recent studies have suggested that the pure form of AD may be rare and that the coexisting brain lesions could tip the scale to clinical diagnosis of dementia.[2–4] A report reviewing the Nun Study (NS) and Honolulu-Asia Aging Study (HAAS) concluded that the total burden of comorbid brain abnormalities was the main determinant of cognitive deficits in clinically diagnosed AD.[2] The combination rather than the type of lesions played a major role. This study also leads to the understanding that there can be a broader opportunity to treat dementia. Pharmacological interventions targeting the comorbidities have improved survival from life-threatening complications. However silent neurodegenerative pathways that proceed during decades could contribute to cognitive decline. Although Alzheimer's transgenic mice expressing human mutant APP, presenilin and tau have advanced our knowledge of AD pathogenesis, studies of AD mouse models with mixed pathology are needed to recapitulate the molecular events of human AD. The comorbidities including brain hypoperfusion, silent ministrokes, diabetes, and cardiovascular dysfunction need to be incorporated into the current AD transgenic models to recapitulate CNS pathology in the human disease.[5] The boundaries that distinctively separated AD from other forms of dementias are slowly disappearing, suggesting that dementia is a confluent syndrome with contributions from multiple pathologies.[3] Comorbidities of dementia include obesity, diabetes, hypertension, and cardiovascular diseases (Fig. 1). The risk factors for these comorbidities are collectively referred to as metabolic syndrome (MetS).

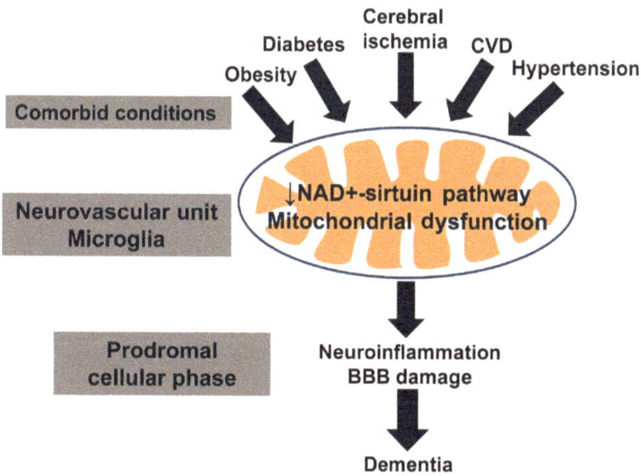

Fig. 1 Comorbidities of Alzheimer's disease. There are several comorbid conditions including obesity, diabetes, cerebral ischemia, cardiovascular diseases, and hypertension that can potentially increase the susceptibility to Alzheimer's disease. The mechanism appears to involve mitochondrial dysfunction in the neurovascular unit and in the microglia. The resulting blood brain damage and neuroinflammation during the prodromal stage of Alzheimer's disease could influence the progression of cognitive decline.

2. CELLULAR PHASE OF AD

Sporadic late-onset AD, the most common form of dementia, is characterized by slow progression over several decades. Cognitive reserve and the ability of brain cells to cope with stress can delay the onset of clinical dementia. There are multiple factors that drive the cellular phase of AD. For example, impaired brain metabolism in early stages appears to play a significant role in cognitive decline.[6] Specifically, defects in frontal and temporoparietal glucose metabolism could contribute to disease progression.[7] Mitochondrial dysfunction is another early event during the prodromal stage of AD[8,9] and it plays an important role in the initiation of neuroinflammation. Linking of these two pathways has provided new insights through the generation of inflammasome,[10–12] a multiprotein cytosolic complex that is generated in response to infection, cellular damage, and metabolic dysregulation.[13] Inflammasome formation leads to the activation of caspase-1 and to the proteolytic cleavage and secretion of the cytokines IL-1β and IL-18.[14] Sterile inflammasomes in response to cellular stress causes neuronal injury.[15] During the disease progression, inflammation gets exacerbated as a result of feed-forward loops and synergistic actions of transcription factors.

For example, secreted inflammatory mediators support astrogliosis and cytokine-activated transcription factors including NF-κB, STAT-1, and c-jun (AP-1) act synergistically to induce more cytokines and chemokines. Many of these events during the presymptomatic phase of this complex disease can become independent self-sustaining pathways later. Presence of comorbidities during the cellular phase of AD can potentially facilitate the progression toward clinical AD. Comorbidities can significantly influence the trajectory of prodromal stage to symptomatic AD. It is being increasingly recognized that the therapeutic targeting of AD needs to start at the prodromal cellular phase.[16] For example, although epidemiological studies have linked the use of antiinflammatory drugs with reduced risk of AD,[17] clinical trials with NSAIDs have failed (reviewed in Ref. [18]), suggesting that the interventions need to start early. Advances in biomarker-based diagnostic criteria can facilitate early interventions.[19,20]

3. CROSS TALK BETWEEN MetS AND THE CELLULAR PHASE OF AD

The major challenge in understanding the complexity of AD pathogenesis is its long cellular phase.[21] This is the stage at which comorbidities can potentially cross talk with AD pathogenesis in mid-life. MetS is a combination of five risk factors including abdominal obesity, hypertriglyceridemia, insulin resistance, high blood pressure, and low levels of good cholesterol (HDL). Current reports are suggesting that around 35% of adults have MetS.[22] The role of comorbidities needs to be examined during the prodromal stage rather than at the time of clinical AD diagnosis, because aged population with comorbidities takes diverse paths in terms of disease management and the type of medications used. Although genetic risk factors play significant roles in susceptibility to AD the role of modifiable risk factors cannot be ignored. A combination of genetic predisposition with unhealthy life styles can dramatically affect the susceptibility to cognitive decline. Consumption of Western diet and lack of physical activities could play important roles during the cellular phase of AD. Although MetS is a known risk factor for cardiovascular disease, diabetes, and stroke, MetS as a risk factor for dementia has received less attention because of mixed results from epidemiological studies.[23–26] An Italian longitudinal study in MCI patients reported that MetS independently predicted an increased risk of progression to dementia in a 3.5-year follow-up.[27] The French three-city study reported association between MetS and vascular dementia (VaD) but not with AD.[28] MetS late in life was found to be not associated as a risk factor for dementia.[25] The mechanism appears to be microvascular damage leading to disrupted cortical connectivity. Insulin resistance has been suggested to be an important link between MetS and cognitive dysfunction. Visceral fat during MetS is characterized by infiltration of macrophages which produce proinflammatory cytokines. The increased levels of circulating cytokines can cross BBB and produce sustained chronic inflammation through an inflammatory loop the mechanism of which we have described in a recent study.[29]

4. TARGETING SIRT3 TO IMPROVE METABOLIC ADAPTATION DURING THE CELLULAR PHASE OF AD

Mitochondrial dysfunction is an early event during the prodromal stage of AD[9] and it plays an important role in the initiation of neuroinflammation. For the therapeutic targeting of these defects, sirtuins appear to show promise.[30] The silent information regulator (SIRT) genes (sirtuins) comprise a highly conserved family of seven proteins that use NAD^+ as a cosubstrate to catalyze the deacetylation and/or the mono-ADP ribosylation of target proteins.[31] They regulate diverse biological mechanisms including longevity, genomic stability, and inflammation. Among the seven members, SIRT3 is exclusively present in mitochondria, where it plays a central role in metabolic regulation[32] (Fig. 2). Acetylation is an

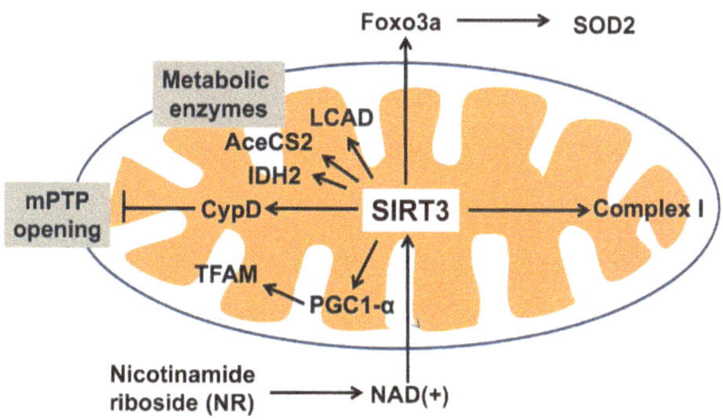

Fig. 2 SIRT3 and metabolic adaptation. SIRT3 deacetylates and activates metabolic enzymes, transcription factors, and other critical proteins in mitochondria. The metabolic enzymes include long chain fatty acid acyl-coA dehydrogenase (LCAD), acetyl CoA synthetase 2 (AceCS2), and isocitrate dehydrogenase (IDH). Overall, SIRT3 mediates adaptive response to metabolic stress especially during the aging process. SIRT3 can be targeted therapeutically by supplementation with nicotinamide riboside, a precursor of NAD^+.

important posttranslational modification that plays a critical role in metabolic regulation.[33] Around 300 acetylation sites have been identified in mitochondrial proteins.[34] SIRT3 is essential for adaptive response to metabolic stress. Targets of SIRT3 deacetylation include metabolic enzymes including long chain fatty acid acyl-CoA dehydrogenase (LCAD), acetyl CoA synthetase 2 (AceCS2), and isocitrate dehydrogenase (IDH), the transcription factor FOXO3a, transcriptional coactivator PGC1-α, antioxidant enzyme SOD2, mitochondrial OPA1 and complex1 proteins.[35] SIRT3 mediates adaptive response to metabolic stress, which is critical during aging. SIRT3 is transcriptionally upregulated by dietary restriction and fasting.[35] Homozygous SIRT3$^{-/-}$ mice are viable and do not display any gross physical or behavioral abnormalities.[36] However, when fed with energy-rich diet, they develop MetS due to impaired mitochondrial metabolism.[37] Single-nucleotide polymorphism of human SIRT3 is associated with susceptibility for MetS.[38] Nicotinamide adenine dinucleotide (NAD^+) is a coenzyme for metabolic pathways and it is also a cosubstrate for many enzymes including sirtuins.[39,40] Depletion of NAD^+ plays a critical role in neurodegeneration.[40–42] Replacing the NAD^+ levels is emerging as an important therapeutic approach.[43] Increasing the cellular level of NAD^+ by administration of nicotinamide riboside (NR), a precursor of NAD^+, is an effective strategy to activate the sirtuin pathway.[44] Other approaches to increase NAD^+ with nicotinamide mononucleotide (NAM), NAD^+, and nicotinic acid have undesirable effects.[45–47]

5. MICROGLIAL PRIMING DURING MetS

Microglia, the resident immune cells of the brain, constitute 5%–10% of the brain cells with region-specific variations. Microglia originate from erythromyeloid precursors in the embryonic yolk sac and migrate to the brain before the blood–brain barrier (BBB) is formed.[48] Microglial synaptic pruning by a complement-dependent pathway plays an important role in the establishment of neuronal network during development.[49] Genome-wide association studies (GWAS) of AD patients have shown that a large number of genetic polymorphisms of risk factor genes are involved in immune regulatory pathways, especially in microglia.[50] Microglia are known to be activated in the vicinity of amyloid plaques in the Alzheimer's brain and they are believed to reduce Aβ burden by phagocytosis. Landreth and coworkers[51] demonstrated that phagocytosis of β amyloid by microglia can be significantly improved with the use of RXR agonist bexarotene, leading to decrease in β amyloid load in AD mouse models. However, uncontrolled chronic inflammation results in the release of neurotoxic factors including proinflammatory cytokines and reactive oxygen species by glial cells, resulting in the neurodegenerative process. In response to injury, microglia change their phenotype and response. Recent reports have suggested that M1/M2 polarization of microglia is an oversimplification. Deep sequencing studies have revealed unique molecular signatures of microglia when compared to other immune cells as well as other brain cells.[52–55] Microglial gene expression patterns are important markers because they reflect the neurodegenerative environment and the detrimental cues sent by MetS from the periphery. Microglia also play crucial intermediary roles in the CNS effects of gut microbiota.[56] For example, mice in germ-free environment with less developed microbiota have immature microglia. Microbiota-generated short chain fatty acids (SCFA) act on GPR34, a SCFA receptor on microglia, leading to its maturation.[56] SCFAR KO mice have microglia with immature phenotype. Western diet causes significant decreases in SCFA and GPR34.[57]

6. PERIPHERAL AND CENTRAL INFLAMMATION CONNECTION

Bidirectional cross talk between peripheral and central inflammation is an important component of AD pathogenesis.[58] Aging-associated chronic low-grade inflammation has been referred to as "inflammaging."[59] The expression of genes in the inflammatory pathways is significantly elevated even during cognitively normal aging.[60] The expression patterns in this study suggest activation of microglia and perivascular macrophages. The progression of neurodegenerative diseases is known to be exacerbated by systemic infection and inflammation.[61] Villeda *et al.* made an interesting observation that exposure of aged animal to young blood reverses the effects of aging at the molecular and functional levels.[62] Microglia in their entire life span, do not directly come in contact with the systemic circulation.[48] Induction of cytokines and chemokines in hippocampus is observed, following systemic challenge with IL-1β and TNFα in mice.[63] Higher peripheral concentrations of proinflammatory cytokines have been reported in Alzheimer's patients.[64] Framingham study has reported elevated circulating IL-1β and TNF-α as markers for the risk of AD.[65] Elevated levels of circulating TNF-α, associated with acute and chronic systemic inflammation, have been shown to contribute cognitive decline in AD.[66] Proinflammatory cytokines are known to pass through BBB.[67–69] Microglia are known to be primed in the aging brain and they respond to peripheral inflammation with greater severity and duration.[70] BBB damage observed during aging further adds to the exacerbation of CNS inflammation with the entry of immune cells into the brain. Activated microglia in the perivascular region can induce the expression of the adhesion molecules through secreted proinflammatory cytokines. Vascular adhesion molecules play important roles in immune cell entry. The cascade involves, rolling adhesion with E-selectin and P-selectin and firm adhesion with ICAM1 and VCAM1, followed by the entry of immune cells. Availability of FDA-approved drugs that can modulate microglial activation and improve brain microvascular function are promising.

7. OVERLAP OF VaD WITH AD

Because 20% of total energy consumption is in the brain, it is highly vascularized to facilitate the uptake of oxygen and nutrients. VaD is the second most common form of dementia after AD. However, significant overlap between these two forms is being recognized. The overlap ranges from AD with vascular dysfunction to mixed type of dementia.[71] When cerebrovascular lesions are often observed in aged brains, it is difficult to consider VaD as a distinct type.[72] Deteriorating vascular function and the progressive neurodegenerative process need to be viewed as converging pathogenic mechanisms. Two-hit vascular hypothesis suggests that defective brain microvascular circulation (first hit) acts as a trigger for the pathological events leading to the second hit of Aβ accumulation.[73] In line with this hypothesis, primary vascular events caused by the comorbidities could trigger a chain of events leading to neurodegeneration. Both VaD and AD share common risk factors including obesity, diabetes, hypertension, and smoking. Dementia could result from combined burden of vascular and neurodegenerative pathology. Cerebral amyloid angiopathy (CAA), observed in majority of AD patients, can cause intracerebral hemorrhage and microbleeds.[74] Thus additive and synergistic effects between VaD and AD can be expected. Understanding the contribution of vascular dysfunction to AD pathogenesis is critical for the development of effective therapeutic targets. Promoting the vascular health in the aging brain can be an important therapeutic strategy.

8. NEUROVASCULAR UNIT FACILITATES MetS–AD CROSS TALK

Comorbidities of AD can exert their deleterious CNS effects through neurovascular unit (NVU) (Fig. 3). NVU contributes to the development of VaD as well as its progression. A recent MRI study in human subjects has reported age-dependent breakdown of BBB.[75] Studies in rodents have shown that feeding of energy-rich diet leads to compromised BBB integrity.[76–78] BBB damage in the aging brain leads to accumulation of blood-derived proteins including immunoglobulins, albumin, fibrinogen, and thrombin.[73] Bien-Ly *et al.* reported lack of BBB permeability in AD mouse models.[79] Essentially this study raises doubt regarding the plasma Aβ-mediated BBB disruption. It appears that BBB damage could be a feature of AD with mixed pathologies. NVU consists of brain microvascular endothelial cells (BMECs), end feet of astrocytes, and pericytes. To meet the high energy demand of active transport across BBB, endothelial cells contain high number of mitochondria. Studies with BMEC have revealed that their susceptibility to oxidative stress.[80] Silencing of SIRT3 leads to decreased viability of endothelial cells.[81] BMECs are uniquely different from other

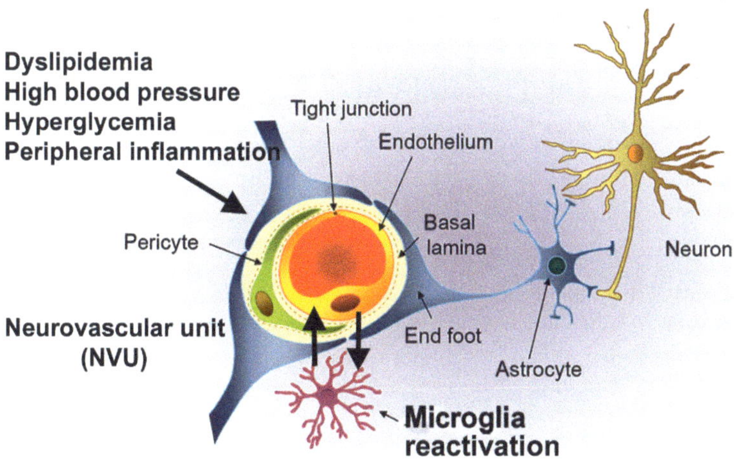

Fig. 3 Metabolic syndrome and the neurovascular unit (NVU). NVU consists of brain microvascular endothelial cells, end feet of astrocytes, and pericytes. Cerebrovascular endothelial cells are critical sensors of dyslipidemia, hyperglycemia, and peripheral inflammation and play critical roles as mediators of microglial activation. Two-way communications between these cell types are critical to maintain homeostasis.

vascular endothelial cells because they are glued together by tight-junction (TJ) proteins including occludin and claudins.[82] As they line the luminal side, they are in constant contact with circulating factors and in communication with circulating immune cells. Therefore, cerebrovascular endothelial cells are critical sensors of peripheral inflammation and mediators of microglial activation. Microglia act as sensors of these signals leading to it reactivation. Microglia not only responds to the cues on the environment in the parenchyma but also to the signals generated by NVU. Microglia play biphasic role in terms of BBB integrity in a context-dependent manner. Following BBB injury, juxtavascular microglia migrate to the site and close the leak through their processes with P2RY12 receptor.[83] However, proinflammatory cytokines released from activated microglia are also known to decrease the expression of TJs and increase the expression of matrix metalloproteinase (MMP-9) which degrades TJ proteins.[84] Higher levels of circulating MMP-9 caused by MMP-9 gene variant are associated with a higher risk for MetS.[85] TNF-α causes microvascular endothelial permeability by activation of MMP-9.[84] Individuals with history of hypertension and high plasma levels of MMP-9 develop white matter hyperintensities.[86] Hyperglycemia-mediated induction of MMP-9 causes astrocyte migration.[87] Circulating MMP-9 levels are higher in children with diabetic ketoacidosis.[88]

9. CEREBRAL ISCHEMIA AND AD

The progression of cognitive decline in AD patients is faster with coexisting cerebral infarction.[89] Cerebral ischemia by tMCAO in CX3CR1/GFP mouse model with the loss of function of microglia showed decreased stroke size.[90] Biphasic functions of microglia after stroke have been reported, suggesting that suppressing microglial activation may not be an effective therapeutic strategy.[91] Microinfarcts are commonly observed in the aging brain.[92,93] The incidence of microinfarcts increases further in VaD patients.[94] Silent infarcts have been shown to be associated with MetS.[95,96] These microinfarcts are generally microscopic in nature. These silent infarcts are typically identified in postmortem examination. Compared to global cerebral ischemia, less information is available with experimental microinfarcts models. A mouse microinfarct model has been developed by Nedergaard and colleagues.[97] This model is generated by unilateral injection of cholesterol crystals. Unlike the classic MCAO model in which neuronal loss is irreversible after 3 h, in the microinfarct model, neuronal loss is delayed over a 24-day period. The chronic effects of microinfarcts could be due to hypoxia resulting from diffuse hypoperfusion, oxidative stress, and inflammation resulting from glial activation. Overall, microinfarcts are considered to contribute independently to cognitive decline. Even in the absence of dementia, they are associated with decreased cognitive function score. These asymptomatic brain lesions can collectively contribute to the progression of AD pathology in additive or synergistic manner.

REFERENCES

1. Selkoe DJ. Alzheimer's disease: genes, proteins and therapy. *Physiol Rev.* 2001;81: 741–766.
2. White LR, Edland SD, Hemmy LS, Montine KS, Zarow C, et al. Neuropathologic comorbidity and cognitive impairment in the Nun and Honolulu-Asia Aging Studies. *Neurology.* 2016;86:1000–1008.
3. Montine TJ, Sonnen JA, Montine KS, Crane PK, Larson EB. Adult Changes in Thought study: dementia is an individually varying convergent syndrome with prevalent clinically silent diseases that may be modified by some commonly used therapeutics. *Curr Alzheimer Res.* 2012;9:718–723.
4. Kawas CH, Kim RC, Sonnen JA, Bullain SS, Trieu T, et al. Multiple pathologies are common and related to dementia in the oldest-old: the 90+ study. *Neurology.* 2015;85:535–542.
5. Snyder HM, Hendrix J, Bain LJ, Carrillo MC. Alzheimer's disease research in the context of the national plan to address Alzheimer's disease. *Mol Aspects Med.* 2015;43–44: 16–24.
6. Habeck C, Risacher S, Lee GJ, Glymour MM, Mormino E, et al. Relationship between baseline brain metabolism measured using [(1)(8)F]FDG PET and memory and executive function in prodromal and early Alzheimer's disease. *Brain Imaging Behav.* 2012;6:568–583.
7. Herholz K. Cerebral glucose metabolism in preclinical and prodromal Alzheimer's disease. *Expert Rev Neurother.* 2010;10:1667–1673.
8. Caldwell CC, Yao J, Brinton RD. Targeting the prodromal stage of Alzheimer's disease: bioenergetic and mitochondrial opportunities. *Neurotherapeutics.* 2015;12:66–80.
9. Du H, Guo L, Yan S, Sosunov AA, McKhann GM, et al. Early deficits in synaptic mitochondria in an Alzheimer's disease mouse model. *Proc Natl Acad Sci USA.* 2010;107:18670–18675.
10. Gurung P, Lukens JR, Kanneganti TD. Mitochondria: diversity in the regulation of the NLRP3 inflammasome. *Trends Mol Med.* 2015;21:193–201.
11. Sorbara MT, Girardin SE. Mitochondrial ROS fuel the inflammasome. *Cell Res.* 2011;21:558–560.
12. West AP, Khoury-Hanold W, Staron M, Tal MC, Pineda CM, et al. Mitochondrial DNA stress primes the antiviral innate immune response. *Nature.* 2015;520(7548): 553–557.
13. Lamkanfi M, Dixit VM. Mechanisms and functions of inflammasomes. *Cell.* 2014;157:1013–1022.
14. Laudisi F, Spreafico R, Evrard M, Hughes TR, Mandriani B, et al. Cutting edge: the NLRP3 inflammasome links complement-mediated inflammation and IL-1beta release. *J Immunol.* 2013;191:1006–1010.

15. Kaushal V, Dye R, Pakavathkumar P, Foveau B, Flores J, et al. Neuronal NLRP1 inflammasome activation of Caspase-1 coordinately regulates inflammatory interleukin-1-beta production and axonal degeneration-associated Caspase-6 activation. *Cell Death Differ.* 2015;22(10):1676–1686.
16. Dubois B, Hampel H, Feldman HH, Scheltens P, Aisen P, et al. Preclinical Alzheimer's disease: definition, natural history, and diagnostic criteria. *Alzheimers Dement.* 2016;12:292–323.
17. McGeer PL, McGeer E, Rogers J, Sibley J. Anti-inflammatory drugs and Alzheimer disease. *Lancet.* 1990;335:1037.
18. Hoozemans JJ, O'Banion MK. The role of COX-1 and COX-2 in Alzheimer's disease pathology and the therapeutic potentials of non-steroidal anti-inflammatory drugs. *Curr Drug Targets CNS Neurol Disord.* 2005;4:307–315.
19. Jack Jr CR, Holtzman DM. Biomarker modeling of Alzheimer's disease. *Neuron.* 2013;80:1347–1358.
20. Jack Jr CR, Knopman DS, Jagust WJ, Shaw LM, Aisen PS, et al. Hypothetical model of dynamic biomarkers of the Alzheimer's pathological cascade. *Lancet Neurol.* 2010;9:119–128.
21. De Strooper B, Karran E. The cellular phase of Alzheimer's disease. *Cell.* 2016;164:603–615.
22. Aguilar M, Bhuket T, Torres S, Liu B, Wong RJ. Prevalence of the metabolic syndrome in the United States, 2003–2012. *JAMA.* 2015;313:1973–1974.
23. Vanhanen M, Koivisto K, Moilanen L, Helkala EL, Hanninen T, et al. Association of metabolic syndrome with Alzheimer disease: a population-based study. *Neurology.* 2006;67:843–847.
24. Razay G, Vreugdenhil A, Wilcock G. The metabolic syndrome and Alzheimer disease. *Arch Neurol.* 2007;64:93–96.
25. Forti P, Pisacane N, Rietti E, Lucicesare A, Olivelli V, et al. Metabolic syndrome and risk of dementia in older adults. *J Am Geriatr Soc.* 2010;58:487–492.
26. Watts AS, Loskutova N, Burns JM, Johnson DK. Metabolic syndrome and cognitive decline in early Alzheimer's disease and healthy older adults. *J Alzheimers Dis.* 2013;35:253–265.
27. Solfrizzi V, Scafato E, Capurso C, D'Introno A, Colacicco AM, et al. Metabolic syndrome and the risk of vascular dementia: the Italian Longitudinal Study on Ageing. *J Neurol Neurosurg Psychiatry.* 2010;81:433–440.
28. Raffaitin C, Gin H, Empana JP, Helmer C, Berr C, et al. Metabolic syndrome and risk for incident Alzheimer's disease or vascular dementia: the Three-City Study. *Diabetes Care.* 2009;32:169–174.
29. Pugazhenthi S, Zhang Y, Bouchard R, Mahaffey G. Induction of an inflammatory loop by interleukin-1beta and tumor necrosis factor-alpha involves NF-kB and STAT-1 in differentiated human neuroprogenitor cells. *PLoS One.* 2013;8, e69585.
30. Min SW, Sohn PD, Cho SH, Swanson RA, Gan L. Sirtuins in neurodegenerative diseases: an update on potential mechanisms. *Front Aging Neurosci.* 2013;5:53.
31. Gan L, Mucke L. Paths of convergence: sirtuins in aging and neurodegeneration. *Neuron.* 2008;58:10–14.
32. Onyango P, Celic I, McCaffery JM, Boeke JD, Feinberg AP. SIRT3, a human SIR2 homologue, is an NAD-dependent deacetylase localized to mitochondria. *Proc Natl Acad Sci USA.* 2002;99:13653–13658.
33. Sol EM, Wagner SA, Weinert BT, Kumar A, Kim HS, et al. Proteomic investigations of lysine acetylation identify diverse substrates of mitochondrial deacetylase sirt3. *PLoS One.* 2012;7, e50545.
34. Kim HS, Patel K, Muldoon-Jacobs K, Bisht KS, Aykin-Burns N, et al. SIRT3 is a mitochondria-localized tumor suppressor required for maintenance of mitochondrial integrity and metabolism during stress. *Cancer Cell.* 2010;17:41–52.

35. Nogueiras R, Habegger KM, Chaudhary N, Finan B, Banks AS, et al. Sirtuin 1 and sirtuin 3: physiological modulators of metabolism. *Physiol Rev*. 2012;92:1479–1514.
36. Ahn BH, Kim HS, Song S, Lee IH, Liu J, et al. A role for the mitochondrial deacetylase Sirt3 in regulating energy homeostasis. *Proc Natl Acad Sci USA*. 2008;105: 14447–14452.
37. Lantier L, Williams AS, Williams IM, Yang KK, Bracy DP, et al. SIRT3 is crucial for maintaining skeletal muscle insulin action and protects against severe insulin resistance in high-fat-fed mice. *Diabetes*. 2015;64:3081–3092.
38. Hirschey MD, Shimazu T, Jing E, Grueter CA, Collins AM, et al. SIRT3 deficiency and mitochondrial protein hyperacetylation accelerate the development of the metabolic syndrome. *Mol Cell*. 2011;44:177–190.
39. Imai S, Guarente L. NAD+ and sirtuins in aging and disease. *Trends Cell Biol*. 2014;24: 464–471.
40. Verdin E. NAD(+) in aging, metabolism, and neurodegeneration. *Science*. 2015;350: 1208–1213.
41. Zhu XH, Lu M, Lee BY, Ugurbil K, Chen W. In vivo NAD assay reveals the intracellular NAD contents and redox state in healthy human brain and their age dependences. *Proc Natl Acad Sci USA*. 2015;112:2876–2881.
42. Zhang H, Ryu D, Wu Y, Gariani K, Wang X, et al. NAD(+) repletion improves mitochondrial and stem cell function and enhances life span in mice. *Science*. 2016;352:1436–1443.
43. Imai S. A possibility of nutriceuticals as an anti-aging intervention: activation of sirtuins by promoting mammalian NAD biosynthesis. *Pharmacol Res*. 2010;62:42–47.
44. Gong B, Pan Y, Vempati P, Zhao W, Knable L, et al. Nicotinamide riboside restores cognition through an upregulation of proliferator-activated receptor-gamma coactivator 1alpha regulated beta-secretase 1 degradation and mitochondrial gene expression in Alzheimer's mouse models. *Neurobiol Aging*. 2013;34:1581–1588.
45. Sauve AA, Wolberger C, Schramm VL, Boeke JD. The biochemistry of sirtuins. *Annu Rev Biochem*. 2006;75:435–465.
46. Bogan KL, Brenner C. Nicotinic acid, nicotinamide, and nicotinamide riboside: a molecular evaluation of NAD+ precursor vitamins in human nutrition. *Annu Rev Nutr*. 2008;28:115–130.
47. Benyo Z, Gille A, Kero J, Csiky M, Suchankova MC, et al. GPR109A (PUMA-G/HM74A) mediates nicotinic acid-induced flushing. *J Clin Invest*. 2005;115:3634–3640.
48. Crotti A, Ransohoff RM. Microglial physiology and pathophysiology: insights from genome-wide transcriptional profiling. *Immunity*. 2016;44:505–515.
49. Hong S, Beja-Glasser VF, Nfonoyim BM, Frouin A, Li S, et al. Complement and microglia mediate early synapse loss in Alzheimer mouse models. *Science*. 2016;352: 712–716.
50. Mhatre SD, Tsai CA, Rubin AJ, James ML, Andreasson KI. Microglial malfunction: the third rail in the development of Alzheimer's disease. *Trends Neurosci*. 2015;38:621–636.
51. Cramer PE, Cirrito JR, Wesson DW, Lee CY, Karlo JC, et al. ApoE-directed therapeutics rapidly clear beta-amyloid and reverse deficits in AD mouse models. *Science*. 2012;335(6075):1503–1506.
52. Beutner C, Linnartz-Gerlach B, Schmidt SV, Beyer M, Mallmann MR, et al. Unique transcriptome signature of mouse microglia. *Glia*. 2013;61:1429–1442.
53. Chiu IM, Morimoto ET, Goodarzi H, Liao JT, O'Keeffe S, et al. A neurodegeneration-specific gene-expression signature of acutely isolated microglia from an amyotrophic lateral sclerosis mouse model. *Cell Rep*. 2013;4:385–401.
54. Hickman SE, Kingery ND, Ohsumi TK, Borowsky ML, Wang LC, et al. The microglial sensome revealed by direct RNA sequencing. *Nat Neurosci*. 2013;16:1896–1905.
55. Butovsky O, Jedrychowski MP, Moore CS, Cialic R, Lanser AJ, et al. Identification of a unique TGF-beta-dependent molecular and functional signature in microglia. *Nat Neurosci*. 2014;17:131–143.

56. Erny D, Hrabe de Angelis AL, Jaitin D, Wieghofer P, Staszewski O, et al. Host microbiota constantly control maturation and function of microglia in the CNS. *Nat Neurosci.* 2015;18:965–977.
57. Agus A, Denizot J, Thevenot J, Martinez-Medina M, Massier S, et al. Western diet induces a shift in microbiota composition enhancing susceptibility to Adherent-Invasive E. coli infection and intestinal inflammation. *Sci Rep.* 2016;6:19032.
58. Di Filippo M, Chiasserini D, Gardoni F, Viviani B, Tozzi A, et al. Effects of central and peripheral inflammation on hippocampal synaptic plasticity. *Neurobiol Dis.* 2013;52: 229–236.
59. Franceschi C. Inflammaging as a major characteristic of old people: can it be prevented or cured? *Nutr Rev.* 2007;65:S173–S176.
60. Cribbs DH, Berchtold NC, Perreau V, Coleman PD, Rogers J, et al. Extensive innate immune gene activation accompanies brain aging, increasing vulnerability to cognitive decline and neurodegeneration: a microarray study. *J Neuroinflammation.* 2012;9:179.
61. Perry VH, Cunningham C, Holmes C. Systemic infections and inflammation affect chronic neurodegeneration. *Nat Rev Immunol.* 2007;7:161–167.
62. Villeda SA, Plambeck KE, Middeldorp J, Castellano JM, Mosher KI, et al. Young blood reverses age-related impairments in cognitive function and synaptic plasticity in mice. *Nat Med.* 2014;20:659–663.
63. Skelly DT, Hennessy E, Dansereau MA, Cunningham C. A systematic analysis of the peripheral and CNS effects of systemic LPS, IL-1beta, [corrected] TNF-alpha and IL-6 challenges in C57BL/6 mice. *PLoS One.* 2013;8, e69123.
64. Swardfager W, Lanctot K, Rothenburg L, Wong A, Cappell J, et al. A meta-analysis of cytokines in Alzheimer's disease. *Biol Psychiatry.* 2010;68:930–941.
65. Tan ZS, Beiser AS, Vasan RS, Roubenoff R, Dinarello CA, et al. Inflammatory markers and the risk of Alzheimer disease: the Framingham Study. *Neurology.* 2007;68: 1902–1908.
66. Holmes C, Cunningham C, Zotova E, Woolford J, Dean C, et al. Systemic inflammation and disease progression in Alzheimer disease. *Neurology.* 2009;73:768–774.
67. Banks WA, Kastin AJ. Blood to brain transport of interleukin links the immune and central nervous systems. *Life Sci.* 1991;48:L117–L121.
68. Banks WA, Kastin AJ, Broadwell RD. Passage of cytokines across the blood–brain barrier. *Neuroimmunomodulation.* 1995;2:241–248.
69. Gutierrez EG, Banks WA, Kastin AJ. Murine tumor necrosis factor alpha is transported from blood to brain in the mouse. *J Neuroimmunol.* 1993;47:169–176.
70. Dilger RN, Johnson RW. Aging, microglial cell priming, and the discordant central inflammatory response to signals from the peripheral immune system. *J Leukoc Biol.* 2008;84:932–939.
71. Iadecola C. The pathobiology of vascular dementia. *Neuron.* 2013;80:844–866.
72. Gorelick PB, Scuteri A, Black SE, Decarli C, Greenberg SM, et al. Vascular contributions to cognitive impairment and dementia: a statement for healthcare professionals from the American Heart Association/American Stroke Association. *Stroke.* 2011;42: 2672–2713.
73. Zlokovic BV. Neurovascular mechanisms of Alzheimer's neurodegeneration. *Trends Neurosci.* 2005;28:202–208.
74. Brenowitz WD, Nelson PT, Besser LM, Heller KB, Kukull WA. Cerebral amyloid angiopathy and its co-occurrence with Alzheimer's disease and other cerebrovascular neuropathologic changes. *Neurobiol Aging.* 2015;36:2702–2708.
75. Montagne A, Barnes SR, Sweeney MD, Halliday MR, Sagare AP, et al. Blood–brain barrier breakdown in the aging human hippocampus. *Neuron.* 2015;85:296–302.
76. Freeman LR, Granholm AC. Vascular changes in rat hippocampus following a high saturated fat and cholesterol diet. *J Cereb Blood Flow Metab.* 2012;32:643–653.

77. Davidson TL, Monnot A, Neal AU, Martin AA, Horton JJ, et al. The effects of a high-energy diet on hippocampal-dependent discrimination performance and blood–brain barrier integrity differ for diet-induced obese and diet-resistant rats. *Physiol Behav.* 2012;107:26–33.
78. Tucsek Z, Toth P, Sosnowska D, Gautam T, Mitschelen M, et al. Obesity in aging exacerbates blood–brain barrier disruption, neuroinflammation, and oxidative stress in the mouse hippocampus: effects on expression of genes involved in beta-amyloid generation and Alzheimer's disease. *J Gerontol A Biol Sci Med Sci.* 2014;69:1212–1226.
79. Bien-Ly N, Boswell CA, Jeet S, Beach TG, Hoyte K, et al. Lack of widespread BBB disruption in Alzheimer's disease models: focus on therapeutic antibodies. *Neuron.* 2015;88:289–297.
80. Navaratna D, Fan X, Leung W, Lok J, Guo S, et al. Cerebrovascular degradation of TRKB by MMP9 in the diabetic brain. *J Clin Invest.* 2013;123:3373–3377.
81. Liu G, Cao M, Xu Y, Li Y. SIRT3 protects endothelial cells from high glucose-induced cytotoxicity. *Int J Clin Exp Pathol.* 2015;8:353–360.
82. Haseloff RF, Dithmer S, Winkler L, Wolburg H, Blasig IE. Transmembrane proteins of the tight junctions at the blood–brain barrier: structural and functional aspects. *Semin Cell Dev Biol.* 2015;38:16–25.
83. Lou N, Takano T, Pei Y, Xavier AL, Goldman SA, et al. Purinergic receptor P2RY12-dependent microglial closure of the injured blood–brain barrier. *Proc Natl Acad Sci USA.* 2016;113:1074–1079.
84. Wiggins-Dohlvik K, Merriman M, Shaji CA, Alluri H, Grimsley M, et al. Tumor necrosis factor-alpha disruption of brain endothelial cell barrier is mediated through matrix metalloproteinase-9. *Am J Surg.* 2014;208:954–960. discussion 960.
85. Yadav SS, Mandal RK, Singh MK, Verma A, Dwivedi P, et al. High serum level of matrix metalloproteinase 9 and promoter polymorphism—1562 C:T as a new risk factor for metabolic syndrome. *DNA Cell Biol.* 2014;33:816–822.
86. Kim Y, Kim YK, Kim NK, Kim SH, Kim OJ, et al. Circulating matrix metalloproteinase-9 level is associated with cerebral white matter hyperintensities in non-stroke individuals. *Eur Neurol.* 2014;72:234–240.
87. Hsieh HL, Lin CC, Hsiao LD, Yang CM. High glucose induces reactive oxygen species-dependent matrix metalloproteinase-9 expression and cell migration in brain astrocytes. *Mol Neurobiol.* 2013;48:601–614.
88. Garro A, Chodobski A, Szmydynger-Chodobska J, Shan R, Bialo SR, et al. Circulating matrix metalloproteinases in children with diabetic ketoacidosis. *Pediatr Diabetes.* 2016. http://dx.doi.org/10.1111/pedi.12359[Epub ahead of print].
89. Sheng B, Cheng LF, Law CB, Li HL, Yeung KM, et al. Coexisting cerebral infarction in Alzheimer's disease is associated with fast dementia progression: applying the National Institute for Neurological Disorders and Stroke/Association Internationale pour la Recherche et l'Enseignement en Neurosciences Neuroimaging Criteria in Alzheimer's Disease with Concomitant Cerebral Infarction. *J Am Geriatr Soc.* 2007;55:918–922.
90. Jolivel V, Bicker F, Biname F, Ploen R, Keller S, et al. Perivascular microglia promote blood vessel disintegration in the ischemic penumbra. *Acta Neuropathol.* 2015;129: 279–295.
91. Ma Y, Wang J, Wang Y, Yang GY. The biphasic function of microglia in ischemic stroke. *Prog Neurobiol.* 2016. pii: S0301-0082(15)30070-8.
92. Brundel M, de Bresser J, van Dillen JJ, Kappelle LJ, Biessels GJ. Cerebral microinfarcts: a systematic review of neuropathological studies. *J Cereb Blood Flow Metab.* 2012;32: 425–436.
93. Vinters HV, Ellis WG, Zarow C, Zaias BW, Jagust WJ, et al. Neuropathologic substrates of ischemic vascular dementia. *J Neuropathol Exp Neurol.* 2000;59:931–945.

94. Haglund M, Passant U, Sjobeck M, Ghebremedhin E, Englund E. Cerebral amyloid angiopathy and cortical microinfarcts as putative substrates of vascular dementia. *Int J Geriatr Psychiatry*. 2006;21:681–687.
95. Park K, Yasuda N, Toyonaga S, Tsubosaki E, Nakabayashi H, et al. Significant associations of metabolic syndrome and its components with silent lacunar infarction in middle aged subjects. *J Neurol Neurosurg Psychiatry*. 2008;79:719–721.
96. Bokura H, Yamaguchi S, Iijima K, Nagai A, Oguro H. Metabolic syndrome is associated with silent ischemic brain lesions. *Stroke*. 2008;39:1607–1609.
97. Wang M, Iliff JJ, Liao Y, Chen MJ, Shinseki MS, et al. Cognitive deficits and delayed neuronal loss in a mouse model of multiple microinfarcts. *J Neurosci*. 2012;32: 17948–17960.

CHAPTER NINE

Mitochondria, Cybrids, Aging, and Alzheimer's Disease

R.H. Swerdlow[1], S. Koppel, I. Weidling, C. Hayley, Y. Ji, H.M. Wilkins
University of Kansas Alzheimer's Disease Center, University of Kansas School of Medicine, Landon Center on Aging, Kansas City, KS, United States
[1]Corresponding author: e-mail address: rswerdlow@kumc.edu

Contents

Abstract

Mitochondrial and bioenergetic function change with advancing age and may drive aging phenotypes. Mitochondrial and bioenergetic changes are also documented in various age-related neurodegenerative diseases, including Alzheimer's disease (AD). In some instances AD mitochondrial and bioenergetic changes are reminiscent of those observed with advancing age but are greater in magnitude. Mitochondrial and bioenergetic dysfunction could, therefore, link neurodegeneration to brain aging.

Progress in Molecular Biology and Translational Science, Volume 146
ISSN 1877-1173
http://dx.doi.org/10.1016/bs.pmbts.2016.12.017

Interestingly, mitochondrial defects in AD patients are not brain-limited, and mitochondrial function can be linked to classic AD histologic changes including amyloid precursor protein processing to beta amyloid. Also, transferring mitochondria from AD subjects to cell lines depleted of endogenous mitochondrial DNA (mtDNA) creates cytoplasmic hybrid (cybrid) cell lines that recapitulate specific biochemical, molecular, and histologic AD features. Such findings have led to the formulation of a "mitochondrial cascade hypothesis" that places mitochondrial dysfunction at the apex of the AD pathology pyramid. Data pertinent to this premise are reviewed.

1. INTRODUCTION

Mitochondria were identified as cell organelles over 100 years ago.[1] Considerable time elapsed before their functions were fully appreciated. During the 1960s it was determined that mitochondria contained their own genome, the mitochondrial DNA (mtDNA),[2,3] and that they generated ATP according to a process defined as the chemiosmotic hypothesis.[4] The membranes that delineated these organelles were also identified, and the fact that these membrane boundaries created compartments that allowed for particular chemical reactions and indeed even pathways to reside was appreciated.

In the second half of the 20th century, a potential role for mitochondria in aging was widely postulated.[5] While mitochondria are neither central nor essential components of all aging hypotheses,[6] their contribution to the aging process as either a primary or downstream contributor to this phenomenon is suspected under a variety of current paradigms.[7] The idea that mitochondria might also contribute to neurodegenerative diseases followed the emerging appreciation of their putative role in aging. This general concept was fueled by the observation that the more common neurodegenerative diseases are "age-related," such that prevalence and incidence for diseases such as Alzheimer's disease (AD) increase with advancing age.[8] Over the past three decades the contribution of mitochondria to neurodegenerative diseases in general, and to AD specifically, has been hotly debated with views ranging from a potential primary role to a mechanistically irrelevant artifact of cell death that arises due to other factors.[9] During this time, though, the debate has taken a notable turn and at this point the main question seems not so much whether mitochondrial dysfunction is important and relevant in selected neurodegenerative diseases, but rather how critical mitochondrial dysfunction will turn out to be in these selected neurodegenerative diseases.[10]

This chapter will address ways in which perceptions of how mitochondria influence brain aging and neurodegeneration have evolved over time. In particular, it will focus on an evolving appreciation of how mitochondria may contribute to AD, and why modifying mitochondrial function is currently considered a viable AD therapeutic target.

2. MITOCHONDRIA AND AGING

2.1 Overview

In the first half of the 20th century it was observed that caloric restriction enhanced rodent life span.[11–15] This contributed to the emerging view that metabolism in general, and perhaps energy metabolism specifically, could help to regulate life span. An appreciation of the idea that metabolism could influence aging came to form the basis of the "Rate of Living Hypothesis" that evolved through the early 20th century,[16,17] which essentially stated species with higher metabolic rates had shorter life spans than species with lower metabolic rates. For example, rodents with high metabolic rates could survive for only a few years, while some reptiles such as turtles with presumably slow metabolic rates could live for many decades. This hypothesis, of course, could not account for long-lived species with apparent high metabolic rates, such as birds.

The Rate of Living Hypothesis was eventually succeeded by the more mechanistically specific Free Radical Theory of Aging that assumed the production of oxygen radicals as a by-product of biochemical reactions should be elevated in species with higher metabolic rates.[18] These free radicals, in turn, would react with molecules within cells and in doing so alter their structure and function. It was further believed that this accumulation of metabolism-generated oxidative stress would drive an aging phenotype.

By the 1970s it was appreciated that within cells, mitochondria were a leading site of free radical generation. With this realization some circles rebranded the Free Radical Theory of Aging as the "Mitochondrial Theory of Aging."[5] As a corollary of this emerging mitochondria-centric approach to aging theory, some began to speculate mitochondria might possess some sort of aging clock.[5,19] To this point mtDNA comprised a particularly attractive candidate, and by the late 1980s it was proposed that an accumulation of somatic mtDNA mutations over time functions as the suspected aging clock.[20] Essentially, mtDNA mutations would accumulate as a consequence of oxidative damage to mtDNA and a modification of its bases, which would in turn compromise the production of respiratory chain subunits through either a lack of production or else through a production of miscoded peptides. In support of this view, an age-associated accumulation of mtDNA mutations, in either the form of deletions or point mutations, was recognized.[21]

Mitochondrial aging was (and is) not central to all theories of aging,[6] and a positive correlation between mtDNA somatic mutations and advancing age could also potentially reflect a consequence of aging as opposed to a driver of aging. Investigators began to critically assess this possibility in the early 21st century in genetically modified mice that were designed to accumulate mtDNA mutations at an accelerated pace. This was accomplished by creating mice with a mutated mtDNA polymerase gamma (mtPOLG), in which the innate proofreading ability of the enzyme was perturbed.[22,23] This allowed for a rapid accumulation of somatic mtDNA mutations across a range of tissues. Phenotype characterizations of these mice showed changes consistent with accelerated aging and were accepted as support for the view that increasing levels of somatic, heteroplasmic mtDNA mutations could drive an aging phenotype.

2.2 Mitochondrial Function and Homeostasis in Advancing Age

The predominant current view is that the functional capacity of an organism's mitochondria declines with advancing age. Much of this thinking was informed by rodent studies, in which mitochondrial functional endpoints were assessed in mice or rats of different ages. For example, it has been reported that the maximum kinetic function of two mitochondrial respiratory chain enzymes, NADH:ubiquinone oxidoreductase (complex I) and cytochrome oxidase (COX; complex IV), declines with advancing age.[24,25] Age-related declines in mitochondrial enzyme activities may represent a specific rather than generalized phenomenon, as the activities of other enzymes such as succinate dehydrogenase (complex II) appears to be preserved. In this respect it is perhaps of interest that complexes I and IV are partially encoded by mtDNA, while complex II is entirely encoded by nuclear DNA.

An interesting parameter in the relationship between brain aging and mitochondria has to do with the number of mitochondria that are present, also referred to here as mitochondrial mass. Some studies have reported that in brains derived from subjects who were free of a neurodegenerative disease prior to death, mtDNA copy number increased with advancing age.[26] This increase in mtDNA was observed despite the fact that mRNA levels were reduced. Increased mtDNA content, therefore, was interpreted by the authors as potentially reflecting a compensatory response to a reduction in mtDNA transcription efficiency. As part of a related finding, another study reported protein levels of an mtDNA-encoded COX protein subunit, COX2, were increased in the brains of aged individuals when compared to the brains of young individuals.[27] These findings in humans were essentially reflected in a more recent study from 5-, 12-, and 24-month old C57Bl/6 mice, in which synaptic mitochondria were found to demonstrate apparent adaptive changes at the protein level, which were arguably compensating for overall detrimental changes including an increase in mtDNA damage.[28]

In general, relative to young organisms, mitochondria from aged organisms have been reported to show decreased ATP production, increased free radical production, depolarization of the mitochondrial membrane potential, and a reduced ability to buffer calcium.[29] Not all studies, though, have uniformly detected such changes, and to some extent attribute a possible preservation of mitochondrial functional indices to compensatory responses.[30]

2.3 Mitochondria and Free Radical Production

Oxidative modifications of cell proteins are seen at increased levels in the aging brain, which could potentially reflect increased mitochondrial free radical production.[31] In corollary to this, levels of at least some mitochondrial antioxidant enzymes (for example, manganese superoxide dismutase, mitochondrial catalase, and periredoxin) increase with advancing age, presumably in response to increased mitochondrial free radical production.[28]

When considering the significance of oxidative stress in aging, it must be kept in mind that oxidative stress is not uniformly a toxic event. Up to certain levels, it appears, oxidative stress functions as a signal transducer.[32,33] For example, it has been reported that oxidative stress can facilitate retrograde signaling from the mitochondria to the nucleus.[34,35] Free radicals, in fact, may promote mitochondrial biogenesis in situations where compensatory increases in mitochondrial mass could prove beneficial.[36–38]

2.4 Mitochondrial DNA and Somatic Mutation

MtDNA differs from nuclear DNA in several notable ways. One unique characteristic is heteroplasmy, which is in some ways the mitochondrial correlate of nuclear heterozygosity. Nuclear heterozygosity infers one copy of a gene contains a particular variant while the other does not. MtDNA heteroplasmy is more complex because cells can each have hundreds to thousands of mtDNA. A state of homoplasmy exists when all mtDNA copies within a cell or organism are identical. Heteroplasmy is present when this is not the case and mtDNA copies diverge at a particular nucleotide or nucleotides. Heteroplasmy can be difficult to absolutely rule out, as doing so depends on the sensitivity of its ascertainment. For example, it appears that individual low-abundance mtDNA sequence deviations, or microheteroplasmies, are relatively common at the 1%–3% level.[39,40] They can be detected at levels even lower than 1%, although at extremely low percentages the reliability of the observed sequence deviation can become questionable. In other words, does an extremely low-abundance deviation represent a bona fide sequence deviation, or could it perhaps represent a sequencing artifact?

Regardless, mtDNA reportedly accumulates mutations at approximately 10 times the rate of nuclear DNA.[41] This has been attributed to a number of factors, including the lack of protective histone proteins, but also to the fact that mtDNA resides in close proximity to electron transport chain-derived free radical production.[42] In one scenario, cytosine can undergo an oxidative deamination to uracil, which pairs with an adenine rather than guanosine upon replication and results in a G-C base pair undergoing conversion to an A-T base pair (Fig. 1A).[43,44] In another scenario, 2-deoxyguanosine is oxidized to 8-hydroxy-2-deoxyguanosine and then to 8-oxo-2-deoxyguanosine. 8-Oxo-2-deoxyguanosine mismatches to an adenine nucleotide, eventually leading to the conversion of a G-C base pair to a T-A base pair (Fig. 1B). 8-Hyroxy-2-deoxyguanosine modifications increase with advancing age.[45] While substitutions in the mtDNA that are consistent with these patterns do appear to accumulate with advancing age, it is interesting to note that in one study of mtDNA POLG mutator mice, which over time accumulate such substitutions at an accelerated pace and show an accelerated aging phenotype, there was no evidence of an age-associated concomitant increase in oxidative stress.[23]

As somatic mutations begin to accumulate they create heteroplasmies whose levels will likely initially reside below the level of detection. If they become fixed in the genome and are replicated over time, their percent of

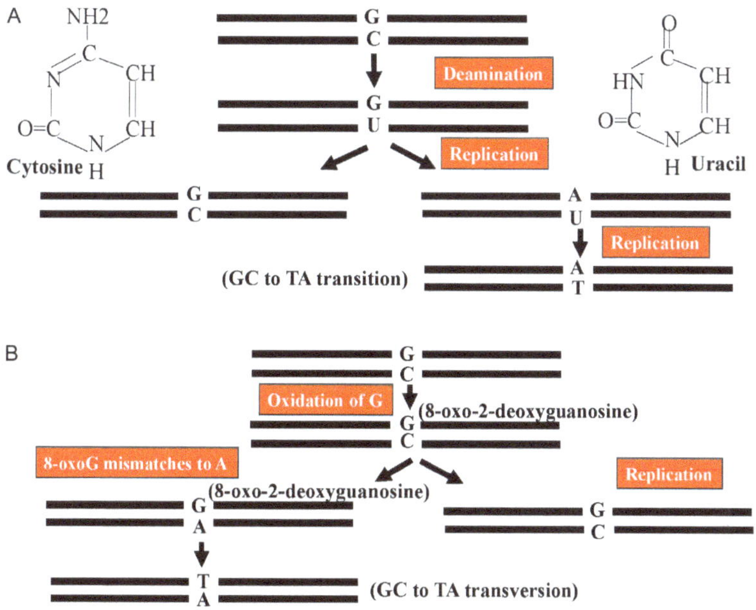

Fig. 1 Oxidation-mediated mtDNA mutation. (A) A G-C pair is converted to a T-A pair following the oxidative deamination of a cytosine nucleotide to a uracil nucleotide. (B) A G-C-pair is converted to a T-A pair following the oxidative conversion of a guanosine to 8-hydroxy-2-deoxyguanosine and then to 8-oxo-2-deoxyguanosine.

the total mtDNA copies increases and the percent heteroplasmy increases. Classically, it has been easier to detect somatic deletion mutations than it has been to detect somatic point mutations. Levels of some deletions, such as the ~5 kDa common deletion, appear to increase in the brains of aging humans.[46]

2.5 Could mtDNA Inheritance Affect Longevity?

It has been speculated that inherited mtDNA variations may influence aging and longevity. One prominent aging study, the Framingham Longevity Study, found that how many years either of an individual's parents lived correlated with how long that individual would live, but the mother's age at death correlated better with the age at death of the child.[47] One possible interpretation of this study is that a maternally inherited genetic factor, perhaps mtDNA, influences aging.

More specific mtDNA studies report particular mtDNA sequences also associate with life expectancy. An ND2 C5178A substitution is reportedly overrepresented in Japanese centenarians,[48] while an ATPase6 G9055A substitution is reportedly overrepresented in French and Irish centenarians.[49,50] Positive haplogroup association studies have also been published. MtDNA haplogroups are defined as patterns of nucleotide substitutions that tend to occur together within individuals, which have arisen over the course of humanity and become fixed in different populations at different frequencies over the course of human migration and history.[51,52] For example, the frequency of haplogroup J in Italian centenarians is reportedly higher than it is in the overall Italian population,[53] and the frequency of haplogroups U and J is reportedly higher in Finnish centenarians than it is in the overall Finnish population.[54]

2.6 Critical Questions About the Role Mitochondria Play in Aging

While there seems to be consensus that an organism's mitochondria and bioenergetic performance change over time, concerns about the place and role of mitochondria in aging are frequently raised. For the case of somatic mutations, it can be pointed out that correlation and association do not prove causation; it is possible that the observed accumulation of mtDNA mutations over life spans, including point mutations and deletions, represents a consequence of aging and is not actually driving aging. MtDNA POLG mutator mice experiments were done with the purpose of addressing this question,[22,23] but some have raised questions about how well this system mechanistically models actual human aging, as well as to how rigorously the phenotype changes observed in these mice truly reflect normal physiologic aging.[55,56] For those accepting the mtDNA POLG mutator mice as a good model of aging more nuanced questions have spurred debate, such as whether point mutations or deletions are primarily responsible for driving age-related changes in the mice.[57,58]

In particular, questions have been raised about whether the magnitude of functional mitochondrial changes seen in aging organisms in general, and in humans specifically, is predictably profound enough to induce physiologic changes. As a corollary to this, it has certainly been reported that compensatory mechanisms are initiated in the face of declining mitochondrial function and it must be considered how well these compensations mitigate the potential consequences of age-related changes in mitochondrial function and cell bioenergetics.[30]

Finally, animal experiments have been reported in which the pharmacologic or genetic induction of mitochondrial dysfunction actually associated with increased life span.[59] At the very least, this suggests that even if mitochondria are a major driver of human aging, the overall picture of how and why they drive aging may turn out to be quite complex.

3. MITOCHONDRIA AND ALZHEIMER'S DISEASE

3.1 Overview

During the 1980s fluorodeoxyglucose positron emission tomography (FDG PET) brain scans revealed cortical glucose utilization is reduced in AD subjects.[60,61] Decreased glucose utilization was featured in neuroanatomically discrete regions, including the posterior temporal and parietal cortices, as well as the posterior cingulate–precuneus region (Fig. 2). The underlying basis for this observation has to date remained uncertain. Proposed

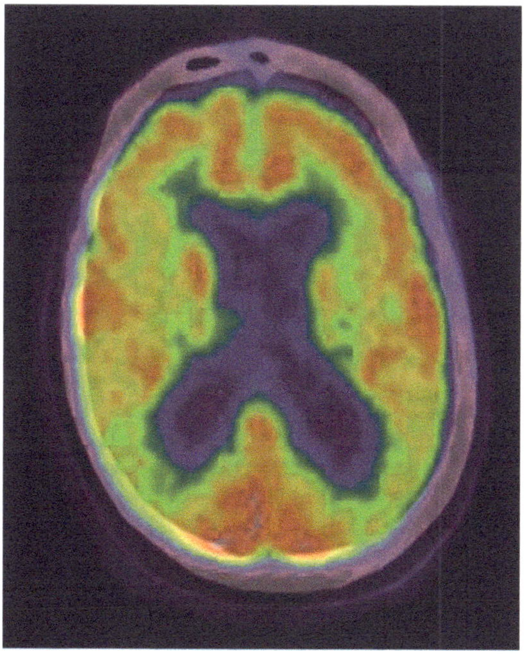

Fig. 2 Fluorodeoxyglucose positron emission tomography (FDG PET) scan from an AD patient. In the normal case the cortical region should show a consistent level of relatively high glucose uptake and, therefore, utilization. In this FDG PET scan from an individual with AD there is attenuation of the high cortical glucose uptake/utilization signal (indicated by a *red-orange color*) in the region of the posterior temporal/inferior parietal cortical regions (indicated by a *yellow-green color*).

possibilities have postulated these declines may reflect an artifact of neuron loss, or a loss of synaptic connectivity. However, when glucose utilization is studied in homogenized brain tissue, where synaptic connectivity is no longer maintained and the amount of material in the assay is standardized, reduced glucose utilization still remains.[62] This raises the possibility that reduced glucose utilization on FDG PET could also possibly reflect perturbed glycolysis flux.

Mitochondria are also different in AD patients than they are in age-matched control subjects. Overall mitochondrial size is reduced, although this is punctuated by the increased presence of overly swollen mitochondria with misshapen cristae.[27,63] Activities of certain mitochondria-localized enzymes are also reduced. This includes a reduction in the activity of the Krebs cycle enzyme α-ketoglutarate dehydrogenase complex and of pyruvate dehydrogenase complex,[64–66] which gates the entry of glycolysis-derived, pyruvate-based carbon into the Krebs cycle. Interestingly, the activity of some Krebs cycle enzymes, specifically the activity of enzymes in the second half of the cycle, has been reported to be increased in brains from AD subjects.[67]

COX activity also tends to be lower in AD subjects than it is in age-matched control subjects.[68] Interestingly, this COX activity reduction is not limited to the brain. In fact, it was first reported to be present in platelet mitochondria derived from AD subjects,[69–74] and only after that was it assayed in the brains of AD subjects, where its activity was similarly and consistently found to be reduced.[70,75–84] In addition to platelet and brain mitochondria, AD COX activity has also been reported in fibroblast cultures derived from sporadic AD subjects.[85]

Determinations of mitochondrial number in AD are to some extent complicated. For the most part, in AD brain the number of normal mitochondria per neuron appears to be reduced, although the amount of mitochondrial material that seems to be undergoing digestion within autophagosomes is increased.[27] Mitochondrial homeostasis is further perturbed in that there is an apparent shift toward increased mitochondrial fission, mitochondria are less likely to radiate from the perikaryon to neurite projections, and messenger RNA and protein levels of the transcriptional coactivator peroxisome proliferator-activated receptor γ coactivator 1α (PGC1α), which facilitates mitochondrial biogenesis, are reduced.[86–90]

By some parameters mtDNA in AD subjects also seems to differ from that of age-matched control subjects. The amount of intact mtDNA is reportedly reduced, even though the amount of autophagosome-localized mtDNA may be increased.[27,91–93] There appears to be increased numbers of some types of presumably somatic mutations, such as deletions and potentially also specific, presumably acquired point mutations.[27,92,94–98] Levels of mtDNA nucleotide oxidative damage are also higher in AD subjects than they are in control subjects.[99]

Some studies claim inherited mtDNA variations are found in higher or lower frequencies in AD subjects as compared to non-AD subjects. Association studies have variably reported particular mtDNA haplogroups are statistically over or underrepresented in AD cohorts and that particular mtDNA single-nucleotide polymorphisms are statistically over or underrepresented in AD cohorts.[100–113] Results across different mtDNA association studies in AD have not been consistent across studies, though, which has led to some concern about the reliability of individual reports.[114–118]

3.2 Could Aβ or APP Account for Differences in AD Mitochondria?

The plaques observed in AD subject brains contain to a large extent beta amyloid (Aβ) protein. A number of investigators have reported that Aβ localizes to mitochondria (in brains from human AD subjects and in amyloid precursor protein (APP) transgenic mice), where it can bind to and interfere with the function of different intramitochondrial proteins including cyclophilin D and the Aβ-binding alcohol dehydrogenase protein.[119–127] Aβ has also been shown in experimental systems to interfere with mitochondrial respiratory chain function, and to specifically inhibit COX activity.[122,128–131]

When added to neuronal NT2 teratocarcinoma cells, Aβ induces increased oxidative stress and cell death. However, when Aβ is added to NT2 teratocarcinoma cells that have been depleted of their endogenous mtDNA (ρ0 cells), that do not produce mtDNA-encoded respiratory chain subunits and are unable to successfully perform oxidative phosphorylation, these toxic Aβ effects are not observed.[132] This suggests Aβ toxicity in in vitro systems is at least to some extent mediated through its effects on cell respiration.

Aβ also impacts other aspects of mitochondrial homeostasis. It appears able to reduce mitochondrial movement,[133] and to shift the mitochondrial fission–fusion balance toward the fission end of the spectrum.[86,87,90]

The Aβ protein derives from a larger parent protein called the APP. An elegant series of experiments has demonstrated APP itself localizes to mitochondria.[134–137] APP in fact contains an N-terminal mitochondrial targeting sequence of strategically placed, positively charged amino acids that lead it to enter mitochondria through the translocase of the outer mitochondrial membrane (TOMM) and translocase of the inner mitochondrial membrane protein import apparatus.[134,136] However, APP also contains a peptide sequence that causes the process of mitochondrial import to prematurely arrest, leaving an intramitochondrial N-terminal end and a long extramitochondrial C-terminal end.[134] The presence of APP at the mitochondria appears to in general interfere with normal mitochondrial function, and to specifically reduce COX activity.[134]

Much of the data showing Aβ-mitochondria and APP-mitochondria physical associations derive from model systems and model organisms, such as transgenic mice that overexpress a mutant human APP transgene.[119–122,124,125,127] However, physical associations have also been demonstrated in brains from deceased AD subjects.[119,126,134] Still, it is not immediately clear how mitochondria-localized APP or Aβ might interfere with the function of mitochondria outside the brains of affected individuals.

3.3 Evidence of a Maternal Inheritance Contribution to AD

Epidemiologic studies suggest that although an individual's AD risk is determined by both parents, maternal influence seems to be more profound than paternal influence.[138–141] This has been shown, for example, by the study of Edland *et al.* who found that among AD probands who also had a demented parent, the demented parent was more than twice as likely to be the mother.[140] Importantly, this relationship was observed even when the age of parental dementia onset was relatively young. This finding, therefore, is unlikely to simply reflect an artifact caused by greater longevity in women vs men, a factor that needs to be considered since survival of mothers to older ages than fathers might also increase their chance of developing a dementing disorder such as AD.

Endophenotype studies also support the presence of an AD maternal inheritance bias. Endophenotypes are incomplete manifestations of a disease; they can be defined by disease-consistent characteristics that are insufficient by themselves to qualify one for a diagnosis, or by biomarkers. AD endophenotype studies have consistently shown that the nondemented adult children of AD mothers are more likely to have AD-like biomarker changes than the nondemented children of AD fathers. The first of these studies was that of Mosconi *et al.*, which reported that FDG PET scans from mostly middle-aged children of AD-affected mothers were more likely to manifest

AD-like glucose utilization patterns than the mostly middle-aged children of AD-affected fathers.[142] Similarly, an arterial spin labeling study found the middle-aged children of AD-affected mothers were more likely to show decreased perfusion patterns than the middle-aged children of AD-affected fathers.[143] Studies have also shown the children of AD-affected mothers tend to have more cerebral amyloid than the children of AD-affected fathers and are also more likely to have increased levels of cerebrospinal fluid oxidative stress markers.[144–147] Several studies have found children of AD mothers also tend to have greater amounts of neuroanatomically specific cerebral atrophy than do the children of AD fathers.[148–152] One study found platelet mitochondria COX activity was lower in the children of AD mothers than it was in the children of AD fathers.[153] Finally, it was reported in the Framingham Longevity study that middle-aged, nondemented APOE4 carriers who also had an AD-affected mother had lower scores on a memory test than did middle-aged, nondemented APOE4 carriers who also had an AD-affected father.[154]

3.4 Could APOE Influence AD Risk by Affecting Mitochondrial Function?

The APOE gene, located at locus 19q13.2, represents the most extensively studied sporadic AD genetic risk factor.[155–157] The APOE gene encodes a protein, apolipoprotein E, that plays a role in lipid and cholesterol transport.[158] There are three relatively common polymorphism-defined APOE alleles, the APOE2, APOE3, and APOE4 variants.[159] The APOE4 version is associated with an increased lifetime risk of developing AD.[155] Several hypotheses have been proposed in an attempt to mechanistically explain the association between APOE variants and AD risk. One hypothesis has arisen from the observation that the apolipoprotein E4 isoform folds differently than the other versions and that this folding difference causes it to be proteolyzed into a smaller peptide that displays a functional mitochondrial targeting sequence.[160,161] This apolipoprotein E4-derived peptide appears to have toxic effects on mitochondria, and in cell culture it even seems to reduce COX activity.[162] Human brains from young deceased APOE4 carriers were also found to have reduced cortical COX activity; this was observed despite an absence of concurrent Aβ accumulation.[163]

As was pointed out in Section 3.3, a cognitive-based endophenotype study of the Framingham Longevity cohort found that among nondemented, middle-aged APOE4 carriers individuals with an AD-affected mother had less robust performance on a test of memory than did subjects with an AD-affected father.[154] If an apolipoprotein E4 degradation product truly functions as a mitochondrial toxin, and inherited mtDNA features do turn out to influence AD risk, it could be the case that APOE4 and particular mtDNA sequences could interact to give rise to a "double" mitochondrial hit that creates a particularly elevated AD risk. Potentially consistent with this possibility are reports that certain mtDNA haplogroups appear to modify the APOE4-associated increase in AD risk.[106,164]

While a reasonably strong case has been made that APOE genotype influences AD risk, it is nevertheless important to point out that the translocase of the outer mitochondrial membrane 40 kDa (TOMM40) subunit gene sits immediately adjacent to the APOE gene.[165] TOMM40 polymorphic variants have also been reported to associate with AD risk,[166–174] although some of these TOMM40 variants are in linkage disequilibrium with the critical APOE isoform-defining variants.[175] This genetic confounding makes it difficult to prove or disprove whether the TOMM40 gene, through the TOMM40 protein, independently contributes to AD risk or contributes at least to some extent to the AD risk currently attributed in most circles to the APOE gene and apolipoprotein E protein.[176]

3.5 Evidence for a Somatic mtDNA Mutation Contribution to AD

Some have hypothesized an accumulation of somatic mtDNA mutations could play a major role in the development of AD, and perhaps even represent a primary driving cause.[21] Data addressing this possibility have been mixed, and studies are limited to some extent because prior to the introduction of next-generation sequencing approaches, it was technically difficult to resolve somatic single-nucleotide changes at the microheteroplasmy level.

Perhaps for this reason initial studies emphasized quantification of large deletions. One early study, that of Corral-Debrinski *et al.*, reported that brains from AD subjects who were less than 75 years of age had higher levels of the ~5 kb "common deletion" than brains from age-matched control subjects.[94] Hamblet *et al.* also found increased levels of the 5 kb common deletion in AD brain.[96] This finding was also essentially corroborated by Cottrell *et al.* and Krishnan *et al.*, who initially found using a histochemical approach that AD brains were more likely to show COX-perturbed neurons in specifically examined areas, and later reported COX-perturbed neurons in AD brains had higher amounts of various large scale mtDNA deletions.[177,178]

Chang *et al.* used a sequential PCR amplification and restriction enzyme digestion approach to screen for the presence of mtDNA point mutations in AD brains.[95] This study did report evidence of increased AD brain mtDNA point mutations, which were felt to likely represent a consequence of mtDNA nucleotide oxidative damage. Interestingly, though, increased levels of the 5 kb common deletion were not detected in the AD brains from this study.

In a study by Coskun *et al.*, the frequencies of several specific D-loop mutations were found to be profoundly increased in the brains of AD subjects.[92] This study further reported expression levels of ND6 were decreased and that the mtDNA to nuclear DNA ratio was reduced in the presence of these mtDNA control region mutations, which was interpreted as support for the view that these mutations interfered with mtDNA transcription and replication. It was also later reported by these investigators that the D-loop mutations that were found to be increased in AD subject brains were also increased in AD subject blood and lymphoblastoid cell line mtDNA.[179]

On the other hand, in their study of AD and control brains Lin *et al.* did not detect a quantitative difference in the burden of microheteroplasmic point mutations.[180] To perform this study, the authors used a clonal sequencing analysis approach of PCR-amplified COX1 amplicons; in addition to analyzing brains from AD and age-matched control subjects, brains from a younger control group were also evaluated. While the microheteroplasmic mutation burdens were comparable between the AD and control groups, this study nevertheless presented several interesting findings: (1) microheteroplasmic mutations were relatively frequent; (2) there was a positive correlation between advancing age and the number of mutations per subject, so that it did appear that the mutation burden did increase with age; (3) there was a negative correlation between advancing age and COX enzyme activity, so that it did appear that COX activity did decline with age; and (4) there was a negative correlation between COX enzyme activity and COX1 mutation burden, so that it did appear that as COX1 mutations accumulated, COX activity fell.

3.6 Evidence of a Mitochondrial Link to Classic AD Histopathology Changes

More than one-cell culture-based study has reported toxin-induced mitochondrial dysfunction, including toxin-mediated COX inhibition, reduces the processing of APP by the nonamyloidogenic α-secretase degradation pathway. This conclusion is based on the finding that cells treated with toxins such as sodium azide (a COX inhibitor) generate reduced levels of soluble APPα (sAPPα).[181,182] It has been further inferred by such studies that decreased α secretase-mediated processing of APP could reflect a shift in APP processing toward its amyloidogenic, β-secretase-mediated processing pathway.[183] The cell culture study of Gabuzda *et al.* to some extent supports this inference, as these authors found that sodium azide-treated cells produced higher levels of an 11 kDa APP cleavage product that was suspected to contain an intact Aβ peptide sequence.[183]

A number of studies utilizing APP transgenic mice more directly suggest pertinent connections between brain energy metabolism and AD histopathology do exist. In the study of Scheffler *et al.*, the investigators used a strategic strain interbreeding approach to create groups of APP transgenic mice that ultimately differed primarily in their mtDNA sequences.[184] It was found that groups of mice with different mtDNA sequences developed profoundly different amounts of amyloid plaques.

Two mouse studies found reducing the amount of COX holoenzyme actually reduced amyloid plaque deposition. For the first of these studies, mice with APP and presenilin 1 (PS1) mutations were crossed with mice engineered for a Cre-loxP-mediated knockout of the cytochrome oxidase 10 (COX10) gene.[185] The COX10 gene encodes a farnesyltransferase that is required for the synthesis of COX heme; eliminating this farnesyltransferase results in a dramatic reduction in COX holoenzyme production. COX activity accordingly declines, as do measureable markers of oxidative stress. In the second of these studies, APP/PS1 mutant mice were crossed with mice designed to express a mitochondria-targeted restriction enzyme that cleaves mtDNA.[186] This led to mtDNA depletion without generating evidence of oxidative stress, and a reduction in levels of the COX1 protein subunit. Increased levels of the APP-derived β-C-terminal fragment (β-CTF) were detected, which did suggest a potential change to APP processing, although no increase in BACE activity was observed, and similar to the findings of Fukui *et al.*, plaque accumulation was reduced.

Kukreja *et al.* evaluated a different mouse model with a predictably different type of respiratory chain defect.[187] The mice in this study were generated by breeding mice expressing a mutant human APP transgene with mice designed to express a dysfunctional mtDNA polymerase γ (PolgA D257A mice) and which accumulate mtDNA mutations at an accelerated rate. In this model of accelerated aging, plaque accumulation was also accelerated. Plaque accumulation differences between the Kukreja *et al.* study and the studies of Fukui *et al.* and Pinto *et al.* are not entirely clear, although it seems reasonable to consider differences in the nature of the induced mitochondrial defects may prove pertinent. One potential distinction between these studies is that the Fukui *et al.* and Pinto *et al.* mice seem to have decreased amounts of respiratory chain enzymes (or at least decreased amounts of COX or COX protein subunits), while the Kukreja *et al.* mice may have made functionally abnormal respiratory chain enzymes rather than less respiratory chain enzymes.

Another relevant mouse study is that of Dumont *et al.*[188] This study crossed transgenic mice that expressed a mutant APP with transgenic mice that overexpressed PGC1α. Contrary to what was perhaps initially expected, the bigenic mice showed increased amyloid plaque deposition. The bigenic mice also demonstrated evidence of perturbed mitochondrial function, as the activities of several mitochondrial-localized enzymes (complex I, succinate dehydrogenase, and citrate synthase) were diminished. Proteosome activity was also diminished in the bigenic mice, and this was felt to play a role in the amyloid deposition increase.

Other data from transgenic mouse studies could be considered potentially consistent with a possible bioenergetics–amyloidosis relationship. It has been shown that in APP transgenic mice, Aβ secretion into brain interstitial fluid is higher when the mice are awake and lower when they are asleep.[189] When APP transgenic mice are manipulated into a state of sleep deprivation, interstitial fluid levels are further elevated.[190] Increasing the whisker stimulation of APP transgenic mice increases, while decreasing the whisker stimulation of these mice decreases, interstitial fluid Aβ levels.[191] When Yamamoto *et al.* used optogenetic stimulation to induce chronic neuronal excitability within the hippocampal perforant pathways of APP transgenic mice, interstitial Aβ and Aβ plaque levels increased.[192] Taken together, studies such as these suggest synaptic activity increases Aβ production, and because synaptic activity creates a state of bioenergetic stress these data at least indirectly argue a link should exist between cell bioenergetics and APP processing/Aβ production.

In humans potential links between bioenergetics and APP processing/Aβ production are harder to establish, although a study of trauma victims did find that emergence from a coma state corresponded temporally with an increase in the brain's interstitial Aβ level.[193] This would seem to be consistent with mouse data that report synaptic activity correlates with Aβ production.

Other relevant human data may be inferred from the study of Vlassenko *et al.*, who reported that areas in which plaques initially present are parts of the brain's default mode network.[194] These regions show a unique bioenergetic pattern that features an increased reliance on aerobic glycolysis, defined by the authors as all glucose utilization that occurs in an adequately oxygenated tissue or adequately oxygenated cell that is not utilized in oxidative phosphorylation. Progressively lower amounts of glucose carbon released as CO_2 in this study were interpreted as being indicative of a progressively increased metabolism of glucose through aerobic glycolysis; non-oxidative phosphorylation uses of glucose include metabolism of glucose to lactate, incorporation into glycogen, a contribution of carbon to fatty acid or cholesterol synthesis, or the entry of glucose into the pentose phosphate shunt. This study stresses that in considering the potential relationship between bioenergetics and APP/Aβ, in addition to considering how much energy metabolism is present, what energy fluxes are present as well as how and why particular fluxes are occurring warrants consideration.

Relationships between neurofibrillary tangles and the tau protein they contain are also reported. Toxic perturbation of cell bioenergetics is certainly recognized to influence the activities of kinases that phosphorylate tau, and to increase tau phosphorylation. This has been demonstrated in both cell culture and animal-based experiments.[195–197] For example, administering the COX inhibitor sodium azide to rats increases tau phosphorylation,[196] as does exposing wild-type mice and mice that express a mutant tau transgene to annonacin, a complex I inhibitor.[198,199] Links between tau phosphorylation and metabolism are also suggested by a study that reports prolonged fasting in mice induces brain tau phosphorylation.[200]

A recent study by Zhao *et al.* demonstrated a potential link between mitochondria and the aggregation of tau into tangles.[201] In this study the authors evaluated the effects of a gene polymorphism in the myelin-associated oligodendrocyte basic protein (MOBP) gene that was previously associated with the risk of developing progressive supranuclear palsy (PSP), a neurodegenerative disease that features tangle accumulation.[202] MOBP is located relatively close to the gene that encodes a protein called appoptosin, a nuclear-encoded protein that localizes to the mitochondrial inner membrane and participates in heme synthesis. The authors found that the MOBP polymorphism influenced appoptosin expression, which increased in the presence of the PSP-associated MOBP polymorphism.[201] Higher amounts of appoptosin lead to increased heme production, which in turn lead to increased production of cytochrome *c*. This resulted in an increase in the amount of cytochrome *c* protein that leaked into the cytoplasm, which in turn activated caspase 3. Caspase 3 then cleaved tau protein at a caspase cleavage site, which generated a tau fragment that aggregated to form tangles and also induced synaptic dysfunction.

Mitochondrial uncoupling has also been demonstrated to induce tau paired helical filament formation.[203] Interestingly, fibroblast cultures prepared from sporadic AD subjects also show altered mitochondrial function[85] and are more likely to bind an antibody that recognizes paired helical filament tau than are fibroblast cultures prepared from non-AD subjects.[204]

4. AD CYTOPLASMIC HYBRID (CYBRID) STUDIES

4.1 Overview

The cytoplasmic hybrid (cybrid) technique makes it possible to transfer mtDNA from one cell to another, and to then perpetuate that transfer (Fig. 3).[205] In some ways it is similar to forming cell hybrids,[206] with an important distinction being that the resulting cell product contains nuclear DNA from only one source. Further, when mtDNA is transferred it is not transferred in its pure form. Rather, it is contained within the mitochondria from the donor source. Whole mitochondria, therefore, actually serve as a transfer vessel. Finally, mitochondrial transfer can be accomplished using different approaches that to some extent determine whether additional cell constituents are also transferred. For example, isolated mitochondria can be injected into the recipient cell, or donor and recipient cells can be mixed in the presence of a detergent that disrupts membrane integrity and allows for a more extensive mixing of cytosolic contents.[207] Ideally, though, how the mtDNA transfer is accomplished should ultimately lead to the same product because as the resultant cell undergoes subsequent growth and division nonperpetuating materials should degrade over time and should dilute over the course of repeated cell divisions. The only obvious transferred component that can replicate and therefore perpetuate is the mtDNA. Conceptually, it could be possible that a templating protein could also be transferred and perpetuate, such as a prion protein, but to date this has not been described in the cybrid literature.

Although simply introducing isolated mitochondria to cultured cells is accompanied by some degree of intracellular mitochondrial internalization, an event referred to as "transformation",[208] the first intentional transfer of mitochondria to recipient cells (in the mid-1970s) featured fusion of enucleated cytoplasts with nucleated cells. The scientific goal of these early studies (which introduced the cybrid term) was to test whether chloramphenicol

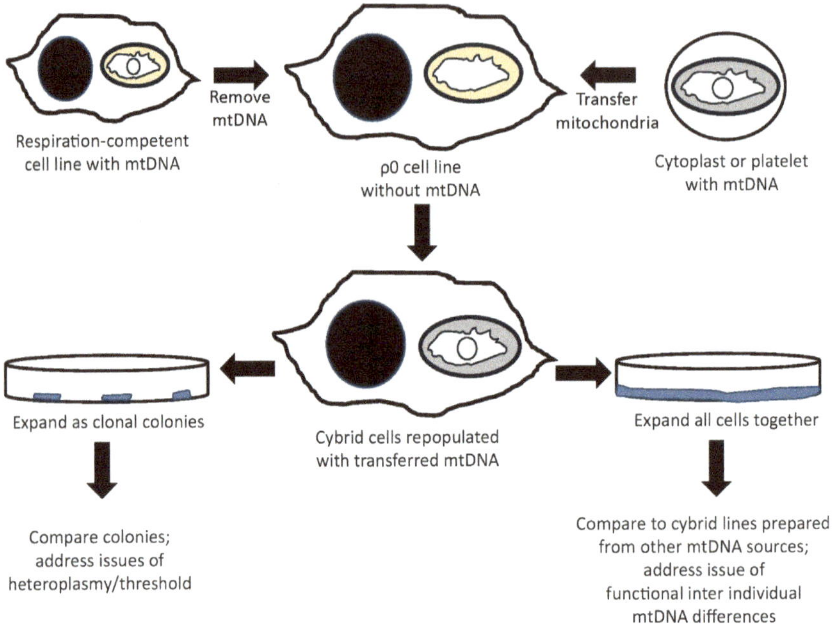

Fig. 3 The cybrid technique. A cell line's endogenous mtDNA is removed to create a ρ0 cell line, which lacks respiratory competence and must be maintained in medium supplemented with pyruvate and uridine. After mixing ρ0 cells with mitochondria-containing cytoplasts or platelets, and facilitating cytosolic mixing by addition of detergent, some ρ0 cells incorporate exogenous mitochondria and by extension their mtDNA. The transferred mtDNA allows for the restoration of respiratory competence, and the newly created cybrid cells can be selected for by removing pyruvate and uridine from the medium (leading to the removal of residual untransformed ρ0 cells). The cybrid cells that result from a single fusion can be grown as separate clonal colonies; in cases where the donor mtDNA carries a heteroplasmic mutation, the individual cybrid clonal lines can be analyzed to address issues of threshold. Alternatively, the cybrid cells that result from a single fusion can be expanded together, creating a single cybrid line that can be compared to other unique cybrid cell lines.

resistance, a characteristic of some cell lines, was an mtDNA-determined trait.[209,210] The investigators found that when mitochondria-containing cytoplasts from a chloramphenicol-resistant cell line were mixed and fused in culture with a chloramphenicol-sensitive cell line, some of the chloramphenicol-sensitive cells acquired chloramphenicol resistance. This lead the investigators to conclude that chloramphenicol resistance was indeed an mtDNA-determined trait.

It is important to note that as a result of this approach, the resulting cybrid cells, at least initially, were presumably heteroplasmic as they should have contained the mtDNA that was endogenous to the accepting cell line, as well as the mtDNA from the donor cytoplast mitochondria. To refine the technique, King and Attardi subsequently developed the idea of using a ρ0 cell line as the accepting cell line.[207] ρ0 cells are cells that have undergone depletion of all detectable mtDNA. The development of ρ0 cell lines, in turn, followed the efforts of several groups to mimic in cultured cell lines the previously observed ability of yeast cells to deplete their mtDNA content under conditions that favored glycolysis.[211–213] These mtDNA-depleting yeast cells were called ρ petites, since prior to its identification as mtDNA cytosolic DNA was initially referred to as ρ DNA.[214] By using ρ0 cells as the accepting cell line, investigators gained the ability to create cell lines that contained only mtDNA from the mitochondrial donor.

Moving forward using ρ0 cell lines as the recipient cells, investigators began to study issues of heteroplasmy, threshold, and in general the biochemical consequences of known mtDNA mutations.[215–220] Mitochondria from human subjects with known homoplasmic or heteroplasmic mtDNA mutations were transferred to ρ0 cells. The resulting cybrid cells were expanded in culture. In instances where heteroplasmic mutations were transferred, the expanding cybrid cells were isolated in order to facilitate the creation of cybrid clones, which ultimately could be shown to contain different ratios of mutant to wild-type mtDNA. These clones with different heteroplasmic ratios were then analyzed biochemically to determine how much of a mutational burden was required for a particular mutation to cause a change in biochemical function, and thereby estimate the percent of mutation that had to be present to reach a phenotypic threshold.

Interest in the cybrid approach to address mtDNA-related questions further developed as more ρ0 cell lines were created, and after it was shown that platelets could serve as mitochondrial donor cells.[219,221] Platelets, which derive from megakaryocytes, lack nuclei and are easily accessed through routine phlebotomy. Through a simple procedure platelet-rich plasma can first be generated from a blood sample, and an enriched platelet fraction can then be prepared through centrifugation of the plasma. The enriched platelet fraction can then be mixed with the ρ0 line of choice to generate cybrid cells.

In the mid-1990s the cybrid approach was first used for a somewhat novel application that involved the utilization of mtDNAs whose sequences were unknown.[222] It was reasoned that biochemical differences between cybrid cell lines prepared from different mitochondrial donors could be used to infer differences existed in their mtDNA sequences. From the perspective of whether mtDNA sequences do indeed vary between individuals applying such an approach did not carry much risk; it was already recognized that mtDNA sequences from individuals that did not derive from a common maternal lineage typically deviate at multiple nucleotide positions. Whether bona fide biochemical differences would be too subtle to detect and produce false-negative results, though, was a concern. A second concern was that variation in biochemical measures could lead to false-positive results or incorrect conclusions about the association of particular mtDNA genomes with a particular characteristic. Regarding these two points it was anticipated that by creating large enough groups of cybrid cell lines from large enough groups of mitochondrial donors with a particular biochemical characteristic, one could accurately ascertain whether the specific biochemical characteristic within that group was influenced by the mitochondrial genome.[205]

At the time this strategy was defined it was felt to also offer particular experimental strengths. In the mid-1990s DNA sequencing was a far more expensive endeavor than it currently is, and the sequencing approaches of that time were not able to reliably detect low-abundance heteroplasmic deviations. Plus, given the high degree of mtDNA polymorphic variation that exists between individuals, unless a particular "smoking gun" sequence mutation that reliably segregated with a cohort was identified, without functional data it would prove difficult to associate individual sequence deviations with a specific group. Doing so would presumably require analyzing very large numbers of individuals.

The first time this approach was utilized was in studies of cybrid cell lines prepared from a group of platelet mitochondria/mtDNA donors with Parkinson's disease (PD).[222] As a group, PD patients were already recognized to have platelet mitochondria complex I activities that were lower than those measured in age-matched control subjects.[223–225] Twenty-four cybrid lines were generated from PD subjects, and 28 cybrid lines were generated from 28 age-matched control subjects. Platelets served as the mitochondria/mtDNA donor source. A human neuroblastoma SH-SY5Y ρ0 cell line that had previously been derived from the standard SH-SY5Y line served as the mitochondria/mtDNA acceptor.[221] The mean complex I activity was found to be ~20% lower in the PD subject-derived cybrid group (simply referred to as the "PD cybrid" group) than it was in the control subject-derived cybrid group (simply referred to as the "control cybrid" group). It was concluded that mtDNA, at least to some extent, contributes to reduced complex I activity in persons with PD.

4.2 AD Cybrid Experiments

The cybrid approach was deemed reasonable to address the specific question of why individuals with AD on average have a lower platelet mitochondria COX activity than age-matched, non-AD subjects.[69–74] Nongenetic explanations included the presence in the circulation of a factor that inhibits COX activity. Because COX contains 13 subunits, 10 of which are encoded by nuclear genes and 3 of which are encoded by mitochondrial genes, genetic explanations could alternatively implicate a nuclear DNA or mtDNA-dependent component. It was a priori hypothesized that mtDNA genes were more likely to contribute to lower COX activity in AD subjects than nuclear DNA genes, since late-onset AD (LOAD) rarely demonstrates recognizable Mendelian inheritance.[226] LOAD is generally considered to show sporadic epidemiology although nevertheless with a genetic influence, and in many ways the unique genetic rules of mtDNA, including heteroplasmy, threshold, mitotic segregation, and maternal inheritance uniquely position it to play a role in otherwise apparent sporadic diseases that also demonstrate altered mitochondrial function.[226]

In terms of applying the cybrid technique to address this question, it was reasoned that if lower mean COX activities were caused by a circulating inhibitory factor, that factor would wash out over the course of expanding the cell lines generated using platelets obtained from AD subjects (herein referred to as "AD cybrids"). Presumably, low AD subject platelet mitochondria COX activity under this scenario would not perpetuate in culture as AD cybrid line COX activities would increase in culture to match that of the cybrid lines generated from platelets obtained from age-matched control subjects (herein referred to as "control cybrids"). It was further reasoned that a nuclear DNA-dependent feature would be unlikely to account for a relative reduction in the AD cybrid COX activity because nuclei are not routinely transferred during the procedure, or if such a transfer did occur the transferred nuclei would be unlikely to perpetuate. Similar to toxin-induced activity reductions, nuclear DNA-dependent reductions in COX activity would predictably wash out after the transfer and selection process was completed.

The first published AD cybrid study was the one by Davis *et al.*[227] This study compared COX data from a group of 20 AD cybrid cell lines to a group of 45 control cybrid cell lines. Transferred mtDNA derived from subject platelet mitochondria, and the acceptor cell line was the SH-SY5Y ρ0 line. COX activity was ~20% lower in the AD cybrid group than it was in the control cybrid group. It is relevant to note that the Davis *et al.*

report was subsequently retracted, although the reasons for the retraction were unrelated to the cybrid data that were presented.[228]

Later in 1997 two other AD cybrid studies were reported. In the study of Sheehan *et al.*, platelets served as the mitochondria/mtDNA donor source, and the acceptor cell line was the SH-SY5Y ρ0 line.[229] An ~50% lower COX activity in the AD cybrids was seen. In the other study, that of Swerdlow *et al.*, platelets served as the mitochondria/mtDNA donor source and the acceptor cell line was an NT2 teratocarcinoma-derived ρ0 line.[230] Fifteen AD cybrid lines were compared to 9 control cybrid lines, and a relative 16% reduction in the AD cybrid group COX activity was observed.

Other studies of unique AD cybrid series have focused in particular on COX activity. In the study of Cardoso *et al.*, the authors used platelet mitochondria to generate AD and control cybrid lines on an NT2 ρ0 nuclear background and found that COX activity in the AD cybrid cell line ($n=6$) group was 22% lower than it was in the control cybrid cell line ($n=5$) group.[231] In the study of Silva *et al.*, the authors used platelet mitochondria to generate AD and control cybrid lines on an SH-SY5Y ρ0 nuclear background and found that COX activity in the AD cybrid cell line ($n=8$) group was ~30% lower than it was in the control cybrid cell line ($n=7$) group.[232] On the other hand, the study of Ito *et al.* also used COX activity as a primary endpoint and found COX activity was comparable between the AD and control cybrid groups.[233] However, there are a number of notable methodologic differences between the Ito *et al.* study and the positive studies thus far mentioned. The Ito *et al.* group used a HeLa cell ρ0 cell line to generate their cybrids, and the mitochondria/mtDNA donor source was mixed; four AD cybrid lines were prepared from platelet mitochondria, three control cybrid lines were prepared from platelet mitochondria, and two control cybrid lines were prepared from fibroblast mitochondria. Also included in the analysis were what were designated as an additional three AD cybrid lines, which were generated by mixing HeLa ρ0 cells with synaptosomes prepared from a brain that was acquired from a deceased AD subject after a 20-h postmortem interval. The authors reported they were able to identify three cell colonies from this fusion that contained mtDNA, and COX activity data ascertained from each of these three colonies were individually included in the analysis. Due to these substantial methodologic differences, it is arguably difficult to conclude that the Ito *et al.* negative study contradicts the positive studies.

A number of AD cybrid studies have evaluated various other aspects of mitochondrial function as well as parameters influenced by mitochondrial function.[227,229–232,234–252] In many cases changes are reported that recapitulate changes that are seen in the brains of AD subjects themselves.[68] Relative to control cybrid cell lines, in AD cybrid cell lines oxidative stress markers are increased,[229–231,234,236,240,241,245,248,249] inflammatory and stress signaling pathways are activated,[234,236,239–241,245,250] Aβ levels are increased,[238,241] glucose utilization is decreased,[232] oxygen consumption is decreased,[232] there is a shift toward mitochondrial fission and a smaller average mitochondrial size,[243,247] numbers of ultrastructurally perturbed mitochondrial are increased,[243,252] PGC1α mRNA levels are reduced,[232] HIF1α protein is reduced,[232] mTOR protein is reduced,[232] SIRT1 protein is reduced,[232] and apoptotic markers are increased.[231,238–241,245]

AD cybrids have also been used to model aspects of AD-specific, mitochondria-related function that are difficult or impossible to study in autopsy brain tissue.[68] For example, mitochondrial membrane potential analyses of AD cybrid lines show a relative degree of depolarization,[235,238,242,252] and AD cybrid mitochondria appear to internalize less calcium and are less able to buffer calcium-mediated intracellular signaling activity than control cybrid lines.[229] AD cybrid ATP levels are reduced.[231,232] It has been shown using differentiated AD cybrid cell lines that mitochondrial movement is relatively reduced.[244] AD cybrid cells are more sensitive to Aβ toxicity than are control cybrid cells.[231,250] AD cybrids have also been used to screen the molecular effects of potential therapeutic interventions; pharmacologic inhibition of mitochondrial fission activity and antioxidants has been shown to benefit certain mitochondria-related functional parameters.[241,245,247,248]

Three studies, one performed using an NT2 ρ0 cell background and two performed using an SH-SY5Y ρ0 cell background, have reported mitochondria-relevant functional changes (including a reduction in COX activity) between cybrid lines generated from human subjects diagnosed with mild cognitive impairment (MCI; a frequent AD precursor state) and cybrids generated from age-matched control subjects.[232,246,248] To date, over 20 cybrid studies have been published that report at least one biochemical or molecular parameter that differs between groups of AD/MCI and control cybrid cell lines.[227,229–232,234–252] Most of these studies have in fact reported multiple divergent parameters. The only categorically negative AD cybrid study was that of Ito *et al.*,[233] which evaluated just one biochemical parameter (COX activity), and which is notable for a variety of distinct methodologic differences that may have caused that study to differ from the other positive studies.

4.3 Implications and Limitations of AD Cybrid Studies

Data from AD cybrid studies support three important assumptions. First, they argue mtDNA accounts for, or at least to some extent contributes to, differences in mitochondrial function that reportedly exist between AD and non-AD subjects. Second, they argue that mitochondrial function can contribute to hallmark extramitochondrial histopathology changes, such as increases in oxidative stress markers[253] and Aβ production. Third, since cybrid cell lines in these studies have been almost exclusively generated through transfer of platelet mitochondria/mtDNA, the AD cybrid literature argues that at least at a molecular or biochemical level, AD changes are not strictly limited to the brain.

Cybrid studies also have limitations.[205,254] Because mitochondria/mtDNA is transferred from platelets, it is possible that what drives mitochondrial dysfunction in AD cybrids is different from what drives mitochondrial dysfunction in AD brains. Given that the nature of mitochondrial dysfunction and its consequences seem to recapitulate so many biochemical, molecular, and physiologic features observed in AD brains, though,[68] it would seem a stretch to propose that mitochondrial dysfunction in AD cybrids is entirely unrelated to mitochondrial dysfunction in AD brain.

The acceptor ρ0 cell lines are derived from tumor cell lines, which limits their ability to rigorously model the characteristics of a human brain. Therefore, when using cybrid cell lines to model AD mitochondrial functional defects, it is probably best not to extrapolate interpretations too far beyond the level of fundamental observation.

One general criticism of the cybrid approach is that although the approach was initially adapted to address questions about the contribution of mtDNA to mitochondrial function in AD, and subsequently used to model aspects of AD subject-specific mitochondrial function, cytosolic components other than mtDNA are transferred during the procedure. This is relevant because any transferred perpetuating component could theoretically lead to sustained changes in mitochondrial function. To date, though, no such component other than mtDNA has been identified. Also, since whole mitochondria are transferred from platelets to acceptor cells, and platelet mitochondria in AD show unique biochemical characteristics, the possibility that a preexisting mitochondrial defect simply did not have adequate time to wash out requires consideration. This possibility was empirically addressed in one of the early AD cybrid studies, which found that the 6-week selection and expansion period later used in most of the AD cybrid studies appears to provide an adequate wash-out period.[230] Moreover, one study reported that with extended time in active culture, a number of altered mitochondrial functional parameters appeared to become more profound.[252] This occurred despite an apparent activation of compensatory adaptations.

While cybrid studies implicate an mtDNA contribution to AD-associated mitochondrial dysfunction, they do not indicate what specific mtDNA features, characteristics, or sequences contribute to that dysfunction. They could reflect an accumulation of somatic mutations; to date data pertinent to the question of whether putative somatic mutations accumulate in circulating blood cells is mixed. One study did not find an obvious age-related accumulation of large deletions in a general population sample,[255] which suggests large mtDNA deletions do not commonly accumulate in circulating cells as they do in the brain. Data addressing the question of whether somatic point mutations accumulate in AD subject blood cells do, however, suggest this may indeed occur.[97,98,179]

Alternatively, it is possible that inherited mtDNA sequence differences may partly or exclusively contribute to AD cybrid line differences in mitochondrial function. Currently, the nature of such potential inherited mtDNA changes, if they in fact exist, is not clear. MtDNA sequence studies performed on AD thus far have not identified a single mutation or variant that distinguishes AD affected from AD-unaffected individuals, although various studies have claimed common inherited mtDNA signatures, such as those defined by the different mtDNA haplogroups, influence AD risk.[113] No mtDNA study to date has definitively addressed the question of whether rare mtDNA mutations or sequence variants associate with AD; such studies would require extensive mtDNA sequence data from large numbers of AD and control subjects.

Finally, a considerable degree of polymorphic variation has been demonstrated to exist within human nuclear respiratory chain subunits.[256] Because the cybrid approach neutralizes the potential functional contributions of a subject's nuclear DNA makeup, valuable information regarding the interplay between that individual's nuclear and mitochondrial genomes is likely to be lost. Substituting the nuclear background of a ρ0 cell for the nuclear background of the mtDNA donor could theoretically magnify or mitigate a particular mtDNA sequence-associated functional consequence.

4.4 The Mitochondrial Cascade Hypothesis

Despite the aforementioned limitations of AD cybrid studies, it is tempting to try and integrate AD cybrid data with data generated from AD tissue, epidemiology, endophenotype, and genetic studies in order to define an overarching hypothesis of AD etiology. Because advancing age is the single greatest sporadic AD risk factor,[8] it is recommended that any resulting hypothesis take into account applicable data and conceptual constructs generated through primary studies of human and animal aging.

A number of investigators have proposed mitochondrial dysfunction could play an important role in AD.[21,27,62,64,69,257–268] Some have proposed mtDNA inheritance or an age-related somatic accumulation of mtDNA mutation could contribute.[21,226] When these conceptual constructs are considered in conjunction with the observation that specific differences in non-brain mitochondrial function seem to exist between AD and non-AD subjects and that differences in mitochondrial function and cell bioenergetics can influence AD histopathology changes, the case can be made that perhaps mitochondria and bioenergetic function play an upstream if not primary role in AD.

The "mitochondrial cascade hypothesis" represents one attempt to comprehensively synthesize data discussed throughout this chapter into a comprehensive hypothesis that tries to account for the development and progression of AD, as well as the relationship of AD to brain aging (Fig. 4).[269–272] The premise is that individuals inherit a baseline level of mitochondrial function and durability; this baseline is determined by genetic contributions from both parents, although the mother makes a greater contribution and has a greater influence because she contributes the mitochondrial genome.[273] Then, as the individual ages certain tissues, and especially the brain, either accumulate somatic mutations or else drift toward increased levels of inherited microheteroplasmic mutations. The rate at which mutations accumulate would presumably be influenced by the individual's genetic makeup, and also by lifestyle factors. These age-related changes would result in declining mitochondrial function, which up to a point could probably be accommodated and addressed through adaptive molecular changes, and thereby define a period of compensated aging. Eventually, though, a threshold of mitochondrial/bioenergetic dysfunction could be reached in which accommodation and compensation are no longer adequate, thus introducing a period of uncompensated brain aging. At this point, an AD phenotype would begin to emerge.

The mitochondrial cascade hypothesis further presumes the classic histologic features of AD, Aβ plaque and tau neurofibrillary tangle deposition, are downstream consequences of changing mitochondrial and bioenergetic function. In its original form the hypothesis speculated Aβ plaques accumulated only after the state of uncompensated brain aging was reached.[271]

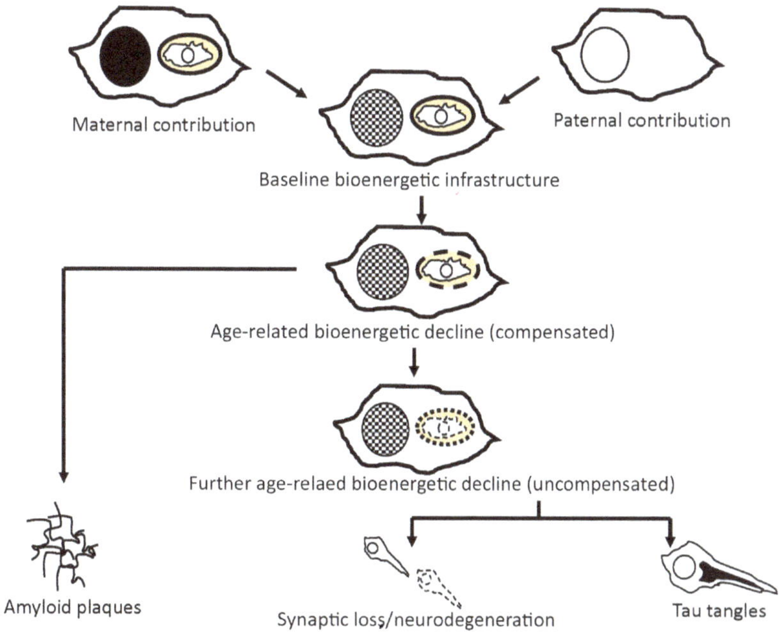

Fig. 4 The AD mitochondrial cascade hypothesis. Inheritance from both parents determines an individual's bioenergetic set-point and durability, with the mother having the greater input due to her contribution of the mtDNA. Over time mitochondrial efficiency declines, likely due to accumulating damage to the mtDNA. At relatively low levels, it is possible to compensate for this change (compensated brain aging), although the compensatory process may itself have consequences. More profound declines in mitochondria function, which may occur as further damage accumulates, can lead to a stage of uncompensated brain aging, which associates with other consequences as well as symptomatic AD.

Recent neuroimaging data from a human cohort longitudinal study now shows, though, that most Aβ plaque deposition occurs during the run-up to the AD clinical syndrome and dramatically slows during the symptomatic stages.[274,275] In conjunction with experimental studies that show Aβ is generated as a by-product of synaptic activity,[191,192] this suggests the possibility that Aβ plaque deposition may primarily represent a by-product of the compensatory changes that accompany age-related mitochondrial functional declines. If correct, this modification to the mitochondrial cascade hypothesis could potentially account for why a substantial percentage of older adults develop Aβ plaques but do not develop the clinical disease[8]; such individuals may avoid making the transition from bioenergetically compensated to uncompensated brain aging.

5. CONCLUSIONS

Decades ago the aging field began to specifically postulate mitochondria and bioenergetic function contributed to aging.[5,19–21] This was originally predicated on correlative and descriptive data. More recent experimental data have emerged, though, that are consistent with this possibility.[22,23]

Over an almost five-decade period it has become increasingly clear that bioenergetic and mitochondrial structural and functional changes also occur in AD.[21,27,62,64,69,257–268] In many cases changes observed in AD are reminiscent of those seen in aging, and in some ways differ primarily in their magnitude.[68] While mitochondrial and bioenergetic changes in AD were initially felt to represent a consequence of the disease, their potential relevance to disease progression has increasingly been considered, and the view that such changes represent valid therapeutic targets has emerged.[276–279]

When considering the hierarchy of biochemical, molecular, and physiologic events that result in AD, some have pointed out that in AD subjects bioenergetic and mitochondrial differences are found outside the brain and that changes in bioenergetic and mitochondrial function can alter how cells and tissues handle other AD phenomena, including how APP is processed to Aβ and Aβ plaque deposition.[9,280] Additional data pertinent to these points have been reported from studies of cybrid cell lines generated through the transfer of AD subject platelet mitochondria/mtDNA to ρ0 cell lines; results from these studies are consistent with the view that mtDNA contributes at least in part to AD mitochondrial and bioenergetic changes and that these changes can drive or at least contribute to a variety of biochemical, molecular, and histologic phenomena observed in AD subject brains.[68]

Synthesizing a spectrum of data from the aging, AD, and cybrid literature supports a conceptual construct that places mitochondrial function and bioenergetics at the apex of AD-associated molecular changes.[269–272] MtDNA would to some extent influence relevant mitochondrial and bioenergetic functional parameters. These molecular changes would similarly play out during the basic process of aging, and in some cases differences observed in both aging and AD would differ mostly by degree, with deficits being more prominent when clinical AD is present. Under this scenario some of the key histologic changes we now associate with AD, such as processing of APP to Aβ and Aβ plaque deposition, would represent downstream consequences of altered mitochondrial and bioenergetic function. Some of these histologic changes may arise during stages where declining brain mitochondrial and bioenergetic function could still be accommodated and compensated for, or could arise after declining brain mitochondrial and bioenergetic function have surpassed a critical level at which point successful compensation is no longer possible. In the brain, classic AD histology changes initiated by mitochondrial and bioenergetic dysfunction could in turn exacerbate failing mitochondrial and bioenergetic function. This mitochondrial cascade hypothesis makes testable predictions and suggests particular therapeutic strategies may be worth pursuing. It will be interesting to see how well the mitochondrial cascade hypothesis absorbs new current and future data generated by the AD research field.

ACKNOWLEDGMENT

Supported in part by the University of Kansas Alzheimer's Disease Center (P30 AG035982).

REFERENCES

1. Lewis MR, Lewis WH. Mitochondria in tissue culture. *Science*. 1914;39:330–333.
2. Nass MM, Nass S. Intramitochondrial fibers with DNA characteristics. I. Fixation and electron staining reactions. *J Cell Biol*. 1963;19:593–611.
3. Nass S, Nass MM. Intramitochondrial fibers with DNA characteristics. II. Enzymatic and other hydrolytic treatments. *J Cell Biol*. 1963;19:613–629.
4. Mitchell P, Moyle J. Chemiosmotic hypothesis of oxidative phosphorylation. *Nature*. 1967;213:137–139.
5. Harman D. The biologic clock: the mitochondria? *J Am Geriatr Soc*. 1972;20:145–147.
6. Lopez-Otin C, Blasco MA, Partridge L, Serrano M, Kroemer G. The hallmarks of aging. *Cell*. 2013;153:1194–1217.
7. Muller FL, Lustgarten MS, Jang Y, Richardson A, Van Remmen H. Trends in oxidative aging theories. *Free Radic Biol Med*. 2007;43:477–503.
8. Swerdlow RH. Is aging part of Alzheimer's disease, or is Alzheimer's disease part of aging? *Neurobiol Aging*. 2007;28:1465–1480.
9. Swerdlow RH, Kish SJ. Mitochondria in Alzheimer's disease. *Int Rev Neurobiol*. 2002;53:341–385.
10. Swerdlow RH. The neurodegenerative mitochondriopathies. *J Alzheimers Dis*. 2009;17:737–751.
11. Osborne TB, Mendel LB, Ferry EL. The effect of retardation of growth upon the breeding period and duration of life of rats. *Science*. 1917;45:294–295.
12. McCay CM, Crowell MF, Maynard LA. The effect of retarded growth upon the length of life and upon the ultimate body size. *J Nutr*. 1935;10:63–79.
13. McCay CM, Maynard LA, Sperling G, Barnes LL. Retarded growth, lifespan, ultimate body size, and age changes in the albino rat after feeding diets restricted in calories. *J Nutr*. 1937;18:1–13.
14. Carlson AJ, Hoelzel F. Apparent prolongation of the life span of rats by intermittent fasting. *J Nutr*. 1946;31:363–375.
15. Masoro EJ. Subfield history: caloric restriction, slowing aging, and extending life. *Sci Aging Knowl Environ*. 2003;2003:Re2.
16. Rubner M. *Das Problem der Lebensdauer und seine Bezieung zu Wachstum und Ernehrung [The problem of longevity and its relation to growth and nutrition]*: 1908. Oldenburg, Munich.

17. Pearl R. *The Rate of Living, Being an Account of Some Experimental Studies on the Biology of Life Duration*. New York: Alfred A. Knopf; 1928.
18. Harman D. Aging: a theory based on free radical and radiation chemistry. *J Gerontol.* 1956;11:298–300.
19. Miquel J, Economos AC, Fleming J, Johnson Jr JE. Mitochondrial role in cell aging. *Exp Gerontol.* 1980;15:575–591.
20. Linnane AW, Marzuki S, Ozawa T, Tanaka M. Mitochondrial DNA mutations as an important contributor to ageing and degenerative diseases. *Lancet.* 1989;1:642–645.
21. Wallace DC. Mitochondrial genetics: a paradigm for aging and degenerative diseases? *Science.* 1992;256:628–632.
22. Trifunovic A, Wredenberg A, Falkenberg M, et al. Premature ageing in mice expressing defective mitochondrial DNA polymerase. *Nature.* 2004;429:417–423.
23. Kujoth GC, Hiona A, Pugh TD, et al. Mitochondrial DNA mutations, oxidative stress, and apoptosis in mammalian aging. *Science.* 2005;309:481–484.
24. Navarro A, Boveris A. The mitochondrial energy transduction system and the aging process. *Am J Physiol Cell Physiol.* 2007;292:C670–C686.
25. Boveris A, Navarro A. Brain mitochondrial dysfunction in aging. *IUBMB Life.* 2008;60:308–314.
26. Barrientos A, Casademont J, Cardellach F, Estivill X, Urbano-Marquez A, Nunes V. Reduced steady-state levels of mitochondrial RNA and increased mitochondrial DNA amount in human brain with aging. *Brain Res Mol Brain Res.* 1997;52:284–289.
27. Hirai K, Aliev G, Nunomura A, et al. Mitochondrial abnormalities in Alzheimer's disease. *J Neurosci.* 2001;21:3017–3023.
28. Stauch KL, Purnell PR, Fox HS. Aging synaptic mitochondria exhibit dynamic proteomic changes while maintaining bioenergetic function. *Aging.* 2014;6:320–334.
29. Green DR, Galluzzi L, Kroemer G. Mitochondria and the autophagy-inflammation-cell death axis in organismal aging. *Science.* 2011;333:1109–1112.
30. Stauch KL, Purnell PR, Villeneuve LM, Fox HS. Proteomic analysis and functional characterization of mouse brain mitochondria during aging reveal alterations in energy metabolism. *Proteomics.* 2015;15:1574–1586.
31. Poon HF, Vaishnav RA, Getchell TV, Getchell ML, Butterfield DA. Quantitative proteomics analysis of differential protein expression and oxidative modification of specific proteins in the brains of old mice. *Neurobiol Aging.* 2006;27:1010–1019.
32. Finkel T. Oxidant signals and oxidative stress. *Curr Opin Cell Biol.* 2003;15:247–254.
33. Finkel T. Reactive oxygen species and signal transduction. *IUBMB Life.* 2001;52:3–6.
34. Haynes CM, Fiorese CJ, Lin YF. Evaluating and responding to mitochondrial dysfunction: the mitochondrial unfolded-protein response and beyond. *Trends Cell Biol.* 2013;23:311–318.
35. Kotiadis VN, Duchen MR, Osellame LD. Mitochondrial quality control and communications with the nucleus are important in maintaining mitochondrial function and cell health. *Biochim Biophys Acta.* 2014;1840:1254–1265.
36. Schulz TJ, Zarse K, Voigt A, Urban N, Birringer M, Ristow M. Glucose restriction extends Caenorhabditis elegans life span by inducing mitochondrial respiration and increasing oxidative stress. *Cell Metab.* 2007;6:280–293.
37. Gomez-Cabrera MC, Domenech E, Romagnoli M, et al. Oral administration of vitamin C decreases muscle mitochondrial biogenesis and hampers training-induced adaptations in endurance performance. *Am J Clin Nutr.* 2008;87:142–149.
38. Ristow M, Zarse K, Oberbach A, et al. Antioxidants prevent health-promoting effects of physical exercise in humans. *Proc Natl Acad Sci USA.* 2009;106:8665–8670.
39. Payne BA, Wilson IJ, Yu-Wai-Man P, et al. Universal heteroplasmy of human mitochondrial DNA. *Hum Mol Genet.* 2013;22:384–390.

40. Ye K, Lu J, Ma F, Keinan A, Gu Z. Extensive pathogenicity of mitochondrial heteroplasmy in healthy human individuals. *Proc Natl Acad Sci USA*. 2014;111:10654–10659.
41. Mecocci P, MacGarvey U, Kaufman AE, et al. Oxidative damage to mitochondrial DNA shows marked age-dependent increases in human brain. *Ann Neurol*. 1993;34:609–616.
42. Bohr VA, Dianov GL. Oxidative DNA damage processing in nuclear and mitochondrial DNA. *Biochimie*. 1999;81:155–160.
43. Hudson EK, Hogue BA, Souza-Pinto NC, et al. Age-associated change in mitochondrial DNA damage. *Free Radic Res*. 1998;29:573–579.
44. Richter C. Reactive oxygen and DNA damage in mitochondria. *Mutat Res*. 1992;275:249–255.
45. Nie B, Gan W, Shi F, et al. Age-dependent accumulation of 8-oxoguanine in the DNA and RNA in various rat tissues. *Oxid Med Cell Longev*. 2013;2013:303181.
46. Corral-Debrinski M, Horton T, Lott MT, Shoffner JM, Beal MF, Wallace DC. Mitochondrial DNA deletions in human brain: regional variability and increase with advanced age. *Nat Genet*. 1992;2:324–329.
47. Brand FN, Kiely DK, Kannel WB, Myers RH. Family patterns of coronary heart disease mortality: the Framingham Longevity Study. *J Clin Epidemiol*. 1992;45:169–174.
48. Tanaka M, Gong JS, Zhang J, Yoneda M, Yagi K. Mitochondrial genotype associated with longevity. *Lancet*. 1998;351:185–186.
49. Ross OA, McCormack R, Curran MD, et al. Mitochondrial DNA polymorphism: its role in longevity of the Irish population. *Exp Gerontol*. 2001;36:1161–1178.
50. Ivanova R, Lepage V, Charron D, Schachter F. Mitochondrial genotype associated with French Caucasian centenarians. *Gerontology*. 1998;44:349.
51. Torroni A, Huoponen K, Francalacci P, et al. Classification of European mtDNAs from an analysis of three European populations. *Genetics*. 1996;144:1835–1850.
52. Ruiz-Pesini E, Mishmar D, Brandon M, Procaccio V, Wallace DC. Effects of purifying and adaptive selection on regional variation in human mtDNA. *Science*. 2004;303:223–226.
53. De Benedictis G, Rose G, Carrieri G, et al. Mitochondrial DNA inherited variants are associated with successful aging and longevity in humans. *FASEB J*. 1999;13:1532–1536.
54. Niemi AK, Hervonen A, Hurme M, Karhunen PJ, Jylha M, Majamaa K. Mitochondrial DNA polymorphisms associated with longevity in a Finnish population. *Hum Genet*. 2003;112:29–33.
55. Khrapko K, Vijg J. Mitochondrial DNA mutations and aging: devils in the details? *Trends Genet*. 2009;25:91–98.
56. Vermulst M, Bielas JH, Kujoth GC, et al. Mitochondrial point mutations do not limit the natural lifespan of mice. *Nat Genet*. 2007;39:540–543.
57. Vermulst M, Wanagat J, Kujoth GC, et al. DNA deletions and clonal mutations drive premature aging in mitochondrial mutator mice. *Nat Genet*. 2008;40:392–394.
58. Kukat A, Trifunovic A. Somatic mtDNA mutations and aging—facts and fancies. *Exp Gerontol*. 2009;44:101–105.
59. Hughes BG, Hekimi S. A mild impairment of mitochondrial electron transport has sex-specific effects on lifespan and aging in mice. *PLoS One*. 2011;6: e26116
60. Foster NL, Chase TN, Fedio P, Patronas NJ, Brooks RA, Di Chiro G. Alzheimer's disease: focal cortical changes shown by positron emission tomography. *Neurology*. 1983;33:961–965.
61. de Leon MJ, Ferris SH, George AE, et al. Positron emission tomographic studies of aging and Alzheimer disease. *AJNR Am J Neuroradiol*. 1983;4:568–571.

62. Swerdlow R, Marcus DL, Landman J, Kooby D, Frey 2nd W, Freedman ML. Brain glucose metabolism in Alzheimer's disease. *Am J Med Sci*. 1994;308:141–144.
63. Baloyannis SJ. Mitochondrial alterations in Alzheimer's disease. *J Alzheimers Dis*. 2006;9:119–126.
64. Gibson GE, Sheu KF, Blass JP, et al. Reduced activities of thiamine-dependent enzymes in the brains and peripheral tissues of patients with Alzheimer's disease. *Arch Neurol*. 1988;45:836–840.
65. Gibson GE, Sheu KF, Blass JP. Abnormalities of mitochondrial enzymes in Alzheimer disease. *J Neural Transm*. 1998;105:855–870.
66. Sorbi S, Bird ED, Blass JP. Decreased pyruvate dehydrogenase complex activity in Huntington and Alzheimer brain. *Ann Neurol*. 1983;13:72–78.
67. Gibson GE, Starkov A, Blass JP, Ratan RR, Beal MF. Cause and consequence: mitochondrial dysfunction initiates and propagates neuronal dysfunction, neuronal death and behavioral abnormalities in age-associated neurodegenerative diseases. *Biochim Biophys Acta*. 2010;1802:122–134.
68. Swerdlow RH. Mitochondria and cell bioenergetics: increasingly recognized components and a possible etiologic cause of Alzheimer's disease. *Antioxid Redox Signal*. 2012;16:1434–1455.
69. Parker Jr WD, Filley CM, Parks JK. Cytochrome oxidase deficiency in Alzheimer's disease. *Neurology*. 1990;40:1302–1303.
70. Bosetti F, Brizzi F, Barogi S, et al. Cytochrome c oxidase and mitochondrial F1F0-ATPase (ATP synthase) activities in platelets and brain from patients with Alzheimer's disease. *Neurobiol Aging*. 2002;23:371–376.
71. Mancuso M, Filosto M, Bosetti F, et al. Decreased platelet cytochrome c oxidase activity is accompanied by increased blood lactate concentration during exercise in patients with Alzheimer disease. *Exp Neurol*. 2003;182:421–426.
72. Cardoso SM, Proenca MT, Santos S, Santana I, Oliveira CR. Cytochrome c oxidase is decreased in Alzheimer's disease platelets. *Neurobiol Aging*. 2004;25:105–110.
73. Parker Jr WD, Mahr NJ, Filley CM, et al. Reduced platelet cytochrome c oxidase activity in Alzheimer's disease. *Neurology*. 1994;44:1086–1090.
74. Valla J, Schneider L, Niedzielko T, et al. Impaired platelet mitochondrial activity in Alzheimer's disease and mild cognitive impairment. *Mitochondrion*. 2006;6:323–330.
75. Kish SJ, Bergeron C, Rajput A, et al. Brain cytochrome oxidase in Alzheimer's disease. *J Neurochem*. 1992;59:776–779.
76. Parker Jr WD, Parks J, Filley CM, Kleinschmidt-DeMasters BK. Electron transport chain defects in Alzheimer's disease brain. *Neurology*. 1994;44:1090–1096.
77. Mutisya EM, Bowling AC, Beal MF. Cortical cytochrome oxidase activity is reduced in Alzheimer's disease. *J Neurochem*. 1994;63:2179–2184.
78. Valla J, Berndt JD, Gonzalez-Lima F. Energy hypometabolism in posterior cingulate cortex of Alzheimer's patients: superficial laminar cytochrome oxidase associated with disease duration. *J Neurosci*. 2001;21:4923–4930.
79. Chagnon P, Betard C, Robitaille Y, Cholette A, Gauvreau D. Distribution of brain cytochrome oxidase activity in various neurodegenerative diseases. *Neuroreport*. 1995;6:711–715.
80. Maurer I, Zierz S, Moller HJ. A selective defect of cytochrome c oxidase is present in brain of Alzheimer disease patients. *Neurobiol Aging*. 2000;21:455–462.
81. Simonian NA, Hyman BT. Functional alterations in Alzheimer's disease: diminution of cytochrome oxidase in the hippocampal formation. *J Neuropathol Exp Neurol*. 1993;52:580–585.
82. Verwer RW, Jansen KA, Sluiter AA, Pool CW, Kamphorst W, Swaab DF. Decreased hippocampal metabolic activity in Alzheimer patients is not reflected in the immunoreactivity of cytochrome oxidase subunits. *Exp Neurol*. 2000;163:440–451.

83. Wong-Riley M, Antuono P, Ho KC, et al. Cytochrome oxidase in Alzheimer's disease: biochemical, histochemical, and immunohistochemical analyses of the visual and other systems. *Vision Res*. 1997;37:3593–3608.
84. Parker Jr WD, Parks JK. Cytochrome c oxidase in Alzheimer's disease brain: purification and characterization. *Neurology*. 1995;45:482–486.
85. Curti D, Rognoni F, Gasparini L, et al. Oxidative metabolism in cultured fibroblasts derived from sporadic Alzheimer's disease (AD) patients. *Neurosci Lett*. 1997; 236:13–16.
86. Wang X, Su B, Lee HG, et al. Impaired balance of mitochondrial fission and fusion in Alzheimer's disease. *J Neurosci*. 2009;29:9090–9103.
87. Manczak M, Calkins MJ, Reddy PH. Impaired mitochondrial dynamics and abnormal interaction of amyloid beta with mitochondrial protein Drp1 in neurons from patients with Alzheimer's disease: implications for neuronal damage. *Hum Mol Genet*. 2011;20:2495–2509.
88. Sheng B, Wang X, Su B, et al. Impaired mitochondrial biogenesis contributes to mitochondrial dysfunction in Alzheimer's disease. *J Neurochem*. 2012;120:419–429.
89. Qin W, Haroutunian V, Katsel P, et al. PGC-1alpha expression decreases in the Alzheimer disease brain as a function of dementia. *Arch Neurol*. 2009;66:352–361.
90. Wang X, Su B, Fujioka H, Zhu X. Dynamin-like protein 1 reduction underlies mitochondrial morphology and distribution abnormalities in fibroblasts from sporadic Alzheimer's disease patients. *Am J Pathol*. 2008;173:470–482.
91. de la Monte SM, Luong T, Neely TR, Robinson D, Wands JR. Mitochondrial DNA damage as a mechanism of cell loss in Alzheimer's disease. *Lab Invest*. 2000;80:1323–1335.
92. Coskun PE, Beal MF, Wallace DC. Alzheimer's brains harbor somatic mtDNA control-region mutations that suppress mitochondrial transcription and replication. *Proc Natl Acad Sci USA*. 2004;101:10726–10731.
93. Rodriguez-Santiago B, Casademont J, Nunes V. Is mitochondrial DNA depletion involved in Alzheimer's disease? *Eur J Hum Genet*. 2001;9:279–285.
94. Corral-Debrinski M, Horton T, Lott MT, et al. Marked changes in mitochondrial DNA deletion levels in Alzheimer brains. *Genomics*. 1994;23:471–476.
95. Chang SW, Zhang D, Chung HD, Zassenhaus HP. The frequency of point mutations in mitochondrial DNA is elevated in the Alzheimer's brain. *Biochem Biophys Res Commun*. 2000;273:203–208.
96. Hamblet NS, Castora FJ. Elevated levels of the Kearns-Sayre syndrome mitochondrial DNA deletion in temporal cortex of Alzheimer's patients. *Mutat Res*. 1997;379:253–262.
97. Casoli T, Di Stefano G, Spazzafumo L, et al. Contribution of non-reference alleles in mtDNA of Alzheimer's disease patients. *Ann Clin Transl Neurol*. 2014;1:284–289.
98. Casoli T, Spazzafumo L, Di Stefano G, Conti F. Role of diffuse low-level heteroplasmy of mitochondrial DNA in Alzheimer's disease neurodegeneration. *Front Aging Neurosci*. 2015;7:142.
99. Mecocci P, MacGarvey U, Beal MF. Oxidative damage to mitochondrial DNA is increased in Alzheimer's disease. *Ann Neurol*. 1994;36:747–751.
100. Egensperger R, Kosel S, Schnopp NM, Mehraein P, Graeber MB. Association of the mitochondrial tRNA(A4336G) mutation with Alzheimer's and Parkinson's diseases. *Neuropathol Appl Neurobiol*. 1997;23:315–321.
101. Fesahat F, Houshmand M, Panahi MS, Gharagozli K, Mirzajani F. Do haplogroups H and U act to increase the penetrance of Alzheimer's disease? *Cell Mol Neurobiol*. 2007;27:329–334.
102. Hutchin T, Cortopassi G. A mitochondrial DNA clone is associated with increased risk for Alzheimer disease. *Proc Natl Acad Sci USA*. 1995;92:6892–6895.

103. Ienco EC, Simoncini C, Orsucci D, et al. May "mitochondrial eve" and mitochondrial haplogroups play a role in neurodegeneration and Alzheimer's disease? *Int J Alzheimers Dis*. 2011;2011:709061.
104. Lakatos A, Derbeneva O, Younes D, et al. Association between mitochondrial DNA variations and Alzheimer's disease in the ADNI cohort. *Neurobiol Aging*. 2010;31:1355–1363.
105. Maruszak A, Canter JA, Styczynska M, Zekanowski C, Barcikowska M. Mitochondrial haplogroup H and Alzheimer's disease—is there a connection? *Neurobiol Aging*. 2009;30:1749–1755.
106. Maruszak A, Safranow K, Branicki W, et al. The impact of mitochondrial and nuclear DNA variants on late-onset Alzheimer's disease risk. *J Alzheimers Dis*. 2011;27:197–210.
107. Santoro A, Balbi V, Balducci E, et al. Evidence for sub-haplogroup h5 of mitochondrial DNA as a risk factor for late onset Alzheimer's disease. *PLoS One*. 2010;5: e12037
108. Shoffner JM, Brown MD, Torroni A, et al. Mitochondrial DNA variants observed in Alzheimer disease and Parkinson disease patients. *Genomics*. 1993;17:171–184.
109. Takasaki S. Mitochondrial haplogroups associated with Japanese Alzheimer's patients. *J Bioenerg Biomembr*. 2009;41:407–410.
110. Tanaka N, Goto Y, Akanuma J, et al. Mitochondrial DNA variants in a Japanese population of patients with Alzheimer's disease. *Mitochondrion*. 2010;10:32–37.
111. van der Walt JM, Dementieva YA, Martin ER, et al. Analysis of European mitochondrial haplogroups with Alzheimer disease risk. *Neurosci Lett*. 2004;365:28–32.
112. Ridge PG, Koop A, Maxwell TJ, et al. Mitochondrial haplotypes associated with biomarkers for Alzheimer's disease. *PLoS One*. 2013;8: e74158
113. Wang Y, Brinton RD. Triad of risk for late onset Alzheimer's: mitochondrial haplotype, APOE genotype and chromosomal sex. *Front Aging Neurosci*. 2016;8:232.
114. Chinnery PF, Taylor GA, Howell N, et al. Mitochondrial DNA haplogroups and susceptibility to AD and dementia with Lewy bodies. *Neurology*. 2000;55:302–304.
115. Edland SD, Tobe VO, Rieder MJ, et al. Mitochondrial genetic variants and Alzheimer disease: a case-control study of the T4336C and G5460A variants. *Alzheimer Dis Assoc Disord*. 2002;16:1–7.
116. Rodriguez Santiago B, Casademont J, Nunes V. Is there a relation between Alzheimer s disease and defects of mitochondrial DNA? *Rev Neurol*. 2001;33:301–305.
117. Wragg MA, Talbot CJ, Morris JC, Lendon CL, Goate AM. No association found between Alzheimer's disease and a mitochondrial tRNA glutamine gene variant. *Neurosci Lett*. 1995;201:107–110.
118. Zsurka G, Kalman J, Csaszar A, Rasko I, Janka Z, Venetianer P. No mitochondrial haplotype was found to increase risk for Alzheimer's disease. *Biol Psychiatry*. 1998;44:371–373.
119. Caspersen C, Wang N, Yao J, et al. Mitochondrial Abeta: a potential focal point for neuronal metabolic dysfunction in Alzheimer's disease. *FASEB J*. 2005;19:2040–2041.
120. Lustbader JW, Cirilli M, Lin C, et al. ABAD directly links Abeta to mitochondrial toxicity in Alzheimer's disease. *Science*. 2004;304:448–452.
121. Manczak M, Anekonda TS, Henson E, Park BS, Quinn J, Reddy PH. Mitochondria are a direct site of A beta accumulation in Alzheimer's disease neurons: implications for free radical generation and oxidative damage in disease progression. *Hum Mol Genet*. 2006;15:1437–1449.
122. Crouch PJ, Blake R, Duce JA, et al. Copper-dependent inhibition of human cytochrome c oxidase by a dimeric conformer of amyloid-beta1-42. *J Neurosci*. 2005;25:672–679.
123. Hansson Petersen CA, Alikhani N, Behbahani H, et al. The amyloid beta-peptide is imported into mitochondria via the TOM import machinery and localized to mitochondrial cristae. *Proc Natl Acad Sci USA*. 2008;105:13145–13150.

124. Du H, Guo L, Fang F, et al. Cyclophilin D deficiency attenuates mitochondrial and neuronal perturbation and ameliorates learning and memory in Alzheimer's disease. *Nat Med*. 2008;14:1097–1105.
125. Dragicevic N, Mamcarz M, Zhu Y, et al. Mitochondrial amyloid-beta levels are associated with the extent of mitochondrial dysfunction in different brain regions and the degree of cognitive impairment in Alzheimer's transgenic mice. *J Alzheimers Dis*. 2010; 20(suppl 2):S535–S550.
126. Yamaguchi H, Yamazaki T, Ishiguro K, Shoji M, Nakazato Y, Hirai S. Ultrastructural localization of Alzheimer amyloid beta/A4 protein precursor in the cytoplasm of neurons and senile plaque-associated astrocytes. *Acta Neuropathol*. 1992;85:15–22.
127. Yao J, Du H, Yan S, et al. Inhibition of amyloid-beta (Abeta) peptide-binding alcohol dehydrogenase-Abeta interaction reduces Abeta accumulation and improves mitochondrial function in a mouse model of Alzheimer's disease. *J Neurosci*. 2011;31:2313–2320.
128. Canevari L, Clark JB, Bates TE. Beta-amyloid fragment 25-35 selectively decreases complex IV activity in isolated mitochondria. *FEBS Lett*. 1999;457:131–134.
129. Casley CS, Canevari L, Land JM, Clark JB, Sharpe MA. Beta-amyloid inhibits integrated mitochondrial respiration and key enzyme activities. *J Neurochem*. 2002;80:91–100.
130. Parks JK, Smith TS, Trimmer PA, Bennett Jr JP, Parker Jr WD. Neurotoxic Abeta peptides increase oxidative stress in vivo through NMDA-receptor and nitric-oxide-synthase mechanisms, and inhibit complex IV activity and induce a mitochondrial permeability transition in vitro. *J Neurochem*. 2001;76:1050–1056.
131. Pereira C, Santos MS, Oliveira C. Mitochondrial function impairment induced by amyloid beta-peptide on PC12 cells. *Neuroreport*. 1998;9:1749–1755.
132. Cardoso SM, Santos S, Swerdlow RH, Oliveira CR. Functional mitochondria are required for amyloid beta-mediated neurotoxicity. *FASEB J*. 2001;15:1439–1441.
133. Wang X, Su B, Siedlak SL, et al. Amyloid-beta overproduction causes abnormal mitochondrial dynamics via differential modulation of mitochondrial fission/fusion proteins. *Proc Natl Acad Sci USA*. 2008;105:19318–19323.
134. Anandatheerthavarada HK, Biswas G, Robin MA, Avadhani NG. Mitochondrial targeting and a novel transmembrane arrest of Alzheimer's amyloid precursor protein impairs mitochondrial function in neuronal cells. *J Cell Biol*. 2003;161:41–54.
135. Anandatheerthavarada HK, Devi L. Amyloid precursor protein and mitochondrial dysfunction in Alzheimer's disease. *Neuroscientist*. 2007;13:626–638.
136. Devi L, Prabhu BM, Galati DF, Avadhani NG, Anandatheerthavarada HK. Accumulation of amyloid precursor protein in the mitochondrial import channels of human Alzheimer's disease brain is associated with mitochondrial dysfunction. *J Neurosci*. 2006;26:9057–9068.
137. Lin MT, Beal MF. Alzheimer's APP mangles mitochondria. *Nat Med*. 2006; 12:1241–1243.
138. Bassett SS, Avramopoulos D, Fallin D. Evidence for parent of origin effect in late-onset Alzheimer disease. *Am J Med Genet*. 2002;114:679–686.
139. Duara R, Lopez-Alberola RF, Barker WW, et al. A comparison of familial and sporadic Alzheimer's disease. *Neurology*. 1993;43:1377–1384.
140. Edland SD, Silverman JM, Peskind ER, Tsuang D, Wijsman E, Morris JC. Increased risk of dementia in mothers of Alzheimer's disease cases: evidence for maternal inheritance. *Neurology*. 1996;47:254–256.
141. Mosconi L, Berti V, Swerdlow RH, Pupi A, Duara R, de Leon M. Maternal transmission of Alzheimer's disease: prodromal metabolic phenotype and the search for genes. *Hum Genomics*. 2010;4:170–193.
142. Mosconi L, Brys M, Switalski R, et al. Maternal family history of Alzheimer's disease predisposes to reduced brain glucose metabolism. *Proc Natl Acad Sci USA*. 2007;104:19067–19072.

143. Okonkwo OC, Xu G, Oh JM, et al. Cerebral blood flow is diminished in asymptomatic middle-aged adults with maternal history of Alzheimer's disease. *Cereb Cortex*. 2014;24:978–988.
144. Mosconi L, Glodzik L, Mistur R, et al. Oxidative stress and amyloid-beta pathology in normal individuals with a maternal history of Alzheimer's. *Biol Psychiatry*. 2010;68:913–921.
145. Mosconi L, Rinne JO, Tsui WH, et al. Increased fibrillar amyloid-{beta} burden in normal individuals with a family history of late-onset Alzheimer's. *Proc Natl Acad Sci USA*. 2010;107:5949–5954.
146. Honea RA, Vidoni ED, Swerdlow RH, Burns JM. Maternal family history is associated with Alzheimer's disease biomarkers. *J Alzheimers Dis*. 2012;31:659–668.
147. Liu Z, Chen HH, Li TL, Xu L, Du HQ. A cross-sectional study on cerebrospinal fluid biomarker levels in cognitively normal elderly subjects with or without a family history of Alzheimer's disease. *CNS Neurosci Ther*. 2013;19:38–42.
148. Honea RA, Swerdlow RH, Vidoni E, Burns JM. Progressive regional atrophy in normal adults with a maternal history of Alzheimer disease. *Neurology*. 2011;76:822–829.
149. Honea RA, Swerdlow RH, Vidoni ED, Goodwin J, Burns JM. Reduced gray matter volume in normal adults with a maternal family history of Alzheimer disease. *Neurology*. 2010;74:113–120.
150. Berti V, Mosconi L, Glodzik L, et al. Structural brain changes in normal individuals with a maternal history of Alzheimer's. *Neurobiol Aging*. 2011;32. 2325.e17-26.
151. Andrawis JP, Hwang KS, Green AE, et al. Effects of ApoE4 and maternal history of dementia on hippocampal atrophy. *Neurobiol Aging*. 2012;33:856–866.
152. Reiter K, Alpert KI, Cobia DJ, et al. Cognitively normal individuals with AD parents may be at risk for developing aging-related cortical thinning patterns characteristic of AD. *Neuroimage*. 2012;61:525–532.
153. Mosconi L, de Leon M, Murray J, et al. Reduced mitochondria cytochrome oxidase activity in adult children of mothers with Alzheimer's disease. *J Alzheimers Dis*. 2011;27:483–490.
154. Debette S, Wolf PA, Beiser A, et al. Association of parental dementia with cognitive and brain MRI measures in middle-aged adults. *Neurology*. 2009;73:2071–2078.
155. Corder EH, Saunders AM, Strittmatter WJ, et al. Gene dose of apolipoprotein E type 4 allele and the risk of Alzheimer's disease in late onset families. *Science*. 1993;261:921–923.
156. Strittmatter WJ, Saunders AM, Schmechel D, et al. Apolipoprotein E: high-avidity binding to beta-amyloid and increased frequency of type 4 allele in late-onset familial Alzheimer disease. *Proc Natl Acad Sci USA*. 1993;90:1977–1981.
157. Pericak-Vance MA, Bebout JL, Gaskell Jr PC, et al. Linkage studies in familial Alzheimer disease: evidence for chromosome 19 linkage. *Am J Hum Genet*. 1991;48:1034–1050.
158. Mahley RW. Apolipoprotein E: cholesterol transport protein with expanding role in cell biology. *Science*. 1988;240:622–630.
159. Mahley RW, Rall Jr SC. Apolipoprotein E: far more than a lipid transport protein. *Annu Rev Genomics Hum Genet*. 2000;1:507–537.
160. Mahley RW, Huang Y. Apolipoprotein E sets the stage: response to injury triggers neuropathology. *Neuron*. 2012;76:871–885.
161. Mahley RW, Weisgraber KH, Huang Y. Apolipoprotein E4: a causative factor and therapeutic target in neuropathology, including Alzheimer's disease. *Proc Natl Acad Sci USA*. 2006;103:5644–5651.
162. Chen HK, Ji ZS, Dodson SE, et al. Apolipoprotein E4 domain interaction mediates detrimental effects on mitochondria and is a potential therapeutic target for Alzheimer disease. *J Biol Chem*. 2011;286:5215–5221.

163. Valla J, Yaari R, Wolf AB, et al. Reduced posterior cingulate mitochondrial activity in expired young adult carriers of the APOE epsilon4 allele, the major late-onset Alzheimer's susceptibility gene. *J Alzheimers Dis*. 2010;22:307–313.
164. Carrieri G, Bonafe M, De Luca M, et al. Mitochondrial DNA haplogroups and APOE4 allele are non-independent variables in sporadic Alzheimer's disease. *Hum Genet*. 2001;108:194–198.
165. Bekris LM, Galloway NM, Montine TJ, Schellenberg GD, Yu CE. APOE mRNA and protein expression in postmortem brain are modulated by an extended haplotype structure. *Am J Med Genet B Neuropsychiatr Genet*. 2009;153B:409–417.
166. Cervantes S, Samaranch L, Vidal-Taboada JM, et al. Genetic variation in APOE cluster region and Alzheimer's disease risk. *Neurobiol Aging*. 2011;32. 2107.e7-17.
167. Hayden KM, McEvoy JM, Linnertz C, et al. A homopolymer polymorphism in the TOMM40 gene contributes to cognitive performance in aging. *Alzheimers Dement*. 2012;8:381–388.
168. Johnson SC, La Rue A, Hermann BP, et al. The effect of TOMM40 poly-T length on gray matter volume and cognition in middle-aged persons with APOE epsilon3/epsilon3 genotype. *Alzheimers Dement*. 2011;7:456–465.
169. Li G, Bekris LM, Leong L, et al. TOMM40 intron 6 poly-T length, age at onset, and neuropathology of AD in individuals with APOE varepsilon3/varepsilon3. *Alzheimers Dement*. 2013;9:554–561.
170. Maruszak A, Peplonska B, Safranow K, Chodakowska-Zebrowska M, Barcikowska M, Zekanowski C. TOMM40 rs10524523 polymorphism's role in late-onset Alzheimer's disease and in longevity. *J Alzheimers Dis*. 2012;28:309–322.
171. Potkin SG, Guffanti G, Lakatos A, et al. Hippocampal atrophy as a quantitative trait in a genome-wide association study identifying novel susceptibility genes for Alzheimer's disease. *PLoS One*. 2009;4: e6501.
172. Roses AD. An inherited variable poly-T repeat genotype in TOMM40 in Alzheimer disease. *Arch Neurol*. 2010;67:536–541.
173. Roses AD, Lutz MW, Amrine-Madsen H, et al. A TOMM40 variable-length polymorphism predicts the age of late-onset Alzheimer's disease. *Pharmacogenomics J*. 2010;10:375–384.
174. Takei N, Miyashita A, Tsukie T, et al. Genetic association study on in and around the APOE in late-onset Alzheimer disease in Japanese. *Genomics*. 2009;93:441–448.
175. Guerreiro RJ, Hardy J. TOMM40 association with Alzheimer disease: tales of APOE and linkage disequilibrium. *Arch Neurol*. 2012;69:1243–1244.
176. Cruchaga C, Nowotny P, Kauwe JS, et al. Association and expression analyses with single-nucleotide polymorphisms in TOMM40 in Alzheimer disease. *Arch Neurol*. 2011;68:1013–1019.
177. Cottrell DA, Blakely EL, Johnson MA, Ince PG, Turnbull DM. Mitochondrial enzyme-deficient hippocampal neurons and choroidal cells in AD. *Neurology*. 2001;57:260–264.
178. Krishnan KJ, Ratnaike TE, De Gruyter HL, Jaros E, Turnbull DM. Mitochondrial DNA deletions cause the biochemical defect observed in Alzheimer's disease. *Neurobiol Aging*. 2012;33:2210–2214.
179. Coskun PE, Wyrembak J, Derbereva O, et al. Systemic mitochondrial dysfunction and the etiology of Alzheimer's disease and down syndrome dementia. *J Alzheimers Dis*. 2010;20(suppl 2):S293–S310.
180. Lin MT, Simon DK, Ahn CH, Kim LM, Beal MF. High aggregate burden of somatic mtDNA point mutations in aging and Alzheimer's disease brain. *Hum Mol Genet*. 2002;11:133–145.
181. Webster MT, Pearce BR, Bowen DM, Francis PT. The effects of perturbed energy metabolism on the processing of amyloid precursor protein in PC12 cells. *J Neural Transm*. 1998;105:839–853.

182. Gasparini L, Racchi M, Benussi L, et al. Effect of energy shortage and oxidative stress on amyloid precursor protein metabolism in COS cells. *Neurosci Lett.* 1997; 231:113–117.
183. Gabuzda D, Busciglio J, Chen LB, Matsudaira P, Yankner BA. Inhibition of energy metabolism alters the processing of amyloid precursor protein and induces a potentially amyloidogenic derivative. *J Biol Chem.* 1994;269:13623–13628.
184. Scheffler K, Krohn M, Dunkelmann T, et al. Mitochondrial DNA polymorphisms specifically modify cerebral beta-amyloid proteostasis. *Acta Neuropathol.* 2012;124:199–208.
185. Fukui H, Diaz F, Garcia S, Moraes CT. Cytochrome c oxidase deficiency in neurons decreases both oxidative stress and amyloid formation in a mouse model of Alzheimer's disease. *Proc Natl Acad Sci USA.* 2007;104:14163–14168.
186. Pinto M, Pickrell AM, Fukui H, Moraes CT. Mitochondrial DNA damage in a mouse model of Alzheimer's disease decreases amyloid beta plaque formation. *Neurobiol Aging.* 2013;34:2399–2407.
187. Kukreja L, Kujoth GC, Prolla TA, Van Leuven F, Vassar R. Increased mtDNA mutations with aging promotes amyloid accumulation and brain atrophy in the APP/Ld transgenic mouse model of Alzheimer's disease. *Mol Neurodegener.* 2014;9:16.
188. Dumont M, Stack C, Elipenahli C, et al. PGC-1alpha overexpression exacerbates beta-amyloid and tau deposition in a transgenic mouse model of Alzheimer's disease. *FASEB J.* 2014;28:1745–1755.
189. Kang JE, Lim MM, Bateman RJ, et al. Amyloid-beta dynamics are regulated by orexin and the sleep-wake cycle. *Science.* 2009;326:1005–1007.
190. Roh JH, Jiang H, Finn MB, et al. Potential role of orexin and sleep modulation in the pathogenesis of Alzheimer's disease. *J Exp Med.* 2014;211:2487–2496.
191. Bero AW, Yan P, Roh JH, et al. Neuronal activity regulates the regional vulnerability to amyloid-beta deposition. *Nat Neurosci.* 2011;14:750–756.
192. Yamamoto K, Tanei Z, Hashimoto T, et al. Chronic optogenetic activation augments abeta pathology in a mouse model of Alzheimer disease. *Cell Rep.* 2015;11:859–865.
193. Brody DL, Magnoni S, Schwetye KE, et al. Amyloid-beta dynamics correlate with neurological status in the injured human brain. *Science.* 2008;321:1221–1224.
194. Vlassenko AG, Vaishnavi SN, Couture L, et al. Spatial correlation between brain aerobic glycolysis and amyloid-beta (Abeta) deposition. *Proc Natl Acad Sci USA.* 2010;107:17763–17767.
195. Escobar-Khondiker M, Hollerhage M, Muriel MP, et al. Annonacin, a natural mitochondrial complex I inhibitor, causes tau pathology in cultured neurons. *J Neurosci.* 2007;27:7827–7837.
196. Szabados T, Dul C, Majtenyi K, Hargitai J, Penzes Z, Urbanics R. A chronic Alzheimer's model evoked by mitochondrial poison sodium azide for pharmacological investigations. *Behav Brain Res.* 2004;154:31–40.
197. Hoglinger GU, Lannuzel A, Khondiker ME, et al. The mitochondrial complex I inhibitor rotenone triggers a cerebral tauopathy. *J Neurochem.* 2005;95:930–939.
198. Rottscholl R, Haegele M, Jainsch B, et al. Chronic consumption of Annona muricata juice triggers and aggravates cerebral tau phosphorylation in wild-type and MAPT transgenic mice. *J Neurochem.* 2016;139:624–639.
199. Yamada ES, Respondek G, Mussner S, et al. Annonacin, a natural lipophilic mitochondrial complex I inhibitor, increases phosphorylation of tau in the brain of FTDP-17 transgenic mice. *Exp Neurol.* 2014;253:113–125.
200. Yanagisawa M, Planel E, Ishiguro K, Fujita SC. Starvation induces tau hyperphosphorylation in mouse brain: implications for Alzheimer's disease. *FEBS Lett.* 1999;461:329–333.

201. Zhao Y, Tseng IC, Heyser CJ, et al. Appoptosin-mediated caspase cleavage of tau contributes to progressive supranuclear palsy pathogenesis. *Neuron*. 2015;87:963–975.
202. Hoglinger GU, Melhem NM, Dickson DW, et al. Identification of common variants influencing risk of the tauopathy progressive supranuclear palsy. *Nat Genet*. 2011;43:699–705.
203. Blass JP, Baker AC, Ko L, Black RS. Induction of Alzheimer antigens by an uncoupler of oxidative phosphorylation. *Arch Neurol*. 1990;47:864–869.
204. Blass JP, Baker AC, Ko L, Sheu RK, Black RS. Expression of 'Alzheimer antigens' in cultured skin fibroblasts. *Arch Neurol*. 1991;48:709–717.
205. Swerdlow RH. Mitochondria in cybrids containing mtDNA from persons with mitochondriopathies. *J Neurosci Res*. 2007;85:3416–3428.
206. Poste G, Reeve P. Enucleation of mammalian cells by cytochalasin B. II. Formation of hybrid cells and heterokaryons by fusion of anucleate and nucleated cells. *Exp Cell Res*. 1972;73:287–294.
207. King MP, Attardi G. Human cells lacking mtDNA: repopulation with exogenous mitochondria by complementation. *Science*. 1989;246:500–503.
208. Clark MA, Shay JW. Mitochondrial transformation of mammalian cells. *Nature*. 1982;295:605–607.
209. Bunn CL, Wallace DC, Eisenstadt JM. Cytoplasmic inheritance of chloramphenicol resistance in mouse tissue culture cells. *Proc Natl Acad Sci USA*. 1974;71:1681–1685.
210. Wallace DC, Bunn CL, Eisenstadt JM. Cytoplasmic transfer of chloramphenicol resistance in human tissue culture cells. *J Cell Biol*. 1975;67:174–188.
211. Morais R, Desjardins P, Turmel C, Zinkewich-Peotti K. Development and characterization of continuous avian cell lines depleted of mitochondrial DNA. *In Vitro Cell Dev Biol*. 1988;24:649–658.
212. Desjardins P, de Muys JM, Morais R. An established avian fibroblast cell line without mitochondrial DNA. *Somat Cell Mol Genet*. 1986;12:133–139.
213. Desjardins P, Frost E, Morais R. Ethidium bromide-induced loss of mitochondrial DNA from primary chicken embryo fibroblasts. *Mol Cell Biol*. 1985;5:1163–1169.
214. Ephrussi B, Hottinger H, Chimenes A. Action de l'acriflavine sur les levures, I: la mutation "petite clonie". *Ann Inst Pasteur*. 1949;76:531.
215. King MP, Koga Y, Davidson M, Schon EA. Defects in mitochondrial protein synthesis and respiratory chain activity segregate with the tRNA(Leu(UUR)) mutation associated with mitochondrial myopathy, encephalopathy, lactic acidosis, and strokelike episodes. *Mol Cell Biol*. 1992;12:480–490.
216. Masucci JP, Davidson M, Koga Y, Schon EA, King MP. In vitro analysis of mutations causing myoclonus epilepsy with ragged-red fibers in the mitochondrial tRNA(Lys)gene: two genotypes produce similar phenotypes. *Mol Cell Biol*. 1995; 15:2872–2881.
217. Michikawa Y, Mazzucchelli F, Bresolin N, Scarlato G, Attardi G. Aging-dependent large accumulation of point mutations in the human mtDNA control region for replication. *Science*. 1999;286:774–779.
218. Chomyn A, Martinuzzi A, Yoneda M, et al. MELAS mutation in mtDNA binding site for transcription termination factor causes defects in protein synthesis and in respiration but no change in levels of upstream and downstream mature transcripts. *Proc Natl Acad Sci USA*. 1992;89:4221–4225.
219. Chomyn A, Lai ST, Shakeley R, Bresolin N, Scarlato G, Attardi G. Platelet-mediated transformation of mtDNA-less human cells: analysis of phenotypic variability among clones from normal individuals—and complementation behavior of the tRNALys mutation causing myoclonic epilepsy and ragged red fibers. *Am J Hum Genet*. 1994;54:966–974.

220. Hofhaus G, Johns DR, Hurko O, Attardi G, Chomyn A. Respiration and growth defects in transmitochondrial cell lines carrying the 11778 mutation associated with Leber's hereditary optic neuropathy. *J Biol Chem*. 1996;271:13155–13161.
221. Miller SW, Trimmer PA, Parker Jr WD, Davis RE. Creation and characterization of mitochondrial DNA-depleted cell lines with "neuronal-like" properties. *J Neurochem*. 1996;67:1897–1907.
222. Swerdlow RH, Parks JK, Miller SW, et al. Origin and functional consequences of the complex I defect in Parkinson's disease. *Ann Neurol*. 1996;40:663–671.
223. Parker Jr WD, Boyson SJ, Parks JK. Abnormalities of the electron transport chain in idiopathic Parkinson's disease. *Ann Neurol*. 1989;26:719–723.
224. Krige D, Carroll MT, Cooper JM, Marsden CD, Schapira AH. Platelet mitochondrial function in Parkinson's disease. The Royal Kings and Queens Parkinson Disease Research Group. *Ann Neurol*. 1992;32:782–788.
225. Benecke R, Strumper P, Weiss H. Electron transfer complexes I and IV of platelets are abnormal in Parkinson's disease but normal in Parkinson-plus syndromes. *Brain*. 1993;116(pt 6):1451–1463.
226. Parker WD. Sporadic neurologic disease and the electron transport chain: a hypothesis. In: Pascuzzi RM, ed. *Proceedings of the 1989 Scientific Meeting of the American Society for Neurological Investigation: New Developments in Neuromuscular Disease*, Bloomington, Indiana: Indiana University Printing Services; 1990.
227. Davis RE, Miller S, Herrnstadt C, et al. Mutations in mitochondrial cytochrome c oxidase genes segregate with late-onset Alzheimer disease. *Proc Natl Acad Sci USA*. 1997;94:4526–4531.
228. Davis RE, Miller S, Herrnstadt C, et al. Retraction. *Proc Natl Acad Sci USA*. 1998;95:12069.
229. Sheehan JP, Swerdlow RH, Miller SW, et al. Calcium homeostasis and reactive oxygen species production in cells transformed by mitochondria from individuals with sporadic Alzheimer's disease. *J Neurosci*. 1997;17:4612–4622.
230. Swerdlow RH, Parks JK, Cassarino DS, et al. Cybrids in Alzheimer's disease: a cellular model of the disease? *Neurology*. 1997;49:918–925.
231. Cardoso SM, Santana I, Swerdlow RH, Oliveira CR. Mitochondria dysfunction of Alzheimer's disease cybrids enhances Abeta toxicity. *J Neurochem*. 2004; 89:1417–1426.
232. Silva DF, Selfridge JE, Lu J, et al. Bioenergetic flux, mitochondrial mass and mitochondrial morphology dynamics in AD and MCI cybrid cell lines. *Hum Mol Genet*. 2013;22:3931–3946.
233. Ito S, Ohta S, Nishimaki K, et al. Functional integrity of mitochondrial genomes in human platelets and autopsied brain tissues from elderly patients with Alzheimer's disease. *Proc Natl Acad Sci USA*. 1999;96:2099–2103.
234. Bijur GN, Davis RE, Jope RS. Rapid activation of heat shock factor-1 DNA binding by H2O2 and modulation by glutathione in human neuroblastoma and Alzheimer's disease cybrid cells. *Brain Res Mol Brain Res*. 1999;71:69–77.
235. Cassarino DS, Swerdlow RH, Parks JK, Parker Jr WD, Bennett Jr JP. Cyclosporin A increases resting mitochondrial membrane potential in SY5Y cells and reverses the depressed mitochondrial membrane potential of Alzheimer's disease cybrids. *Biochem Biophys Res Commun*. 1998;248:168–173.
236. De Sarno P, Bijur GN, Lu R, Davis RE, Jope RS. Alterations in muscarinic receptor-coupled phosphoinositide hydrolysis and AP-1 activation in Alzheimer's disease cybrid cells. *Neurobiol Aging*. 2000;21:31–38.
237. Ghosh SS, Swerdlow RH, Miller SW, Sheeman B, Parker Jr WD, Davis RE. Use of cytoplasmic hybrid cell lines for elucidating the role of mitochondrial dysfunction in Alzheimer's disease and Parkinson's disease. *Ann N Y Acad Sci*. 1999;893:176–191.

238. Khan SM, Cassarino DS, Abramova NN, et al. Alzheimer's disease cybrids replicate beta-amyloid abnormalities through cell death pathways. *Ann Neurol.* 2000;48:148–155.
239. Onyango IG, Tuttle JB, Bennett Jr JP. Altered intracellular signaling and reduced viability of Alzheimer's disease neuronal cybrids is reproduced by beta-amyloid peptide acting through receptor for advanced glycation end products (RAGE). *Mol Cell Neurosci.* 2005;29:333–343.
240. Onyango IG, Bennett Jr JP, Tuttle JB. Endogenous oxidative stress in sporadic Alzheimer's disease neuronal cybrids reduces viability by increasing apoptosis through pro-death signaling pathways and is mimicked by oxidant exposure of control cybrids. *Neurobiol Dis.* 2005;19:312–322.
241. Onyango IG, Ahn JY, Tuttle JB, Bennett Jr JP, Swerdlow RH. Nerve growth factor attenuates oxidant-induced beta-amyloid neurotoxicity in sporadic Alzheimer's disease cybrids. *J Neurochem.* 2010;114:1605–1618.
242. Thiffault C, Bennett Jr JP. Cyclical mitochondrial deltapsiM fluctuations linked to electron transport, F0F1 ATP-synthase and mitochondrial Na+/Ca+2 exchange are reduced in Alzheimer's disease cybrids. *Mitochondrion.* 2005;5:109–119.
243. Trimmer PA, Swerdlow RH, Parks JK, et al. Abnormal mitochondrial morphology in sporadic Parkinson's and Alzheimer's disease cybrid cell lines. *Exp Neurol.* 2000;162:37–50.
244. Trimmer PA, Borland MK. Differentiated Alzheimer's disease transmitochondrial cybrid cell lines exhibit reduced organelle movement. *Antioxid Redox Signal.* 2005;7:1101–1109.
245. Zhang H, Liu Y, Lao M, Ma Z, Yi X. Puerarin protects Alzheimer's disease neuronal cybrids from oxidant-stress induced apoptosis by inhibiting pro-death signaling pathways. *Exp Gerontol.* 2011;46:30–37.
246. Silva DF, Santana I, Esteves AR, et al. Prodromal metabolic phenotype in MCI cybrids: implications for Alzheimer's disease. *Curr Alzheimer Res.* 2013;10:180–190.
247. Gan X, Huang S, Wu L, et al. Inhibition of ERK-DLP1 signaling and mitochondrial division alleviates mitochondrial dysfunction in Alzheimer's disease cybrid cell. *Biochim Biophys Acta.* 2014;1842:220–231.
248. Gan X, Wu L, Huang S, et al. Oxidative stress-mediated activation of extracellular signal-regulated kinase contributes to mild cognitive impairment-related mitochondrial dysfunction. *Free Radic Biol Med.* 2014;75:230–240.
249. Yu Q, Fang D, Swerdlow RH, Yu H, Chen JX, Yan SS. Antioxidants rescue mitochondrial transport in differentiated Alzheimer's disease trans-mitochondrial cybrid cells. *J Alzheimers Dis.* 2016;54:679–690.
250. Costa RO, Ferreiro E, Martins I, et al. Amyloid beta-induced ER stress is enhanced under mitochondrial dysfunction conditions. *Neurobiol Aging.* 2012;33. 824.e5-16.
251. Jeong JH, Yum KS, Chang JY, et al. Dose-specific effect of simvastatin on hypoxia-induced HIF-1alpha and BACE expression in Alzheimer's disease cybrid cells. *BMC Neurol.* 2015;15:127.
252. Trimmer PA, Keeney PM, Borland MK, et al. Mitochondrial abnormalities in cybrid cell models of sporadic Alzheimer's disease worsen with passage in culture. *Neurobiol Dis.* 2004;15:29–39.
253. Nunomura A, Perry G, Aliev G, et al. Oxidative damage is the earliest event in Alzheimer disease. *J Neuropathol Exp Neurol.* 2001;60:759–767.
254. Khan SM, Smigrodzki RM, Swerdlow RH. Cell and animal models of mtDNA biology: progress and prospects. *Am J Physiol Cell Physiol.* 2007;292:C658–C669.
255. Meissner C, Mohamed SA, Klueter H, Hamann K, von Wurmb N, Oehmichen M. Quantification of mitochondrial DNA in human blood cells using an automated detection system. *Forensic Sci Int.* 2000;113:109–112.

256. Lu J, Wang K, Rodova M, et al. Polymorphic variation in cytochrome oxidase subunit genes. *J Alzheimers Dis*. 2010;21:141–154.
257. Sims NR, Finegan JM, Blass JP. Altered glucose metabolism in fibroblasts from patients with Alzheimer's disease. *N Engl J Med*. 1985;313:638–639.
258. Sims NR, Finegan JM, Blass JP. Altered metabolic properties of cultured skin fibroblasts in Alzheimer's disease. *Ann Neurol*. 1987;21:451–457.
259. Sims NR, Finegan JM, Blass JP, Bowen DM, Neary D. Mitochondrial function in brain tissue in primary degenerative dementia. *Brain Res*. 1987;436:30–38.
260. Blass JP, Zemcov A. Alzheimer's disease. A metabolic systems degeneration? *Neurochem Pathol*. 1984;2:103–114.
261. Hoyer S. Brain oxidative energy and related metabolism, neuronal stress, and Alzheimer's disease: a speculative synthesis. *J Geriatr Psychiatry Neurol*. 1993;6:3–13.
262. Beal MF. Aging, energy, and oxidative stress in neurodegenerative diseases. *Ann Neurol*. 1995;38:357–366.
263. Reddy PH, Beal MF. Are mitochondria critical in the pathogenesis of Alzheimer's disease? *Brain Res Brain Res Rev*. 2005;49:618–632.
264. Moreira PI, Carvalho C, Zhu X, Smith MA, Perry G. Mitochondrial dysfunction is a trigger of Alzheimer's disease pathophysiology. *Biochim Biophys Acta*. 2010;1802:2–10.
265. Moreira PI, Cardoso SM, Santos MS, Oliveira CR. The key role of mitochondria in Alzheimer's disease. *J Alzheimers Dis*. 2006;9:101–110.
266. Zhu X, Perry G, Moreira PI, et al. Mitochondrial abnormalities and oxidative imbalance in Alzheimer disease. *J Alzheimers Dis*. 2006;9:147–153.
267. Brinton RD. The healthy cell bias of estrogen action: mitochondrial bioenergetics and neurological implications. *Trends Neurosci*. 2008;31:529–537.
268. Mancuso M, Siciliano G, Filosto M, Murri L. Mitochondrial dysfunction and Alzheimer's disease: new developments. *J Alzheimers Dis*. 2006;9:111–117.
269. Swerdlow RH, Burns JM, Khan SM. The Alzheimer's disease mitochondrial cascade hypothesis. *J Alzheimers Dis*. 2010;20(suppl 2):S265–S279.
270. Swerdlow RH, Khan SM. The Alzheimer's disease mitochondrial cascade hypothesis: an update. *Exp Neurol*. 2009;218:308–315.
271. Swerdlow RH, Khan SM. A "mitochondrial cascade hypothesis" for sporadic Alzheimer's disease. *Med Hypotheses*. 2004;63:8–20.
272. Swerdlow RH, Burns JM, Khan SM. The Alzheimer's disease mitochondrial cascade hypothesis: progress and perspectives. *Biochim Biophys Acta*. 2014;1842:1219–1231.
273. Giles RE, Blanc H, Cann HM, Wallace DC. Maternal inheritance of human mitochondrial DNA. *Proc Natl Acad Sci USA*. 1980;77:6715–6719.
274. Jack Jr CR, Wiste HJ, Lesnick TG, et al. Brain beta-amyloid load approaches a plateau. *Neurology*. 2013;80:890–896.
275. Burns JM, Swerdlow RH. Backwaters and rapids on the amyloid river. *Neurology*. 2013;80:878–879.
276. Swerdlow RH. Bioenergetic medicine. *Br J Pharmacol*. 2014;171:1854–1869.
277. Swerdlow RH. Mitochondrial medicine and the neurodegenerative mitochondriopathies. *Pharmaceuticals*. 2009;2:150–167.
278. Swerdlow RH. Role and treatment of mitochondrial DNA-related mitochondrial dysfunction in sporadic neurodegenerative diseases. *Curr Pharm Des*. 2011;17:3356–3373.
279. Swerdlow RH. Treating neurodegeneration by modifying mitochondria: potential solutions to a "complex" problem. *Antioxid Redox Signal*. 2007;9:1591–1603.
280. Swerdlow RH. Alzheimer's disease pathologic cascades: who comes first, what drives what. *Neurotox Res*. 2012;22:182–194.

CHAPTER TEN

The Kidney in Aging: Physiological Changes and Pathological Implications

H. Sobamowo, S.S. Prabhakar[1]
Texas Tech University Health Sciences Center, Lubbock, TX, United States
[1]Corresponding author: e-mail address: sharma.prabhakar@ttuhsc.edu

Contents

Abstract

Aging is associated with progressive decline in renal function along with concurrent morphological changes that ultimately lead to glomerulosclerosis. The mechanisms leading to such changes in the kidney with age as well as the basis of controversies that surround the physiological basis vs pathological nature of aging kidney are the focus of this in-depth review. In addition, the renal functional defects of acid–base homeostasis and electrolyte disturbances in elderly and the physiological basis of such disorders are also discussed.

Progress in Molecular Biology and Translational Science, Volume 146
ISSN 1877-1173
http://dx.doi.org/10.1016/bs.pmbts.2016.12.018

1. INTRODUCTION

The prevalence of chronic kidney disease (CKD) in the US adult population was 11% (19.2 million). By stage, an estimated 5.9 million individuals (3.3%) had stage 1 (persistent albuminuria with a normal glomerular filtration rate (GFR)), 5.3 million (3.0%) had stage 2 (persistent albuminuria with a GFR of 60–89 mL/min/1.73 m^2), 7.6 million (4.3%) had stage 3 (GFR, 30–59 mL/min/1.73 m^2), 400,000 individuals (0.2%) had stage 4 (GFR, 15–29 mL/min/1.73 m^2), and 300,000 individuals (0.2%) had stage 5, or kidney failure. Aside from hypertension and diabetes, age is a key predictor of CKD, and 11% of individuals older than 65 years without hypertension or diabetes had stage 3 or worse CKD. By 2 years of age, the GFR of a child nears adult levels, and it remains there until the fourth decade. Age is associated with a physiological decline in GFR which is almost about slightly under 1 mL/min for year or by about 8 mL/min/1.73 m^2/decade. Such a decline starts from the middle of the fourth decade. There is variation in the rate of decline given gender, race, and burden of comorbid disease. A basic question is whether the 11% of individuals older than 65 years without hypertension or diabetes with CKD stage III or worse CKD stage are due to pathologic or physiologic changes. We will first discuss the evaluation of renal function in general population and then address the changes in the elderly.[1]

2. EVALUATION OF RENAL FUNCTION

A nationally representative sample of 15,625 noninstitutionalized adults aged 20 years and older from the NHANES III was analyzed. Kidney function (GFR), kidney damage (albuminuria), and stages of CKD (GFR and albuminuria) were estimated from calibrated serum creatinine level, spot urine albumin level, age, sex, and race. GFR was estimated using the simplified Modification of Diet in Renal Disease Study equation and compared with the Cockcroft–Gault equation for creatinine clearance (CCr). Normal aging is accompanied by renal functional and structural deterioration.

To examine the hemodynamic and growth-related mechanisms of age-associated nephron loss, as well as the potential beneficial effects of antihypertensive therapy, studies were performed in normal aging Munich-Wistar rats, and in rats, receiving long-term antihypertensive therapy with the angiotensin-converting enzyme inhibitor (ACEI) enalapril.[2] Compared with young rats, untreated old rats studied at 2.5 years of age exhibited normal blood pressure but increased glomerular capillary pressure due to a reduction in afferent arteriolar resistance.[2] Glomerular size increased proportionately to changes in body weight, while kidney weight increased to a lesser degree. Albuminuria rose significantly after 10 months of age and was accompanied by development of modest, but significant, glomerular sclerosis. ACEI therapy from the age of 3 months lowered systemic and glomerular capillary pressures, did not affect glomerular size, and significantly ameliorated development of albuminuria and structural injury.[2] In protocol 2, untreated rats were compared with a treated group in which enalapril therapy was delayed until the age of 1 year, when albuminuria was already rising. Subsequent increases in albuminuria and development of sclerosis were significantly attenuated, although not entirely prevented. These findings suggest that hemodynamic mal-adaptations may contribute to age-related loss of renal function in the rat and that antihypertensive therapy may serve to delay this process.

Age-related changes in GFR, effective renal plasma flow (RPF), and tubular excretory capacity in adult males were evaluated by Davies and Shock.[3] Measurements of inulin clearance, diodrast clearance, and diodrast Tm were made under basal conditions in 70 males between the ages of 20 and 90 years. 9–12 subjects were selected from each decade on the basis of medical history, physical examination, and urine analysis. All subjects were free from history or clinical evidence of renal disease, essential hypertension, cerebrovascular accident, or heart disease. All subjects were ambulatory and afebrile. The average inulin clearance, diodrast clearance, and diodrast Tm decreased linearly beyond the age of 30 years (Fig. 1A and B). The average inulin clearance dropped from 122.8 to 65.3 cc/min/1.73 m^2 between the ages of 20 and 90 years (46%). Diodrast clearance dropped from 613 to 289 cc plasma/min/1.73 m^2 between the ages of 20 and 90 years (53%). Over the same age span, the diodrast Tm dropped from 54.6 to 30.8 mg L/min/1.73 m^2 (43.5%).[3] Lack of experimental evidence in that period precludes the ability to define the mechanisms for the observed changes. Subsequently, these data have been compared with renal function using Cr-based formulae (MDRD and Cockroft and Gault), which established that there is wide variability in the loss of renal function with aging (Fig. 1C).

2.1 Changes in Renal Physiology With Aging

The incidence and prevalence of CKD in elderly patients are continuously increasing worldwide. Loss of renal function is not only considered to be

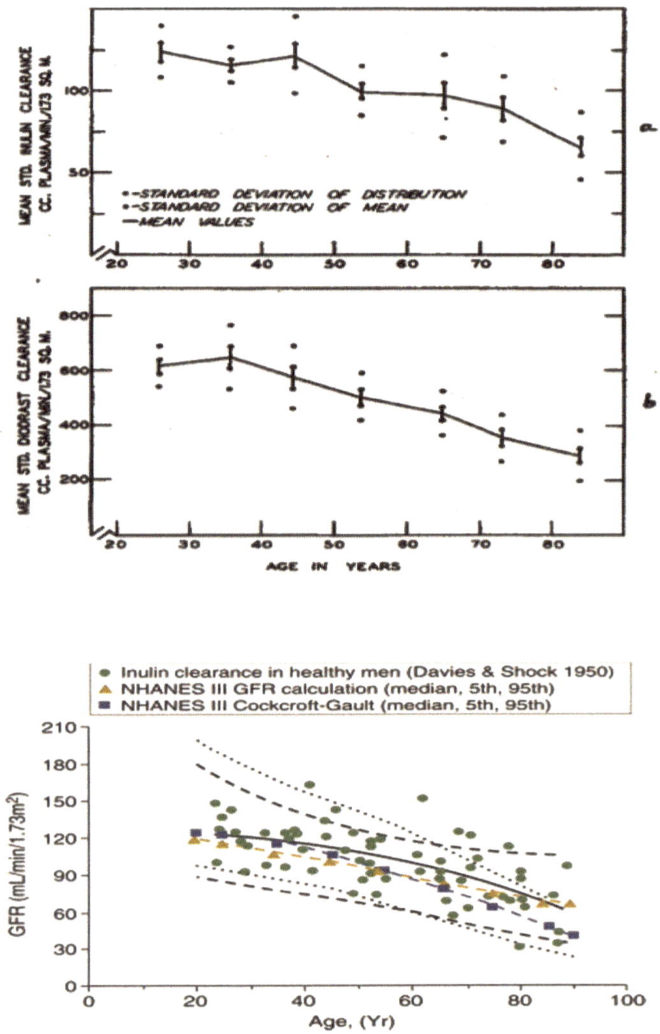

Fig. 1 Effects of age on renal function. GFR as measured by inulin clearance (A) diodrast clearance (B), and calculated creatinine clearance (C).

part of the aging process itself but also reflects the multiple morbid states of many geriatric patients. Chronic renal failure has many clinical consequences and not only results in a delayed excretion of toxins cleared by the kidneys but also affects erythropoiesis, water, and electrolyte balance as well as mineral bone metabolism. Furthermore, CKD directly leads to and aggravates geriatric syndromes especially with regards to the onset of frailty.

The age-related reduction in CrCl is accompanied by a reduction in daily urinary creatinine excretion due to reduced muscle mass.[14] The net effect is near-constancy of serum creatinine while true GFR (and CrCl) declines. The incidence of both microalbuminuria and overt proteinuria increase with advancing age, even in the absence of diabetes, HTN, or elevated SCr. Calculating the GFR using specific algorithms validated for the elderly population and together with measuring the amount of proteinuria will allow an estimation of renal function in elderly patients with high accuracy.[4]

The fawn-hooded hypertensive (FHH) rat serves as a genetic model of spontaneous hypertension associated with glomerular hyperfiltration and proteinuria. However, the knowledge of the natural course of hypertension and kidney disease in FHH rats remains fragmentary and the underlying pathophysiological mechanisms are unclear. In this study, over the animals' lifetime, the survival rate, blood pressure (telemetry), indices of kidney damage, the activity of renin-angiotensin (RAS), nitric oxide (NO) systems, and CYP450-epoxygenase products (EETs) were followed very closely. Compared to normotensive controls, no elevation of plasma and renal RAS was observed in prehypertensive and hypertensive FHH rats; however, RAS inhibition significantly reduced systolic blood pressure (137 ± 9–116 ± 8 and 159 ± 8–126 ± 4 mmHg, respectively) and proteinuria (62 ± 2–37 ± 3 and 132 ± 8–87 ± 5 mg/day, respectively).[5] Pharmacological RAS inhibition reduced angiotensin (ANG) II and increased ANGI–VII in the kidney. This factor may have delayed the progression of kidney disease. It was also noted that renal NO and EETs declined in the aging FHH rats but not in the control strain. The present results, published in "the clinical and experimental hypertension journal," demonstrate that exaggerated vascular responsiveness to ANGII, indicate that RAS may contribute to the development of hypertension and kidney disease in FHH rats.[5]

Therapeutic enhancement of this activity besides RAS inhibition could be attempted in the therapy of human hypertension associated with kidney disease. A study was designed by Cavanaugh *et al.*, to examine the prospective association between kidney function and three outcomes: survival to age 85 with functional independence, survival to age 85 with disability, and death before age 85.[6] This prospective study was conducted at 40 US clinical centers and participants were postmenopausal women who were enrolled between 1993 and 1998 with baseline biomarker assessments who had the potential to reach age 85 before September 2013 ($N = 7178$). Kidney function was measured according to estimated glomerular filtration rate (eGFR) calculated from serum creatinine collected at baseline. Disability was defined as mobility or activity of daily living limitations measured by questionnaire. eGFR was greater than 90 mL/min/1.73 m^2 in 22.7% of women, 60–89 mL/min/1.73 m^2 in 66.5%, 45–59 mL/min/1.73 m^2 in 8.7%, and less than 45 mL/min/1.73 m^2 in 2.0%. Median follow-up was 15 years. Of 4953 survivors, 3155 reported no physical disability at age 85.

Two thousand two hundred twenty-five participants died before age 85. Women with an eGFR of 90 mL/min/1.73 m^2 or greater had 2.71 times greater odds of survival to age 85 with functional independence than of dying before 85 (95% confidence interval (CI) = 1.62–4.51) than those with an eGFR less than 45 mL/min/1.73 m^2. Women with an eGFR of 60–89 mL/min/1.73 m^2 had 3.04 times (95% CI = 1.85–5.00) greater odds, and women with an eGFR of 45–59 mL/min/1.73 m^2 had 2.22 times (95% CI = 1.313.76) greater odds. Better kidney function was not significantly associated with greater likelihood of survival to age 85 with independent function than of surviving with disability.[6] Normal aging is accompanied by renal functional and structural deterioration.

2.2 General Mechanisms of Aging

Some understanding of the general mechanisms of aging can help us better understand the relationship between aging and renal dysfunction. Theories on aging have evolved according to the development of our understanding of biology. General theories that considered aging a global system-wide process have been replaced by the idea that the aging of a particular organism results from the sum of the aging of its individual cells. This approach to the understanding of aging is supported by much experimental evidence. In human beings, the aging-related dysfunction of organs and tissues, such as the brain or subcutaneous fat, is closely related to a reduction in the number of cells. The replicative capacity of cells explained from a variety of mammals is roughly proportional to the lifespan of the animals; this finding suggests a relationship between cellular aging and whole animal aging.

The WRN gene, a gene involved in the development of Werner's syndrome, a disease characterized by the appearance of a precocious aging phenotype in humans, is homologous to a family of DNA helicases of *Escherichia coli*. Aging thus might be a cellular autonomous process. In addition to containing individual cells, organisms also exhibit a complex system of cell-to-cell relationships and derangements in these relationships which could also be involved with aging.[7]

Independently of these considerations, two main theories have been proposed to explain aging: The first hypothesis, an environmental one suggests that aging is the consequence of the repetitive action of exogenous factors in a normal organism, resulting in an accumulated damage that outstrips the normal repair processes. The second, or genetic, theory proposes that aging occurs because of a genetic program that determines the progressive appearance of the different aging-related phenotypic changes. These two ideas are not mutually exclusive, and the accumulated damage could reflect an environment-dependent repetitive injury that triggers a genetic program of aging.

Organisms must obtain nutrients and in the case of aerobic organisms, oxygen from the external media or environment, to maintain cellular function and homeostasis. During the cellular metabolism of nutrients and oxygen, different toxic intermediate molecules arise; but cells have defense systems that are able to clear these molecules. Sometimes, however, production of toxic molecules overrides the protective mechanisms with the resultant damage that leads to progressive aging. The most important evidence supporting this hypothesis is dietary restriction by caloric restriction without compromising essential nutrients which is the most reproducible way of slowing aging. Two main groups of toxic molecules likely are involved in the aging process-reactive active species and advanced glycosylation end products (AGEs).

Reactive oxygen intermediates (ROI) have been the most widely studied. These molecules, including superoxide anion, hydrogen peroxide, and hydroxyl radical, among others, are formed during the progressive reduction of molecular oxygen to form water within the cell, but also as a consequence of the action of different cellular enzymes. ROI can induce chemical changes in many substances essential for normal cell function, including nucleic acids, proteins, and lipids, with subsequent structural and functional cell damage. The levels of macromolecules exhibiting oxidative damage increase in certain tissues of aged organisms. There are at least two recent studies that clearly support the role of ROI in aging. First, drosophila strains bearing extra copies of genes encoding both superoxide dismutase and catalase, the main enzymes involved in ROI removal, have longer lifespans than do drosophila without the extra genes. Second, the age-i mutant of *Caenorhabditis elegans*, characterized by an increased lifespan, also displays higher levels of superoxide dismutase.[7]

AGEs comprise the other group of molecules formed as a consequence of the basic metabolic process that seems to play a pathogenetic role in aging development. This is formed by the long-term interaction of reducing sugars, such as glucose or fructose, with the amino groups of intracellular or extracellular proteins. AGEs are increased in several pathologic situations, particularly diabetes, and in aged organisms. Protein glycosylation or the interaction of AGEs with specific cell receptors is associated with the development of functional and structural changes similar to those characterized by aging. A close relationship exists between AGE and ROI, as it seems that ROI are generated in cells after the interaction of AGEs with their receptors. Glycated proteins can undergo oxidative damage, with the subsequent accumulation of glycoxidation products such as N-epsilon (carboxymethyl) lysine, which are considered good markers of aging-related tissue damage.[7]

The genetic theory of aging proposes that the lifespan of any organism is determined by a specific genetic program. This theory is supported by the observation that Pacific salmon undergo a rapid senescence after spawning. A limited number of genes, including age-i, daf-2, and clk in *C. elegans* and WRN in human beings, have been related to the development of aging-related phenotypic changes. The exact mechanism of this programmed cell death has not been adequately defined; several possibilities have been proposed: the telomere shortening theory proposes that cells do not completely replicate their chromosomes during a cell division cycle with associated loss of very late replicating DNA sequences. The theory of terminal differentiation with programmed cell senescence as the consequence of the activation or inactivation of particular genes after a certain number of cell divisions. Other hypotheses propose that aging is the result of minimal but repetitive DNA injuries or progressive loss of copies of important genes. These theories only partially explain the aging phenomenon, particularly at a cellular level. A universal mechanism of cell aging has not yet been determined. Living organisms are provided with a genetic program, including a specific aging program, which controls their different functions. To maintain these functions, organisms must obtain nutrients and oxygen from the external medium and, in the processing of these metabolites, toxic molecules are formed. Although organisms can disarm most of these molecules, a small number of them can interfere with normal basic functions, or even with the genetic program, thus determining a particular rate of aging. Other external stimuli also might influence the aging process.

2.3 Kidney in Aging—Changes Physiological or Pathological?

2.3.1 Aging-Related Morphologic Changes

Although assessment of specific aging-related morphologic renal changes in the elderly is not easy because of the high prevalence of superimposed

Table 1 Morphologic Changes in the Kidneys of Elderly Individuals

Macroscopic changes	Reduced size and weight
	Relative cortical atrophy
Vascular changes	Hyalinosis of arterial walls
Glomerular changes	Increased number of sclerosed glomeruli
	Hypertrophy of the remnant glomeruli
	Increased thickness of basal membrane
	Mesangial matrix expansion
	Irregular fusion of foot processes
Tubular changes	Reduction in the number of tubules
	Atrophy of the tubular epithelium
	Tubular dilation
	Increased thickness of basal membrane
Interstitial changes	Interstitial fibrosis

vascular or inflammatory diseases, studies in apparently disease-free individuals have provided valuable information (Table 1).

Renal mass increases progressively from about 50 g at birth to over 400 g at the fourth decade, and then declines to fewer than 300 g by the ninth decade.[7] The loss of renal mass mainly depends on progressive atrophy of renal cortex, with relative sparing of the medulla. The cortical atrophy roughly reflects a decreased number of functioning nephrons. Under the age of 40, few glomeruli appear sclerosed; in contrast, by the eighth decade, between 10% and 30% of the glomeruli are completely sclerosed, the glomeruli of the outer cortex being especially affected.[7] The remaining functioning glomeruli appear to increase in size although recent measurements performed by computer-assisted image analysis suggest that after the fourth decade glomerular size declines slightly. The glomerular number decreases and the glomerular shape changes, with decreased lobulation, and reduction in length of the glomerular tuft perimeter relative to total area with glomerular tuft collapse. Degeneration of cortical glomeruli results in atrophy of both afferent and efferent arterioles leading to global sclerosis. Mesangial matrix increases progressively, and glomerular basement membrane undergoes progressive thickening; free intraglomerular anastomoses appear and functioning capillary loops are reduced (so-called glomerular simplification). Eventually, the increased extracellular matrix condenses into hyaline material and collapses the glomerular tuft, finally inducing complete glomerular sclerosis (Fig. 2A and B). The incidence of glomerular sclerosis increases with advancing age but, again, with wide variability.

Degeneration of glomeruli in the renal cortex in turn results in atrophy of the afferent and efferent arterioles; in the juxtamedullary area, glomerulosclerosis seems to cause the formation of a direct channel between these two arterioles. These channels could contribute to the maintenance of medullary blood flow as cortical perfusion declines. Tubular structures also decrease with aging. Although some studies suggested dissociation between glomerular and tubular atrophy, this hypothesis has not been confirmed, and a close relationship appears to exist between degenerative changes in glomeruli and those in tubules. Interstitial changes, with increased fibrosis, also frequently occur in the aging kidney.

Studies of aging-related renal changes in experimental models, performed mostly in albino rats, are consistent with the pathologic findings in humans. Two strains of rats, Fisher 344 and Sprague–Dawley, are especially prone to the development of age-related nephropathy, but other albino strains also develop variable degrees of renal damage with aging. In these animals, glomerular sclerosis is readily demonstrated after 24 months, but increased glomerular basement membrane thickness and progressive expansion of mesangial matrix are detected as early as 3 months. Glomerular size of the intact glomeruli increases with age. Studies also noted an increase in the number of mesangial cells with age.[7] Intratubular casts occur more frequently in old rats, with flattening and atrophy of the tubular epithelia. Interstitial fibrosis, a constant characteristic of these animals has been detected as early as after 8 months in Lewis rats. It is believed that the interstitial changes precede glomerular sclerosis in the renal aging process. Recently, morphologic studies were performed in 24-month-old Wistar rats (Fig. 2B). The morphologic changes observed were similar to those previously described. After staining with Syrius red, using a computer-assisted planimetric procedure, the mesangial matrix expansion and interstitial fibrosis were found to be increased by 273% and 181% were detected, respectively, when compared to 3-month-old animals.

The biochemical nature of the extracellular matrix accumulation in the aging kidney was evaluated, and studies performed both in human beings

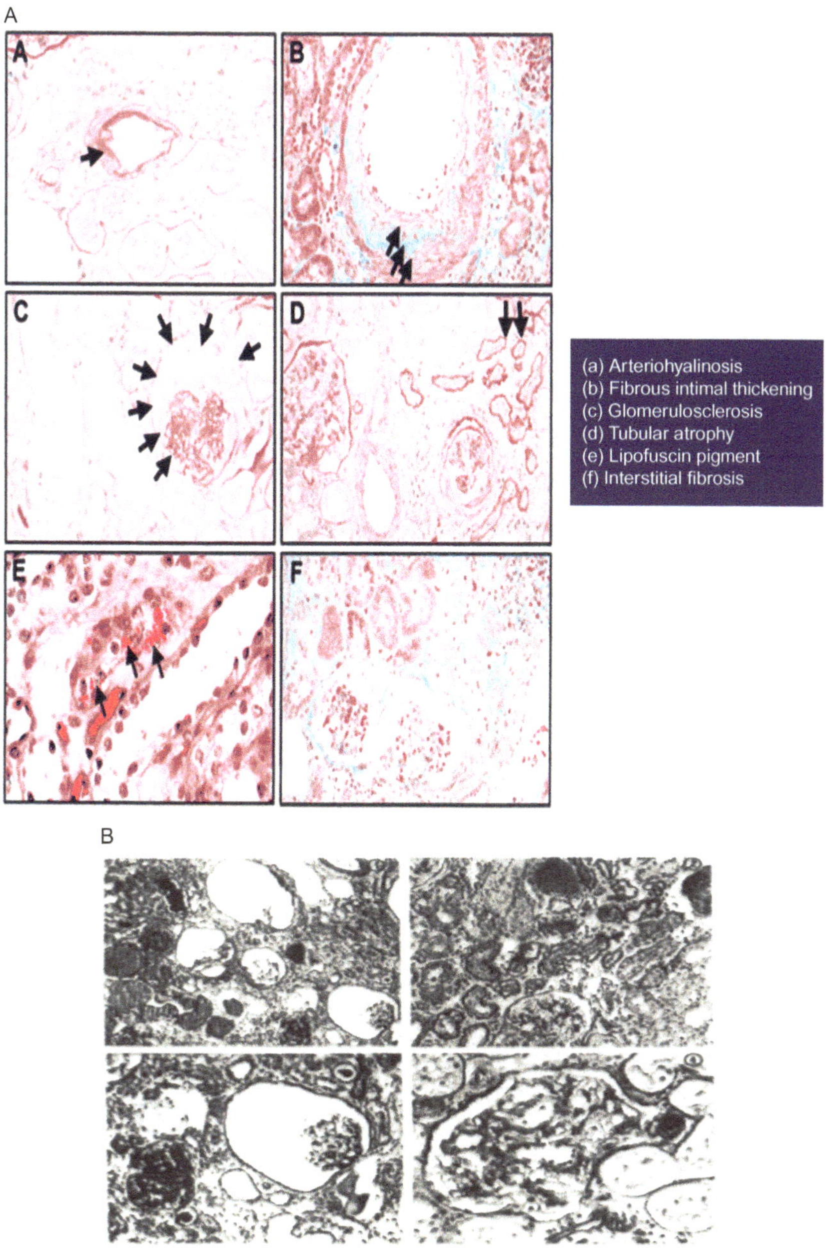

Fig. 2 (A) Histology of renal senescence. (B) Morphologic changes in the renal cortex of 24-month-old Wistar rats. (1) Diffuse glomerular and tubular changes, with cystic appearance and atrophy of the glomerular tuft of some glomeruli, glomerulosclerosis, tubular dilation, and intratubular casts (PAS 100 ×). (2) Tubular atrophy, reduplication of basal membranes, and interstitial expansion (PAS 400 ×). (3) Magnification of a sclerosed glomerulus, near another glomerulas with cystic appearance, in an area with interstitial expansion (PAS 400 ×). (4) Arteriolar hyalinosis (PAS 600 ×).

and in rats have confirmed that the chemical composition of the glomerular basement membrane differs between young and old individuals. Several changes have been detected in the old individuals, including increased non-enzymatic glycosylation of proteins and changes in the degree of sulfation of glycosaminoglycan. The most widely found biochemical change is increased collagen content. Abrass *et al.* recently questioned the hypothesis of collagen accumulation by performing immunofluorescence studies in Fisher 344 rats with a wide panel of antibodies. These authors demonstrated a moderate increase of collagens I and III only in areas with interstitial fibrosis, but detected no changes in collagens I, III, and IV at the glomerular level. The changes observed in the glomerular tuft, particularly in the glomerular basement membrane, were related to an increased content of various laminin isoforms, whereas in the interstitial compartment, a generalized immuno-staining for fibronectin and thrombospondin were observed. The relationship between interstitial fibrosis and collagen I accumulation also seems to be supported by the demonstration of increased levels of type-I collagen mRNA in the cortex of old rats. In contrast to the results from Abrass, preliminary results from this laboratory demonstrated an increased collagen type-IV mRNA (alpha-i chain) in the renal cortex of 24-month-old Wistar rats. This finding suggests that accumulation of this collagen plays a role in the genesis of the morphologic renal changes observed in aged rats. Differences in rat strain, age of the rat at the time of the study, or sensitivity of the techniques might account for the apparent discrepancies detected in the different studies.[7]

2.3.2 Functional Changes

Aging kidneys manifest significant functional (Table 2) as well as morphologic changes. In human beings, renal blood flow (RBF) decreases by about 10% per decade after the maximum level is reached in young adulthood. For example, RPF of approximately 600 mL/min/1.73 m^2 during the third decade decreases to about 300 mL/min/1.73 m^2, a 50% reduction, in the ninth decade. The decrease in RBF is associated with significant increases in afferent and efferent arteriolar resistances and the decline in RBF cannot be explained as a secondary phenomenon associated with the previously described renal mass reduction, as specific studies designed to answer this question demonstrated that the decreased RBF was accompanied by a real decline in blood perfusion per unit of renal tissue mass.[7] Changes in cardiac function can also not account for the reduction in this parameter, as the minimal reduction in the percentage of cardiac output

Table 2 Functional Changes in the Kidneys of Elderly Individuals

Renal blood flow	Decreased[a]
	Relative increase of medullary blood flow
Glomerulus	Decreased glomerular filtration rate[a]
	Increased filtration fraction[a]
	Increased permeability to macromolecules[b]
Tubule	Impaired ability for sodium handling
	Deranged tubular transport
	Impaired concentration and dilution
	Impaired acidification
Other	Decreased synthesis of renin
	Decreased 1α-hydroxylase activity

[a]Generally accepted in humans but not in rats.
[b]Increased prevalence of microalbuminuria in humans. Over proteinuria in rats.

directed to the kidney (that occurs in the elderly) does not explain the observed decline in RBF. Studies utilizing the xenon washout technique have demonstrated that the reduction in RBF is not uniform throughout the kidney. According to the anatomic descriptions, cortical blood flow is preferentially decreased in the elderly, with a relative sparing of the blood flow in juxtamedullary glomeruli. As these glomerular structures have a higher filtration fraction than do the cortical glomeruli, the observation that filtration fraction increases with advancing age could be explained by this observation.

Changes in RBF in experimental animals differ from those observed in human beings. The absolute values of RBF remain stable between 3 and 20–24 months and even slight increase in this parameter have been observed in 15- to 18-month-old Sprague–Dawley rats. When RBF is factored by kidney weight, however, it significantly decreased with aging and these data have been sometimes interpreted as indicative of an aging-related significant derangement of RBF. The analysis of preglomerular and postglomerular resistances by micropuncture has yielded different results, depending on the rat strain. These resistances were increased in old Munich-Wistar rats 1161, but were decreased in Sprague–Dawley animals.[7]

GFR has been studied extensively in the elderly. Cross-sectional studies have demonstrated that GFR decreases progressively after age 30–40 years. This decreased GFR was detected not only in cross-sectional studies but also in longitudinal studies. The Baltimore Longitudinal Study of Aging also found a progressive decline of glomerular filtration with aging. The rate of decline was 0.8 mL/min/1.73 m^2/year, a rate similar to that previously reported in cross-sectional studies. Both diminished glomerular lobulation and sclerosis of glomeruli reduce the surface area available for filtration and contribute to the observed age-related decline in GFR. In addition, age-related changes in CV hemodynamics, such as reduced CO and systemic HTN, are likely to play a role in reducing renal perfusion and filtration. eGFR did not change in approximately one-third of the patients included in this longitudinal study. In contrast to the decline in GFR, plasma creatinine does not change with increasing age. It is important to note that muscle mass from which creatinine is derived decreases with age at approximately the same rate as does GFR. In consequence, the age-related loss of GFR is not reflected by an increased concentration of plasma creatinine. As a result of this factor, this parameter must be used with caution in elderly populations to assess GFR, since it underestimates renal function. The commonly used formula for estimating CCr from plasma creatinine values always take into account the age of the patients and is preferred.

Another aspect of glomerular function that has been extensively studied is the permselectivity of the filtration barrier. Although some reports describe an increased prevalence of proteinuria in a population of persons over 65 years, only a minority of disease-free patients over 80 years show clinical proteinuria.[7] When the glomerular permeability to macromolecules was studied, there was no difference detected between young and old individuals. This therefore seems to suggest that the permselectivity of the glomerular filtration barrier in human beings is only minimally altered in the elderly. Glomerular function in old rats differs from that in older humans. Assessing different reports on GFR is difficult because data that are frequently expressed are corrected for the body and kidney weight; these two variables increase in old animals, but it does seem that GFR remains stable in albino rats until 18–24 months of age and then declines progressively. On the other hand, proteinuria is a constant manifestation of renal dysfunction in old rats in strains that develop glomerulosclerosis (Fig. 3). Although the exact nature of this selectivity defect has not been elucidated, some evidence points to a combined charge and size defect as responsible for the aging-related increased proteinuria.

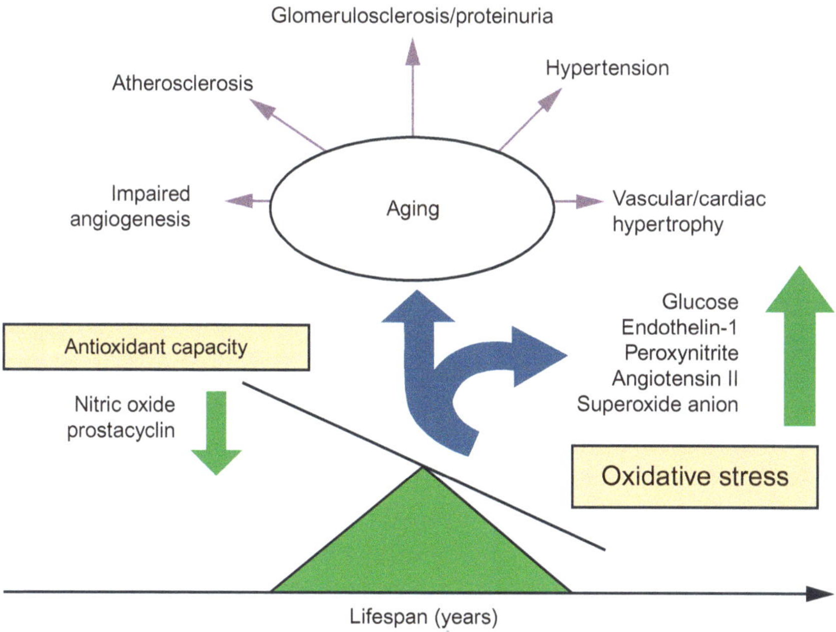

Fig. 3 Proposed mechanisms of the vascular and renal aging process.

2.4 Aging and Tubular/Electrolyte Balance

2.4.1 Sodium and Water Handling

The aging kidney is able to maintain normal electrolyte homeostasis under steady-state conditions, but it has impaired ability to respond to perturbations of fluid and electrolyte balance. When old individuals are sodium-deprived, sodium excretion progressively declines, but it takes longer to achieve equilibrium in comparison to when younger people undergo the same deprivation. The mean half-time for reduction of sodium excretion is 17 h in individuals less than 30 years old, but is prolonged to 31 h in subjects more than 60 years old. Nevertheless, the elderly can achieve sodium equilibrium even when given diets with very low sodium content. Elderly subjects also have problems with sodium overloads, edema, and hypertension which frequently occur in this population. Short-term sodium-loading studies show distinct age-related sodium excretion patterns: after a 2-L normal saline load, individuals older than 40 years show a lower 24-h sodium excretion (with a significantly greater portion of the sodium excreted at night) than do their younger counterparts. One of the best-known aspects of tubular dysfunction in the elderly is their relative inability to adequately concentrate and dilute the urine.

2.4.2 What Accounts for the Relative Inability of Elderly Individuals to Reabsorb Sodium Normally?

A number of factors could explain the aging effect on sodium handling. First, aging-associated structural renal alterations, such as interstitial fibrosis or a decreased number of tubules, might play a part in the homeostasis. Second, decreased GFR and its associated hyperfiltering glomeruli also might explain some aspects of the deranged sodium homeostasis. As a result of this, a decreased GFR could produce, on a short-term basis, a relative inability to excrete a sodium load. On the other hand, the hyperfiltering nephrons excrete a solute load significantly higher than do normal nephrons, with a subsequent osmotic diuresis and natriuresis. These hyperfiltering nephrons are unable to readily reabsorb sodium. Third, hormonal changes also might account for changes in sodium homeostasis. When plasma aldosterone was measured in elderly individuals, values were significantly lower than those in young people. This decreased aldosterone synthesis could contribute to the relative inability of the aging kidney to conserve sodium. Another hormone proposed to account for this deranged sodium excretion is atrial natriuretic peptide (ANP). Plasma levels of. this hormone significantly increase in elderly subjects. However, the natriuretic response after the infusion of exogenous ANP seems to be decreased in healthy elderly men. These data have been interpreted as a relative decreased responsiveness of the aging kidney to ANP. Decreased concentrating ability in the elderly has been attributed to changes in the functional status of the hypothalamic–pituitary axis. However, when the release of arginine vasopressin (AVP) was analyzed under different physiologic stimuli, elderly subjects exhibited increased AVP release with respect to young individuals. Not all studies have had similar results, however, and decreased aging-related AVP release also has been demonstrated. Moreover, nonosmotic AVP release also might be deranged in the elderly. In any case, it is generally accepted that the most important mechanism involved in the elderly's renal concentrating defect is an inadequate renal response to endogenous AVP. In humans, this lack of response has been attributed to aging-related tubulointerstitial structural changes as well as to a derangement in the intrarenal mechanisms responsible for the maintenance of medullary hypertonicity, including solute transport by the thick ascending limb of the loop of Henle and relatively slow medullary blood flow. Studies in rats also suggest that impaired responsiveness of the collecting duct cells to AVP is involved in the concentrating defect in old animals. The basis of this defect could be a decreased number of V2 receptors, but a recent report failed to demonstrate such a decrease.

In consequence, defective coupling of this receptor to the adenylate cyclase system likely explains the lack of response to AVP in older rats. The intrinsic mechanisms of the renal diluting defect in the elderly have been less studied, but they might relate to the aging-related decrease in GFR and decreased solute transport in the thick ascending limb of the loop of Henle.

Studies in the elderly suggest Na handling is fairly normal in the proximal tubule, but the capacity to reabsorb Na in the ascending limb of the loop of Henle is markedly impaired. The reduced loop capacity to reabsorb sodium has two important consequences:

(1) The amount of sodium delivered to the more distal segments increases.
(2) The capacity to concentrate the medullary interstitium is reduced, which further contributes to the inability to concentrate the urine.

Age-related abnormalities in several hormonal systems controlling Na excretion play a role in impaired ability to conserve Na. Levels of plasma renin and of blood and urinary aldosterone fall significantly, and responses to appropriate stimuli such as Na restriction are blunted. The impaired response to Na deprivation (relative salt wasting) makes the elderly patient more susceptible to development of a cumulative Na deficit and its attendant systemic complications. Similarly, the renal response to a sodium load is sluggish in older patients.

2.5 Disorders of Water Balance

Studies comparing the maximal urinary density or osmolality after water deprivation in young and old individuals have clearly demonstrated that kidneys from old subjects do not form urine as concentrated as that of young people. Renal diluting ability also is impaired in elderly individuals. During water diuresis, urine osmolality in old subjects is significantly higher than that in young subjects, and solute-free water clearance is lower. The same defects in urinary concentration and dilution also have been described in laboratory rats.

In response to water deprivation, both the maximal decrease in urine volume and the increase in urine osmolality in healthy elderly subjects are significantly diminished. The maximum urine osmole after dehydration is 1109 mOsm/kg in subjects aged 20–39 years, 1051 mOsm/kg in those aged 40–59 years, and 882 mOsm/kg in those aged 60–79 years. Several conditions may contribute to this defect: the reduced number of functioning nephrons may contribute to an obligatory solute diuresis in the remaining intact nephrons, altered responsiveness to exogenous AVP, and the release of endogenous AVP in response to appropriate stimuli is abnormal and in some cases, there is after rising serum osmolality, so volume depletion or hyperosmolality stimuli are less effective.

Serum Na levels remain within the normal range in healthy elderly individuals but defective Na and water homeostatic mechanisms render this population markedly susceptible to derangement. Hyponatremia is the most common electrolyte disorder in the elderly, occurring in as many as one quarter of all hospitalized elderly patients. The most common underlying mechanisms of geriatric hyponatremia:

(1) Decreased ability to excrete water
(2) Water intoxication in the setting of diuretic therapy
(3) Oversecretion of AVP

Hypernatremia is also prominent in the elderly. At particularly high risk are institutionalized older patients with cognitive impairment, who often manifest failure to recognize thirst and/or physical inability to obtain fluids. Cerebrovascular disease may also inhibit thirst, as well as limiting the physical ability to gain access to fluids.

2.6 Potassium Disorders

Significant abnormalities in cellular and total body potassium occur with advancing age. The erythrocyte K^+ concentration is decreased, and both total body K^+ and total exchangeable body K^+ are reduced by about 20% compared with younger subjects. The mechanisms responsible include decreased muscle mass, alterations of cell membrane characteristics, nutritional deficiencies, and inability of the kidney to conserve potassium. Hypokalemia is the most prominent potassium abnormality in the elderly population. The most prominent cause of hypokalemia in the elderly is probably diuretic therapy.

2.7 Acid–Base Balance

The healthy elderly are generally able to maintain normal values for serum pH, Pco_2, and HCO_3 concentration. There is a modest but significant decrease in serum HCO_3 levels (within the normal range) with aging. Impaired acidification seems to be a consequence of reduced renal mass, although some studies suggest an intrinsic acidification defect, possibly associated with impaired ammonium excretion. These systems adequately dispose of the normal daily acid load. Studies of ammonium loading in elderly patients indicate a reduced ability to excrete an acute exogenous acid load.

3. CALCIUM, PHOSPHORUS, AND MAGNESIUM DISORDERS IN AGING

Other aging-related tubular defects, including defective phosphate management, have been analyzed less extensively. Defective phosphate reabsorption by the proximal tubule partially depends on the increased PTH concentration associated with decreased GFR. But parathyroidectomy only partially prevents this defect, so alternative mechanisms of deranged phosphate reabsorption must be at play. Phosphate transport is decreased in cultured renal tubular cells, and Levi *et al.* have suggested that this decreased transport results from changes in the chemical composition of cell membranes. Moreover, the expression of the Na-phosphate cotransporter decreases with age in tubular cells. Serum levels of total Ca, ionized Ca, phosphorus, magnesium, and PTH usually remain within the normal range in the elderly. There may be a tendency toward increased serum PTH levels with advancing age. A decrease in vitamin D levels is frequently seen in elderly patients who are in poor health due to lack of exposure to sunlight, dietary deficiency, and impaired conversion to calcitriol. Although renal tubular calcium absorption appears to remain relatively intact with aging, calcium metabolism is impaired. This may be due to age-related decreases in intestinal Ca^+ absorption, reduced renal 1α-hydroxylase activity, diminished $1{,}25(OH)_2$ vitamin D3 activity and decreased intestinal adaptation to dietary Ca^+ restriction.

Other disorders of tubular transport widely studied are decreases in sodium-hydrogen exchange and in sodium-coupled phosphorus reabsorption. These defects also have been demonstrated in laboratory animals and even in preparations of brush-border vesicles. This age-related decline in sodium-dependent phosphate transport precedes the effect of age on sodium-hydrogen exchange in brush-border membrane vesicles. These data suggest that all membrane transport functions at the proximal tubule are not similarly affected during the aging process.

3.1 Renal Hormonal Synthesis

The renal aging process is also characterized by decreased renin synthesis. Studies in humans and in rats have demonstrated decreased concentrations and activities of plasma renin despite normal plasma concentration of renin substrate, as well as decreased renal renin content in elderly individuals. In these subjects, maneuvers designed to stimulate renin secretion amplified the differences in plasma renin levels with respect to the young population. Jung *et al.* demonstrated decreased renin mRNA content in renal tissue in 12-month-old Sprague–Dawley rats, even in the absence of significant changes in renal renin. In contrast, Corman *et al.* detected significantly decreased renin content in 30-month-old female WAG/nj rats, without changes in renin mRNA expression. A deficit in 1α-hydroxylase activity is another characteristic of aged subjects. As a consequence of this defect, plasma levels of 1,25-dihydroxycholecalciferol decrease in this population, with a subsequent derangement in calcium homeostasis

3.2 Mechanisms Responsible for Renal Changes During Aging

Analysis of the mechanisms involved in the development of aging-related renal changes has been performed at two levels. Most studies have looked for the immediate reasons that explain the changes detected in renal structure and function in old human beings or animals. However, these studies have not established a relationship between the specific mechanisms studied and the aging process. In consequence, a second set of studies or hypotheses have tried to establish a causal link between the aging process itself and the possible mechanisms involved in the genesis of the renal changes.[7]

Aging-related morphologic renal changes are similar to those detected in renal disease and experimental models characterized by progressive chronic renal failure, including glomerulonephritis, diabetes, and surgical reduction of renal mass. Although the exact biochemical composition of the expanded extracellular matrix in aging kidneys is not fully comparable to that in any of those pathologic situations, one could hypothesize that aging share some of the pathogenetic mechanisms proposed for these diseases. Table 3 lists the most widely accepted mechanisms of extracellular matrix expansion and changes in cell numbers in the kidney in progressive renal diseases. Some of these mechanisms have been widely explored in experimental models of glomerulonephritis or diabetes, as well as in rats with surgical renal mass reduction. Four main aspects of the table must be stressed. First, the number of cells at a particular time in disease progression is regulated by the balance between cell proliferation and apoptosis (programmed cell death). Second, increased cell numbers can precede glomerulosclerosis and interstitial fibrosis, even in situations in which cell proliferation is not readily detected. Third, changes in the complex equilibrium between matrix synthesis and

Table 3 Mechanisms and Factors Involved in the Expansion of Extracellular Matrix and Change in Cell Numbers in Progressive Renal Diseases

General mechanisms
Changes in the proliferation rate of resident or infiltrating cells
Changes in the apoptosis rate of resident or infiltrating cells
Increased synthesis of normal or abnormal extracellular matrix components
Decreased degradation of normal or abnormal extracellular matrix components
Factors involved in the regulation of these mechanisms
Growth factors: PDGF, EGF, TGFβ, FGF, IGF-1[a]
Cytokines Il-1, Il-13, TNF
Vasoactive peptides: AII, ET, ANP
Lipid mediators: PGE_2, PGI_2, TxA_2, PAF
Others: NO, ROI

[a]*Abbreviations*: *AII*, angiotensin II; *ANP*, atrial natriuretic peptide; *EGF*, epidermal growth factor; *ET*, endothelin; *FGF*, fibroblastic growth factor; *Il-1*, interleukin-1; *Il-13*, interleukin-13; *IGF-1*, insulin-like growth factor 1; *NO*, nitric oxide; *PAF*, platelet-activating factor; *PDGF*, platelet-derived growth factor; *PGE_2*, prostaglandin E_2; *PGI_2*, prostacyclin; *ROI*, reactive oxygen intermediates; *TGFβ*, transforming growth factor β; *TNF*, tumor necrosis factor; *TxA_2*, thromboxane A_2.

degradation are, perhaps, the main critical point in fibrosis development. Fourth, from a functional point of view, a wide overlap exists between the classical growth factors and the different vasoactive factors, as the former may induce significant hemodynamic effects, whereas the latter may modify the rate of proliferation and protein synthesis in different cells.[7]

These mechanisms have been inadequately explored in aging. The number of mesangial cells increases in the early stages of aging in rats, but no additional studies have confirmed this finding. It would be very important to assess the rates of proliferation and apoptosis of the different renal cells as a function of age to ascertain the importance of these phenomena in the genesis of the progressive replacement of cells by extracellular matrix. In addition, glomerular hypertrophy frequently occurs in elderly individuals and old rats. Although this hypertrophy has been frequently considered the consequence of intrarenal hemodynamic changes, some authors believe that the hypertrophy depends on the release of local mediators, including growth factors, and that it could be an indirect marker of the increased local production of growth-promoting metabolites. Only one recent report demonstrated an inverse relationship between urinary epidermal growth factor excretion and age, and suggested that a reduced production of this growth factor retards the repair process at the kidney level. In contrast, different vasoactive autacoids have been studied as possible mediators of the functional changes of the aging kidney, and it is now a well-recognized fact that autacoids regulate cell proliferation and/or extracellular matrix synthesis. Thus, ANGII and endothelin have well-defined effects on cell proliferation, whereas nitric oxide (NO) and ANP possibly inhibit cell growth and extracellular matrix synthesis. Changes in these vasoactive, growth-promoting metabolites could be involved in the development of the aging-related morphologic changes. Finally, data about the possible importance of the extracellular matrix degradation in the genesis of the structural changes related to aging correlated with decrease in glomerular and tubular proteinase activities in aging rat kidneys. According to the previously discussed criteria, the possible consequences of this decreased proteinase activity would be increased extracellular matrix and subsequent morphologic changes in the kidney. Some of these mechanisms are likely influenced by gender, an important determinant of the rate at which the kidney is damaged with age. Morphologic alterations are less severe in old women than in old men, and male rats have a higher mortality rate due to renal failure. These gender differences seem to depend on the presence of androgens rather than on the absence of estrogens; castration of older animals prevented the development of renal dysfunction in male rats without modifying the pattern of renal damage in female rats. TGF13 is involved in the development of aging-related morphologic changes, particularly interstitial fibrosis. Expression of TGF13 mRNA in renal cortex of Wistar rats increased progressively with aging (Fig. 4). The blockade of TGF13 expression by long-term treatment with captopril partially prevented the development of interstitial fibrosis but not of glomerulosclerosis. TGF13 modulates the proliferation rate of different renal cell types, as well as matrix synthesis and degradation. Thus, the progressive glomerulosclerosis associated with experimental glomerulonephritis, diabetes, or renal mass reduction could depend on TGF13 mRNA overexpression. Although a direct temporal relationship between aging and TGF13 mRNA expression was demonstrated, only interstitial fibrosis seemed to depend on this overexpression, in contrast with previous results in other pathologic situations. Thus pathogenesis of aging-related renal dysfunction is not fully comparable to those of glomerulonephritis, diabetes, or renal mass reduction.[7]

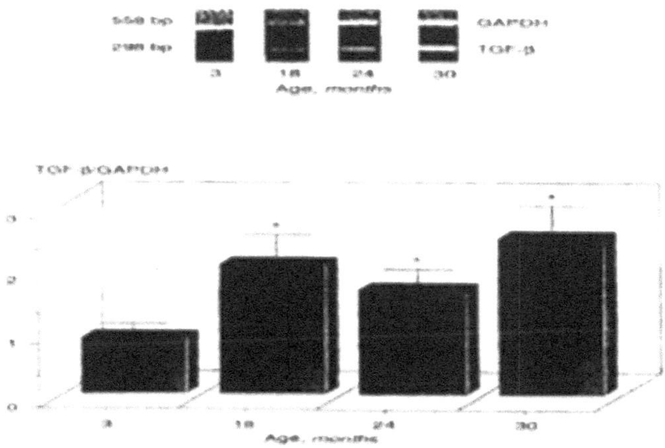

Fig. 4 Expression of the TGFβ mRNA in the renal cortex from 3-month-old, 18-month-old, 24-month-old, and 30-month-old rats. *Upper panel*, the simultaneous amphtication of the TGF-pl and GAPDH (housekeeping gene) mRNAs by using RT-PCR, in samples from rats of different ages. *Lower panel*, the ratio between the two amplification products (TGFf 1/GAPDH) was calculated and the mean SCM of six different rats are given. $^*P < 0.05$ vs 3-month-old rats. *Published with permission of* J Am Soc Nephrol.

3.3 Functional Mechanisms

Aging-related functional changes are closely related to the morphologic alterations previously described. However, they are not only the consequence of structural changes, as a deranged regulation of different aspects of normal renal function also seems to be involved in the genesis of these changes. A complex network of functional relationships between the different renal structures exists, and changes in the function of a particular structure can depend on the dysfunction of others. Renal vessels in healthy elderly humans or rats do not show structural changes significant enough to completely explain the reduction in RBF, except when other pathologic situations such as arteriosclerosis are superimposed on the basic aging process.[15] In consequence, authors have looked for the intrinsic causes of the decreased RPF in the disease-free elderly individual. It is generally accepted that the basic defect in the vessels of these individuals is an impaired ability to relax in the presence of some well-defined vasorelaxant stimuli.[17] From the initial description of decreased renal vasodilation after pyrogen injection in human beings four decades ago, authors have demonstrated, in human beings or in animals that the normal vasodilatory responses induced by acetylcholine, amino acids, glycine, or after food ingestion are blunted in the elderly. However, some discrepancies exist in this area. A normal, unblunted renal vasodilatory response to acetylcholine and L-arginine infusion in older rats has been shown. On the other hand, the agonist induced renal vasoconstrictive response in the aging kidney might not be the same in human beings and rats. The ANGII-induced reduction of RBF in elderly humans was independent of the age of the subjects, and the renal vasodilatory response to acute ACEI persisted even in old people. In contrast, Tank *et al.* found that kidneys of aged rats exhibited an exaggerated response to systemic vasoconstrictor stimuli. Taken together, these data suggest that the combination of defective vasodilation with normal or increased vasoconstriction accounts for the reduced RBF that characterizes aging. The basis for these altered vascular responses in the aging kidney is not understood. The relative impact of age, sex, body build, hypertension, systemic atherosclerosis, intrarenal vascular disease, and interstitial fibrosis on glomerulosclerosis and glomerular size was investigated using multiple linear regression. Both age and intrarenal vascular disease exhibited highly significant, independent associations with glomerulosclerosis.[16]

Pyrogen injection, acetylcholine, and amino acid infusion share a common mechanism of inducing renal vessel relaxation, that is, the local release of NO, one of the most important vasodilator mechanisms. Defective synthesis or activity of the NO system might be involved in the impaired renal vascular vasodilatory response observed in old individuals. Three mechanisms could account for this possible defective response: reduced production of NO in response to different stimuli, increased degradation of the NO released, or increased synthesis of this metabolite with a decreased response of the target cells. The first of these three hypotheses is supported by the fact that the 24-h urinary excretion of nitrites plus nitrates (considered an indirect index of NO synthesis in the kidney) as well as the glomerular synthesis of nitrites is decreased in older rats. However, in studies of aged rats with NO synthesis blockade, the dependence of RBF on nitric oxide is greater in old than in young individuals. They suggested that increased synthesis of this compound is necessary to maintain renal perfusion. As a consequence of the increased and maintained basal NO synthesis, the response to stimuli such as acetylcholine or amino acids would decrease. Unfortunately, neither direct measurements of local NO synthesis in renal vessels nor a detailed analysis of the expression and activity of NO synthases in renal cortex is available at present. However, a report from Hwang *et al.* demonstrated that cultured proximal tubule epithelial cells from kidneys of old donors express significantly higher amounts of constitutive NO synthase and this report opens new perspectives in the study of this problem. The reasons for a possible increase in basal NO synthesis in renal cortex of the aging kidney is unclear. Nitric oxide could act as a counteracting mechanism of some vasoconstrictor mediators that might be increased in aging. High-circulating endothelin levels have been described in plasma of aging men, endothelin mRNA expression is increased in cultured vascular endothelial cells from old compared with young individuals, and increased endothelin-1 secretion was detected in aged cultured human umbilical vein endothelial cells. Endothelin, via its ET-B receptor, might increase NO synthesis. This group, in collaboration with the Department of Pathology of the University of Granada, has found that mRNA expression of preproendothelins-1 and -3 increases in old rats (unpublished data). This work therefore supports a possible role for this peptide in the vascular renal changes that characterize aging.

Nitric oxide and endothelin are not the only vasoactive systems that have been studied as possible mediators of aging-related functional changes. Impaired arterial baroreflex, with a subsequent increased renal sympathetic activity, has been proposed as a possible mechanism for increased renal vascular resistance. The previously mentioned data concerning renin synthesis in elderly human beings and rats would lead us to expect that ANGII decreases with aging, and some studies support this contention. However, one recent report proposing that this peptide actually increases with age, and the significant renal vasodilation with increases in RPF observed in older rats in response to the acute ANGII blockade, suggest that intrarenal ANGII is activated in aged rats.

Synthesis of platelet-activating factor (PAF), a lipid mediator with well-recognized vasoconstrictor ability, seems to be increased in isolated glomeruli of aged rats; this finding suggests a pathogenic role for this autacoid in aging-induced altered vascular responses. The equilibrium between prostacyclin and thromboxane is also deranged with aging, and the ratio of prostaglandin 12 to thromboxane A2 decreases in the urine of older humans, as well as in the glomeruli and inner and outer medulla of older rat kidneys. Finally, impaired ANP-induced relaxation of renal arteries in rats and monkeys provide yet another way of impairing renal vasodilation.

Another possible explanation for the impaired renal vasodilatory response is defective activation of intracellular second messengers that mediate vascular smooth cell relaxation. A blunted cAMP response to adrenergic agonists has been described in blood vessels with aging, and this alteration modifies vascular responses to stress and exercise. Further this response could depend on impaired guanine nucleotide regulatory protein (G protein) function. However, it was not possible to detect changes in the amount of these G proteins in aged rats, particularly Gs and Gi, at the renal level. On the other hand, the defective response to ANP in blood vessels from elderly individuals might be due to accelerated degradation of cGMP by phosphodiesterases.

Additional studies are needed, including evaluations of the different intracellular systems involved in vascular smooth muscle cell relaxation, to clarify the importance of these mechanisms in the development of aging-related decreased RPF. Changes in GFR in the elderly are generally attributed to progressive glomerular sclerosis and decreased RPF. However, formation of the glomerular ultrafiltrate is a finely regulated phenomenon, and the reduction in GFR occurs more slowly than does the fall in RPF, the result being an increased filtration fraction. Two mechanisms might account for the relative GFR maintenance, even in the presence of significant hemodynamic and structural changes. First, the aging process produces a nonhomogeneous derangement of RBF, with a preferentially decreased cortical flow. As the filtration fraction of the juxtamedullary glomeruli seems to be higher than that of the cortical glomeruli, the GFR would diminish less than in the case of a homogeneous reduction in renal perfusion. Second, it is well known that the remaining glomeruli undergo hemodynamic changes to compensate for the lack of function of the sclerosed glomeruli. This phenomenon, known as secondary hyperfiltration, could maintain adequate GFR even in the presence of a significant reduction of functioning nephrons. The mechanisms involved in the development of hyperfiltration have been extensively studied in different experimental models. In most cases, hyperfiltration seems to depend on increased glomerular plasma flow and increased hydrostatic pressure in the glomerular capillary network as a consequence of selective vasodilation of the afferent arteriole. Micropuncture studies in aged rats have analyzed the determinants of glomerular ultrafiltration. Most of these studies demonstrate decreased resistance of the glomerular afferent arteriole with an increased glomerular plasma flow. Controversy exists, however, with respect to changes in the glomerular capillary pressure in these old rats, because normal or increased values have been detected. Moreover, an increased age-related ultrafiltration coefficient also has been described. A recent micropuncture study by Baylis, performed in Munich-Wistar rats, failed to demonstrate the previously described changes in old rats, as glomerular plasma flow decreased and afferent arteriole resistance increased, whereas the ultrafiltration coefficient did not change. This study points to a possible interstrain variability in the mechanisms involved in the development of aging-induced GFR changes. However these studies were performed in rats as old as 2 years, when glomerulosclerosis and proteinuria are readily detected but when absolute values of GFR are maintained.

As in the case of RBF, changes in GFR in the elderly might be due, at least partially, to an imbalance among the autacoids that regulates intraglomerular hemodynamics. However, no studies have analyzed independently these possible local mediators. As regulation of intraglomerular hemodynamics depends on the regulation of the afferent and efferent arterioles, the autacoids involved could be the same in both processes. Additional studies are needed to clarify the specific mediators that lead to the renovascular and glomerular changes of aging. Aging and activation of the mechanisms involved in the renal changes in the elderly have focused on a particular deranged aspect of renal structure or function. However, these studies have not provided enough clues about how aging determines the activation of these mechanisms. For instance, some authors have proposed that reduced RBF in elderly individuals depends on changes in the local synthesis of NO but how aging induces these changes remains unclear. The knowledge of the general mechanisms of aging and of the direct mediators of renal dysfunction may assist to better understand the aging process at the renal level. Alterations in the genetic program, induced by turning on the cell death program (apoptosis) or by exogenous damage to DNA may explain the activation of particular pathogenetic mechanisms in the aging kidney. Thus, changes in the synthesis of certain growth factors, vasoactive mediators, or cell transporters might be associated with the genetic changes that induce aging. Unfortunately, a detailed analysis of the genetic changes in aged kidney cells, unlike in aged flbroblasts have not been performed. Anderson and Brenner suggest that aging-related progressive renal damage is the consequence of a continuous exogenous stimulus, the diet, on renal structure and function. Long-term low-protein feeding and chronic ANGII-converting enzyme inhibition have a protective effect on the glomerulus in the aging rat. Since these two maneuvers lower intraglomerular pressure in other experimental models of renal disease, these data have been interpreted as indicating that glomerular hypertension, induced by a high-protein diet, causes age-induced nephropathy. The mechanisms connecting intraglomerular hypertension with progressive damage of glomerular structures are currently being investigated, and it seems that changes in the mechanical forces acting on the cells might induce significant phenotypic changes in resident glomerular cells, thereby modulating the release of local mediators. The critical point for validating this theory would be the direct demonstration of increased intraglomerular pressure in elderly experimental animals.[7]

Anderson *et al.* reported increased intraglomerular pressure in 24-month-old Munich-Wistar rats. However, Baylis did not find significant changes in glomerular pressure in males up to 20 months of age of the same rat strain, even though the animals had significant glomerular structural damage. Moreover, intraglomerular pressures in castrated male and female Munich-Wistar rats, which do not develop glomerulosclerosis, did not differ from values in intact male rats. Other rat strains that develop glomerulosclerosis earlier, for example, Sprague–Dawley, show small increases of intraglomerular pressure at 13–18 months but not at 20–22 months of age. All these data suggest that intraglomerular hypertension is a relevant, but not the sole, factor in aging-related renal damage, nor is it likely the initial mechanism that triggers progressive renal dysfunction in the elderly.

ROIs also might be a link between aging and renal damage. The first argument supporting a role for ROI in the progressive renal damage of aging comes from the observation that these metabolites likely are involved in the pathogenesis of other renal diseases characterized by progressive extracellular matrix expansion and decreased GFR, such as experimental renal mass reduction or glomerulonephritis. Moreover, the cellular biology of resident glomerular cells is clearly influenced by ROI. Although high concentrations of these metabolites usually induce cell necrosis, under particular experimental conditions, they induce proliferation of mesangial cells. This fact could be related to tyrosine phosphorylation of the platelet-derived growth factor receptor and the pp60c-src protein. Further, ROI promote changes resembling apoptosis in tubular cells. By promoting cell proliferation or apoptosis, ROI might play different pathogenetic roles during different stages of aging. In early stages, increased cell proliferation associated with increased synthesis of extracellular matrix might induce matrix expansion. In more advanced stages, apoptosis could reduce the number of cells in different parts of the nephron as well as in the interstitium. ROI also might modulate synthesis of the vasoactive factors involved in the dysfunction of the aging kidney. ROI increase prostanoid synthesis in varying renal structures. PAF synthesis by mesangial cells also might be increased in the presence of ROI. In addition, ROI stimulate endothelin production in cultured human mesangial cells.

Expression of preproendothelin-1 mRNA increased in cultured bovine aortic endothelial cells incubated with hydrogen peroxide. The relationships between NO and ROI are less well documented. Superoxide anion may inactivate NO, thereby inhibiting this vasodilatory system. However, recent results from the laboratory points to alternative possibilities; messenger RNA expression of one of the enzymes involved in NO synthesis, the endothelial constitutive nitric oxide synthase, as well as its activity, might be increased in cultured bovine aortic endothelial cells incubated with a ROI generating system such as xanthine oxidase. The relationship between these in vitro findings and the in vivo results in aged individuals must be evaluated carefully in the future.

Data concerning the ability of ROI to modulate cytokine synthesis by different glomerular cells are scarce. Synthesis of tumor necrosis factor seems to be stimulated by ROI, and deferoxamine, an iron chelator with well-defined antioxidant properties which might regulate tumor necrosis factor release in mesangial cells. On the other hand, ROI might be an intermediate metabolite in the release of monocyte chemoattractant protein and monocyte colony-stimulating factor induced by tumor necrosis factor. The possible role of these cytokines in the aging kidney has not been studied. Two recent reports stressed the importance of ROI in the pathogenesis of the changes that characterize the aging kidney. ROI synthesis induces well-defined effects in different renal structures and triggers the functional and morphologic changes that characterize aging. The role of AGEs in the development of diabetic complications has been studied extensively. Increased in elderly individuals, AGEs might induce some of the changes involved in the development of aging-induced renal dysfunction. Receptors for AGEs have been described in macrophages and monocytes. These cells, which transiently infiltrate the renal parenchyma, might synthesize interleukin-1, insulin-like growth factor I, tumor necrosis factor, and granulocyte-macrophage colony-stimulating factor in response to receptor binding by AGEs. In addition, AGE receptors have been identified on glomerular mesangial cells, where they seem to play a role in the modulation of PDGF-induced extracellular matrix synthesis. Prolonged administration of AGEs to normal rats induced glomerular hypertrophy and extracellular matrix expansion. This finding underscores the importance of these metabolites in the development of glomerulosclerosis.

Finally, the functional properties of several important matrix components are altered by AGE formation, disrupting the normal matrix-to-matrix and cell-to-matrix interactions, thus favoring the progressive matrix expansion of aging kidneys. The finding that about one-third of the subjects included in the Baltimore Longitudinal Study of Aging did not show any change in the GFR, and the existence of rat strains that do not develop any aging-related renal damage, suggest that the renal dysfunction of the elderly is due to an accumulation of damage induced by minimal, clinically undetected, renal disease, and is not the consequence of the aging process itself. Although it is a well-recognized fact that aged patients with superimposed diseases, such as hypertension or diabetes, show a more rapid decrease of renal function, healthy human beings and laboratory rats with well-controlled disease develop these renal changes in the absence of any detected renal disease. It is likely that a complex relationship between external influences, including diet, and the genetic program of a particular individual determines the variability observed in aging humans.

3.4 Inflammatory and Prothrombotic Markers and the Progression of Renal Disease in Elderly Individuals

It was hypothesized that these markers may also be determinants of the progression of renal disease. The association of six markers: serum C-reactive protein (CRP), white blood cell (WBC) count, fibrinogen, factor VII, albumin, and hemoglobin with subsequent elevations of creatinine and decline in estimated GFR in the Cardiovascular Health Study, a community-based cohort of elderly individuals, was analyzed. Linear regression was used to determine predictors of an annualized change in serum creatinine as the main outcome. Duration of follow-up was 7 years for the original cohort and 4 years for the more recently recruited black cohort. A total of 588 (12.7%) individuals had a decline in estimated GFR of at least 3 mL/min per year per 1.73 m^2. Higher CRP ($P<0.001$), WBC count ($P<0.001$), fibrinogen ($P<0.001$), and factor VII ($P<0.001$) levels and lower albumin ($P<0.001$) and hemoglobin levels ($P<0.001$) were associated with a rise in creatinine, after adjusting for age.[8] With additional adjustments for race, gender, baseline creatinine, systolic and diastolic BP, lipid levels, weight, and pack-years smoking, higher CRP, factor VII, fibrinogen, WBC count, and lower albumin and hemoglobin levels remained associated with a rise in creatinine.

Similar results were found for decline in estimated GFR. The decline in GFR was greater with increasing number of inflammatory or prothrombotic markers that were above the median (below for hemoglobin and albumin). Inflammatory and prothrombotic markers are predictors for a change in kidney function in elderly individuals.[12] Interventions that reduce inflammation might confer significant cardiovascular and renal benefits.[8] The baseline-adjusted predictor of developing CKD included age, glomerular filtration rate, hematuria, hypertension, diabetes, serum lipids, obesity, smoking status, and consumption of alcohol. Treated diabetes in male subjects, and treated hypertension, systolic blood pressure >160 mmHg and/or diastolic blood pressure >100 mmHg, diabetes, and treated diabetes in female subjects were associated with more than a doubling of the HR.

For the development of CKD stage III or higher, proteinuria of ≥ 2+, and proteinuria and hematuria were associated with more than a doubling of the HR in male subjects. The prevalence of newly developed CKD over 10 years was 19.2% in adults. Various studies suggested that not only hypertension and diabetes but also several metabolic abnormalities were independent risk factors for developing CKD.[18]

3.4.1 Sexual Dimorphism

Chris *et al.* reports that with advancing age, kidney function declines, and structural damage develops. It is difficult to dissociate the contribution of "normal aging" from underlying hypertension, atherosclerosis, glucose intolerance/diabetes, obesity, dyslipidemias, and/or undiagnosed CKD.[9] Although the average decline in GFR in men, after age 40, is about 1% per year, longitudinal studies reveal a tremendous variability in the rate of loss of GFR with age. The female kidney is relatively protected and this sex difference is seen across species and persists irrespective of genetic background/race. Although they are leading causes for renal failure, diabetes, and hypertension do not cause racial differences in developing ESRD. Minority women especially are at greater risk for ESRD than white women. Further studies are needed to determine whether earlier initiation of dialysis is a factor in higher ESRD incidence among minorities.[19] Indeed, in a very well-researched "normal" population of potential kidney donors, there was a gradual decline in GFR in men after approximately age 30 while remaining stable in women beyond the age of 50 (Fig. 5).

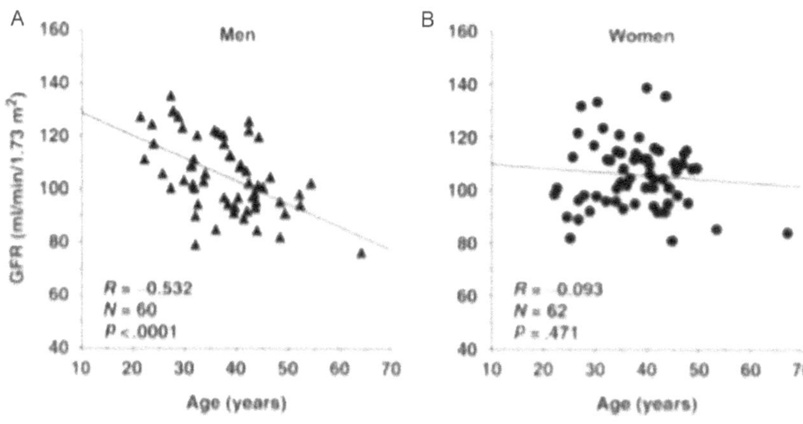

Fig. 5 Glomerular filtration rate (GFR; measured by inulin clearances) in cross-sectional studies in normal men and women of different ages who were evaluated as potential kidney transplant donors.

Nephron endowment at birth determines the propensity to develop later kidney damage.[21] Although nephron number cannot be noninvasively assessed in man, there is a strong association between low birth weight and later development of hypertension and kidney disease, and low birth weight is also likely to predict accelerated age-dependent injury. The occurrence of larger glomeruli in men is solely dependent on their greater body surface area. Similarly, the greater total glomerular volume seen in men as compared to women reflects increased kidney weight in men. Sex is not an independent determinant of total glomerular volume.[22] Nephron endowment is similar in male and female rats, mice, and man, but it is interesting to note that despite similar, low nephron numbers at birth, female rats of the MWF/ZTM strain are protected against age-dependent kidney damage vs males. Chronically increased glomerular blood pressure (BP) is another factor implicated in the pathogenesis of CKD but is not always present in the aging kidney. Rats that exhibit mild kidney damage with age (e.g., the normotensive Munich-Wistar, MW, rat) do not exhibit increased glomerular BP until after age-dependent injury is established, and male rats of the MWF/ZTM strain display normal glomerular BPs despite accelerated age-dependent kidney disease. However, age-dependent injury and functional declines are accelerated by systemic hypertension in man and glomerular hypertension occurs in aging, sclerosis-prone male Sprague–Dawley (SD) rats. In female rats of most strains, there is protection against age-induced renal structural damage.

Surprisingly, although there is little clinical data, what is available suggests that there is no sex difference in the rate of development of age-dependent kidney injury in aging humans. This is an area where additional research is needed and with the enhanced imaging techniques now available, noninvasive "histological" measurements in kidneys of normal aging men and women would greatly enhance our understanding. The GFR is determined both by the number of functioning glomeruli[23] and by the renal hemodynamics, which are controlled by the renal vascular resistance vessels.[10] The kidney vasculature has a complex architecture with resistances both before and after the glomerulus (the afferent and efferent arterioles), which regulate RPF and glomerular BP.[9]

In the aging male Munich-Wistar rats, parallel increases in afferent and efferent resistances cause falls in RPF, whereas glomerular BP is maintained, hence GFR falls. In the aging female Munich-Wistar rats, preservation of GFR is associated with relatively unchanged RPF and glomerular BP, facilitated by mild relaxation of both afferent and efferent renal arteriolar

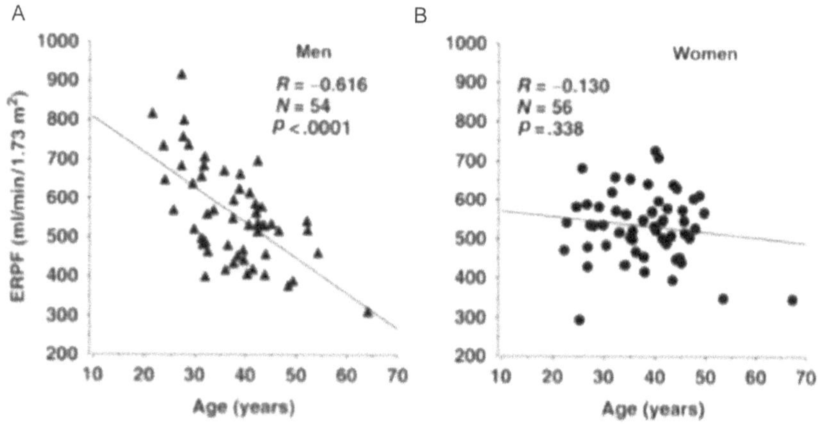

Fig. 6 Effective renal plasma flow (ERPF; measured by para-aminohippurate clearances) in cross-sectional studies in normal men and women of different ages who were evaluated as potential kidney transplant donors.

resistance. In normal aging women, RPF is also maintained, whereas RPF falls in men (Fig. 6). In normal young adult men and women (and male and female rats), the males exhibit higher values of GFR and RPF, whereas by age 70, the values of RPF are similar in normotensive men and women reflecting preservation of GFR and RPF in aging females (Fig. 6).

3.4.2 Causes of Sex Differences

There are likely to be several reasons why men are more vulnerable to age-dependent kidney dysfunction, including differences conferred by the sex chromosomes. The possible roles of the sex steroids were evaluated carefully.

3.4.3 Estrogens

There is strong evidence that estrogens exert kidney/cardiovascular protection. Premenopausal women exhibit a slower rate of progression of nondiabetic CKD compared with men and this sex difference is lost in diabetic CKD, possibly in association with the falls in circulating estrogen.[11] Aging female C57Bl6 mice develop glomerulosclerosis after menopause and estrogen supplementation reverses glomerular damage in female, injury-prone mice. In addition to protecting the kidney by improving cardiovascular health, estrogens suppress vascular smooth muscle and mesangial cell growth and extracellular matrix accumulation, thus inhibiting the development of glomerular sclerosis. However, estrogens can sometimes be associated with worse renal pathology, as in the type 2 diabetic mouse kidney, the stroke-prone spontaneously hypertensive rat and in the presence of severe hypertriglyceridemia. The kidney contains many estrogen receptors (ERs) and has many estrogen-regulated genes, mainly controlled by ERα.[9]

Studies in ER knockout mice suggest that ERα activation contributes to glomerular hypertrophy and sclerosis after uninephrectomy and with diabetes. In most cases, however, ERα is protective and is required for vascular repair from atherosclerosis in mice of both sexes. It also protects the podocyte from apoptotic injury. Mesangial cells from female glomerular sclerosis-prone mice express decreased ERα, and ERα depletion occurs in high salt-induced hypertension and renal damage. The ERα knockout female mouse develops accelerated albuminuria and glomerular damage with age. Stimulation of the ERβ may also be protective since the ERβ knockout mouse develops age-dependent hypertension. There is increasing evidence that stimulation of the membrane ER GPR30 exerts renal and cardiovascular protective actions.[20] Estrogen supplementation in rats and mice is often beneficial, but two large clinical trials report adverse cardiovascular responses to hormone replacement therapy (HRT) in postmenopausal women. Animal studies routinely use 17β-estradiol given subcutaneously, whereas clinical trials often use oral conjugated equine estrogens (containing many estrogens, progestins, androgens, and other substances, which have less predictable actions). Also, late initiation of HRT was associated with less benefit and/or increased cardiovascular risk compared with women in whom HRT was initiated at or close to menopause.[9]

3.4.4 Androgens

In rats, castration of young adult males prevents age-dependent glomerular sclerosis. Androgens are profibrotic, stimulating mesangial extracellular matrix production, and inhibiting matrix degradation. Androgens are also associated with greater kidney damage and higher BP, and chronic antagonism of androgens is protective in several hypertensive rat models. Testosterone also promotes podocyte apoptosis via an androgen receptor-mediated effect.[9] In normal men, however, low androgens correlate with increased cardiovascular risk and insulin resistance. Androgen levels fall in men with hypertension, renal disease, and aging, although whether this contributes to the more rapid progression of nondiabetic CKD and age-dependent renal dysfunction seen in men is unclear.

In women, testosterone levels increase after menopause as cardiovascular risk increases but remain much lower than in age-matched men. Women with polycystic ovary syndrome have elevated androgen levels and increased cardiovascular risk although this increased risk may be more related to insulin insensitivity than androgen level. In fact, a low testosterone to bioavailable estrogen ratio correlates with a proatherogenic adipocytokine profile in both men and women, and a recent review concludes that there is no clear link between elevated testosterone levels and cardiovascular disease in women. Not all actions of estrogens on the kidney are beneficial, in fact, there are two clinical studies which suggest that oral HRT worsens proteinuria and accelerates the age-dependent decline in renal function in postmenopausal women, whereas transvaginal delivery was not associated with the loss of renal function. In contrast, other studies report reductions in proteinuria with estrogen, progesterone, and combination therapy. It seems reasonable to favor transdermal or transvaginal administration of 17β-estradiol, avoiding oral administration and use of conjugated equine estrogens. Also, the timing of HRT is important and initiation of HRT in women who are many years postmenopausal should be avoided. One interesting effect of normal aging is the marked change in the estrogen: androgen ratio that occurs between the sexes, with older men exhibiting approximately 4 × higher estradiol and 20 × higher testosterone than older women. Perhaps, more consideration should be given to this ratio when considering cardiovascular/renal health during aging.

3.5 Aging Kidney and the Interplay Between the Nitric Oxide and ANGII Systems

NO is vasodilatory, inhibits growth of contractile cells as well as extracellular matrix production, inhibits oxidative stress, and also inhibits renal sodium reabsorption. ANGII has opposing actions, since in addition to directly and indirectly promoting renal sodium retention and vasoconstriction, it also promotes cell growth, fibrosis, and oxidative stress and inflammation. Chronic NO deficiency develops in man and experimental animals in many types of CKD causing hypertension and a profibrotic state, which contribute to injury progression. There is also strong animal and clinical evidence that overactivity of intrarenal ANGII is part of the pathogenesis of hypertension and CKD.[24] The possible contribution of NO deficiency/ANGII overactivity to development of age-dependent kidney damage and dysfunction and how this might relate to the sex differences have been discussed by some investigators.[13] Total NO production falls in the aging male Sprague–Dawley rat and kidney injury develops rapidly, whereas in the aging Sprague–Dawley female, there is little CKD and total NO production is

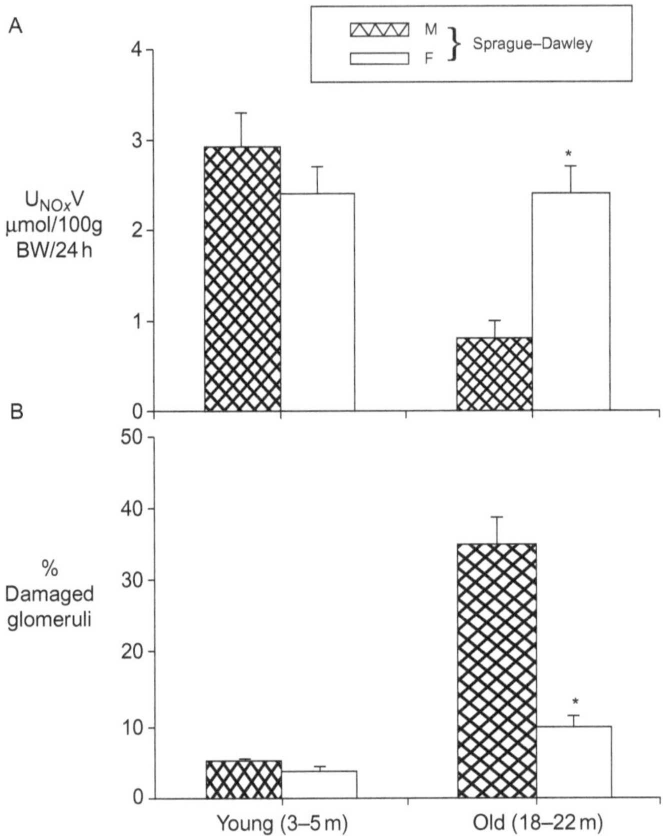

Fig. 7 The 24-h urinary excretion of NO_2+NO_3 (NOX), UNOXV (A), and the percentage of damaged glomeruli (B) (i.e., those showing segmental and global sclerosis) in young adult (3–5 months) and old (18–22 months) male (M) and female (F) Sprague–Dawley rats.

maintained (Fig. 7). Some of these sex differences are due to estrogen that exert multiple direct and indirect NO stimulatory actions.[9]

4. CONCLUSIONS

Renal function starts declining from the fourth decade and often leads to severe renal insufficiency in very elderly humans. While there is wide variability in the rate of such age-related renal functional decline and controversies surround the question of whether such change is physiological or pathological, the mechanisms leading to renal functional decline seem to attract the attention of several investigators as evident from this in-depth analysis. These open-ended questions regarding the pathophysiological basis of morphological and functional changes of aging also complicate treatment strategies in the elderly. Aging population is also susceptible for multiple disorders related to electrolyte, acid–base and water handling by the kidney. Further investigations are warranted to gain better insights into these renal defects and disorders and to optimally manage such geriatric dilemmas.

REFERENCES

1. Coresh J, Astor BC, Greene T, Eknoyan G, Levey AS. Prevalence of chronic kidney disease and decreased kidney function in the adult US population: Third National Health and Nutrition Examination Survey. *Am J Kidney Dis.* 2003;41(1):1–12.
2. Anderson S, Rennke HG, Zatz R. Glomerular adaptations with normal aging and with long-term converting enzyme inhibition in rats. *Am J Physiol.* 1994;267(1 pt 2):F35–F43.
3. Davies DF, Shock NW. Age changes in glomerular filtration rate, effective renal plasma flow, and tubular excretory capacity in adult males. *J Clin Invest.* 1950;29(5):496–507.
4. Brenner BM, Lawler EV, Mackenzie HS. The hyperfiltration theory: a paradigm shift in nephrology. *Kidney Int.* 1996;49:1774–1777.
5. Doleželová Š, Jíchová Š, Husková Z, et al. Progression of hypertension and kidney disease in aging fawn-hooded rats is mediated by enhanced influence of renin-angiotensin system and suppression of nitric oxide system and epoxyeicosanoids. *Clin Exp Hypertens.* 2016;26:1–8[Epub ahead of print].
6. Rodriguez-Puyol D. Nephrology forum. The aging kidney. *Kidney Int.* 1998;54: 2247–2265.
7. Nephrology forum the aging kidney Diego Rodrfguez-Puyol http://dx.doi.org/10.1038/4499994.
8. Fried L, Solomon C, Shlipak M, et al. Inflammatory and prothrombotic markers and the progression of renal disease in elderly individuals. *J Am Soc Nephrol.* 2004;15(12): 3184–3191.
9. Braun F, Brinkkötter PT. Decline in renal function in old age. *J Gerontol Geriat.* 2016;49:469. http://dx.doi.org/10.1007/s00391-016-1109-y.
10. Remuzzi A, Puntorieri S, Mazzoleni A, Remuzzi G. Sex related differences in glomerular ultrafiltration and proteinuria in Munich-Wistar rats. *Kidney Int.* 1988;34:481–486.
11. Braun F, Brinkkötter PT. Decline in renal function in old age: part of physiological aging versus age-related disease.
12. Fried L, Solomon C, Shlipak M, et al. Inflammatory and prothrombotic markers and the progression of renal disease in elderly individuals. *J Am Soc Nephrol.* 2004;15(12): 3184–3191.
13. Baylis C. Sexual dimorphism: the aging kidney, involvement of nitric oxide deficiency, and angiotensin II overactivity. *J Gerontol A Biol Sci Med Sci.* 2012;67(12):1365–1372. http://dx.doi.org/10.1093/gerona/gls171. Published online 2012 Sep 7.
14. Levi M, Rowe JW. Renal function and dysfunction in aging. In: Seldin DW, Giebisch G, eds. *The Kidney: Physiology and Pathophysiology.* New York: Raven Press; 1992:3433–3456.
15. Baylis C, Corman B. The aging kidney: insights from experimental studies. *J Am Soc Nephrol.* 1998;9(4):699–709.
16. Bleyer AJ, Shemanski LR, Burke GL, Hansen KJ, Appel RG. Tobacco, hypertension, and vascular disease: risk factors for renal functional decline in an older population. *Kidney Int.* 2000;57:2072–2079.

17. Hemmelgarn BR, Zhang J, Manns BJ, et al. Progression of kidney dysfunction in the community-dwelling elderly. *Kidney Int.* 2006;69:2155–2161.
18. Kasiske BL. Relationship between vascular disease and age-associated changes in the human kidney. *Kidney Int.* 1987;31:1153–1159.
19. Weinstein JR, Anderson S. The aging kidney: physiological changes. *Adv Chronic Kidney Dis.* 2010;17(4):302–307. http://dx.doi.org/10.1053/j.ackd.2010.05.002. PMCID: PMC2901622.
20. Yamagata K, Ishida K, Sairenchi T, et al. Risk factors for chronic kidney disease in a community-based population: a 10-year follow-up study. *Kidney Int.* 2007;71:159–166.
21. Xue JL, Eggers PW, Agodoa LY, Foley RN, Collins AJ. Longitudinal study of racial and ethnic differences in developing end-stage renal disease among aged medicare beneficiaries. *J Am Soc Nephrol.* 2007;18:1299–1306.
22. Neugarten J, Gallo G, Silbiger S, Kasiske B. Glomerulosclerosis in aging humans is not influenced by gender. *Am J Kidney Dis.* 1999;34:884–888.
23. Bertram JF, Douglas-Denton RN, Diouf B, Hughson MD, Hoy WE. Human nephron number: implications for health and disease. *Pediatr Nephrol.* 2011;26:1529–1533.
24. Neugarten J, Kasiske B, Silbiger SR, Nyengaard JR. Effects of sex on renal structure. *Nephron.* 2002;90:139–144.

CHAPTER ELEVEN

Mitochondrial Perturbation in Alzheimer's Disease and Diabetes

F. Akhter, D. Chen, S.F. Yan, S.S. Yan[1]
School of Pharmacy, Higuchi Bioscience Center, University of Kansas, Lawrence, KS, United States
[1]Corresponding author: e-mail address: shidu@ku.edu

Contents

Abstract

Mitochondria are well-known cellular organelles that play a vital role in cellular bioenergetics, heme biosynthesis, thermogenesis, calcium homeostasis, lipid catabolism, and other metabolic activities. Given the extensive role of mitochondria in cell function, mitochondrial dysfunction plays a part in many diseases, including diabetes and Alzheimer's disease (AD). In most cases, there is overwhelming evidence that impaired mitochondrial function is a causative factor in these diseases. Studying mitochondrial function in diseased cells vs healthy cells may reveal the modified mechanisms and molecular components involved in specific disease states. In this chapter, we provide a concise overview of the major recent findings on mitochondrial abnormalities and their link to synaptic dysfunction relevant to neurodegeneration and cognitive decline in AD and diabetes. Our increased understanding of the role of mitochondrial perturbation indicates that the development of specific small molecules targeting aberrant mitochondrial function could provide therapeutic benefits for the brain in combating

Progress in Molecular Biology and Translational Science, Volume 146
ISSN 1877-1173
http://dx.doi.org/10.1016/bs.pmbts.2016.12.019

aging-related dementia and neurodegenerative diseases by powering up brain energy and improving synaptic function and transmission.

1. INTRODUCTION

Emerged evidence suggests that the deleterious and advanced cellular changes in aging and diabetes are linked to mitochondrial dysfunction.[1,2] Brain aging is often characterized by neuronal loss and synaptic alteration, which are associated with mitochondrial abnormalities, energy failure, respiratory chain impairment, generation of reactive oxygen species (ROS), and neuronal perturbation.[3] Further, various evidences suggest that mitochondrial dysfunction is a prominent and early oxidative stress-associated factor that produces neuronal abnormalities in aging and diabetes, resulting in susceptibility to aging-related neurodegenerative diseases.[4] In the neurons, mitochondria are distributed throughout the length of the axons, presynaptic terminals, and dendrites. Mitochondria play active roles in regulating synaptogenesis and morphological/functional responses to synaptic activity; thus, mitochondrial dysfunction can lead to a stark neuronal energy deficit and, in the long run, to modifications in neuronal synapses and neurodegeneration in the aging brain.[1]

Alzheimer's disease (AD) is a chronic aging-related disease with two pathological features: abnormal accumulations of amyloid beta peptide (Aβ) and phosphorylation of tau protein in the brain. Increased evidence indicates that mitochondrial and synaptic dysfunction is an early pathological feature of AD.[5] Aβ has deleterious effects on mitochondrial function and structure and contributes to energy failure, respiratory chain impairment, ROS generation, induction of mitochondrial permeability transition pore (mPTP), imbalance of calcium homeostasis, disruption of mitochondrial dynamics, and mitochondrial DNA/RNA mutations.[6] Although Aβ directly and indirectly causes abnormal mitochondrial and neuronal function, recent studies have highlighted the association between early mitochondrial dysfunction and the accumulation of Aβ in mitochondria, implicating mitochondrial Aβ in AD pathogenesis.[7–28] These observations provide a better understanding of the relationship between mitochondria and AD pathogenesis.

Mitochondrial malfunction, synaptic damage, and the resultant impairment in cognitive function are pathological features of diabetes-affected brains.[2] Diabetes adversely affects the brain and increases the risk for depression and dementia.[29–39] In neurons, synaptic mitochondria are vital for the maintenance of synaptic function and transmission through normal mitochondrial dynamics, distribution, and trafficking as well as energy metabolism and synaptic calcium modulation. Imbalance of mitochondrial dynamics contributes to oxidative stress and hyperglycemia-induced alterations in mitochondrial morphology and function.[38,40,41] Diabetes elicits AD-like brain changes linked with cognitive decline and neurodegeneration, such as elevated tau expression and phosphorylation and accumulation of Aβ,[42–46] mitochondrial dysfunction, disruption of mitochondrial dynamics,[37,38,41,47–51] oxidative stress,[40,49] neuroinflammation, loss of synapses, impaired learning and memory, and synaptic plasticity deficits.[29,35,36,44,52–55] The underlying mechanisms and strategies to rescue such injury and dysfunction are not well understood. Studies have identified several cellular and mitochondrial cofactors that are directly or indirectly involved in AD- and diabetes-mediated alterations in mitochondrial and synaptic structure and function. Such factors include cyclophilin D (CypD), presequence protease (PreP), Aβ, mPTP, *N*-methyl-D-aspartate, and the receptor for advanced glycation endproducts (RAGE).

This chapter addresses several aspects of AD- and diabetes-induced mitochondrial dysfunction with a special focus on mitochondrial molecular mechanisms underlying synaptic pathology and cognitive dysfunction.

2. MITOCHONDRIAL FUNCTION

Mitochondria are essential organelles for cell survival, playing a crucial role in calcium homeostasis, energy metabolism, detoxification of ROS generation, and induction of cell death, including apoptosis and necrosis. Mitochondria in different types of cells or in different subcompartments of one cell differ significantly in their morphology and function and can be divided into multiple subgroups within one cell.[56] The recent recognition of mitochondrial heterogeneity facilitates our understanding of mitochondrial biology.

Mitochondria are the major site of ATP synthesis and are also the site of amino acid biosynthesis, fatty acid oxidation, steroid metabolism, calcium homeostasis, and ROS production and detoxification. The inner mitochondrial membrane is largely impermeable and contains a variety of enzymes, including those responsible for making ATP, and forms the major barrier between the cytosol and the mitochondrial matrix. The five complexes of the respiratory chain [complex I (NADH ubiquinone oxidoreductase), complex II (succinate ubiquinone oxidoreductase), complex III (ubiquinone-cytochrome *c* reductase), complex IV (cytochrome oxidase), and complex V (ATP synthase)] are embedded in the inner mitochondrial membrane. The transmission of electrons along the respiratory chain provides the energy to pump protons from the matrix into the intermembrane space, thereby generating the electrochemical gradient required to drive ATP synthesis.[56]

3. SYNAPTIC MITOCHONDRIAL PATHOLOGY IN AD

Synapses are the neuronal contact sites through which neurons receive and send information.[57,58] Energy provision and calcium fluctuation in synapses are prerequisite for interneuronal communication.[59] To meet the high energy demands and to cope with constant calcium flux, synapses are enriched with mitochondria for on-site energy provision and calcium modulation.[60]

Although the detrimental impacts of Aβ on synapses and synaptic function are extensive, multiple studies demonstrate that mitochondrial structure and function are particularly susceptible to the effects of mitochondrial Aβ accumulation.[7–28] Further, synaptic mitochondria serve as a reservoir for Aβ accumulation in aging and AD[1,5,61–64]; thus mitochondrial dysfunction is a major player in the synaptic alterations seen in AD and diabetes.[3,56,65]

First, mitochondrial and neuronal malfunction in AD is linked to the progress accumulation of Aβ in the mitochondria of both human AD and transgenic AD mouse brains.[1,7–9,11,15–18,66–68] Aβ can directly import into mitochondria via the translocase of the outer membrane machinery,[67] RAGE,[69] or other unknown mechanisms. Aβ may also be locally produced in mitochondria via gamma-secretase that is localized in mitochondria.[70–72] Notably, accumulation of mitochondrial Aβ precedes extracellular Aβ deposition in AD brains, increases with age, and associates with early onset synaptic loss, synaptic damage, and mitochondrial oxidative damage,[5,7,10,11,22,73–83] suggesting that early accumulation of Aβ in mitochondria may be an initiating pathological event, leading to mitochondrial and neuronal perturbation. Second, interaction of Aβ with mitochondrial matrix proteins such as amyloid-binding alcohol dehydrogenase (ABAD)[7,10,11,84] and CypD[23,85–87] exacerbates Aβ-induced mitochondrial and neuronal stress. Increasing PreP activity by antagonizing the Aβ-ABAD interaction decreases mitochondrial and cerebral Aβ accumulation in AD mice overexpressing Aβ and improves mitochondrial function.[88] Third, increasing neuronal PreP expression and activity in Aβ-enriched synaptic mitochondria of mAPP mice greatly reduces mitochondrial accumulation. Accordingly, synaptic function and learning and memory are significantly improved in PreP-overexpressed mitochondria.[28] These data strongly indicate that PreP is critical for maintaining mitochondrial integrity and function by clearance of mitochondrial Aβ. Strategies that reduce Aβ levels in mitochondria in addition to the brain by increasing PreP expression and activity are critical to consider as new avenues for both preventing and halting AD progression at the early stage. One such therapeutic strategy involves the development of a small-molecule agonist of PreP in order to safely decrease mitochondrial and cerebral Aβ accumulation by accelerating Aβ clearance.

These recent studies highlight the significant role of Aβ in synaptic mitochondrial pathology and significantly advance our understanding of the mechanisms underlying mitochondrial dysfunction in AD, especially in the early stage when the presence of Aβ has not yet set in motion the devastating cognitive impairments often associated with AD. Ameliorating alterations in mitochondrial function could improve synaptic function and reverse cognitive decline in AD.

In AD and non-AD cell and animal models, treatment of mitochondria-targeted molecules mitoQ and SS31 significantly reverses Aβ-induced CypD elevation, mitochondrial fusion/fission proteins imbalance, and neurite growth.[89] MitoQ and SS31 also reduce mutant huntingtin-induced mitochondrial toxicity and synaptic damage.[90] Additionally, antioxidants attenuate mitochondrial transport and function in cybrid cells containing AD-derived mitochondria.[4,91] These results suggest a close relationship between neuronal mitochondrial dysfunction and synaptic perturbation and the value of eliminating neuronal mitochondrial oxidative stress in the treatment of neuronal/synaptic alterations in AD.[1,90]

4. IMPACT OF CypD-DEPENDENT mPTP ON MITOCHONDRIAL DEFECTS

CypD is a crucial component of the mPTP. CypD released from matrix can bind to the adenine nucleotide translocase in the inner mitochondrial membrane and the voltage-dependent anion channel in the outer mitochondrial membrane to trigger the opening of mPTP, a nonselective, high conductance pore allowing the transport not only calcium but any solute below the pore size. The opening of mPTP results in osmotic swelling, dissipation of the mitochondrial membrane potential, reduced mitochondrial calcium retention capacity, decreased membrane potential, increased ROS production, and eventually, cell death (Fig. 1).[92] Increased expression of CypD occurs in neurodegenerative diseases including AD, Parkinson's disease (PD), Huntington's disease (HD),[23,87,93–97] and diabetes, and contributes to mitochondrial perturbation.[5,23,65,98]

Studies from in vitro cellular and in vivo animal models have demonstrated that blockade of CypD significantly attenuates mPTP-related mitochondrial dysfunction and cell death, which are relevant to the pathogenesis

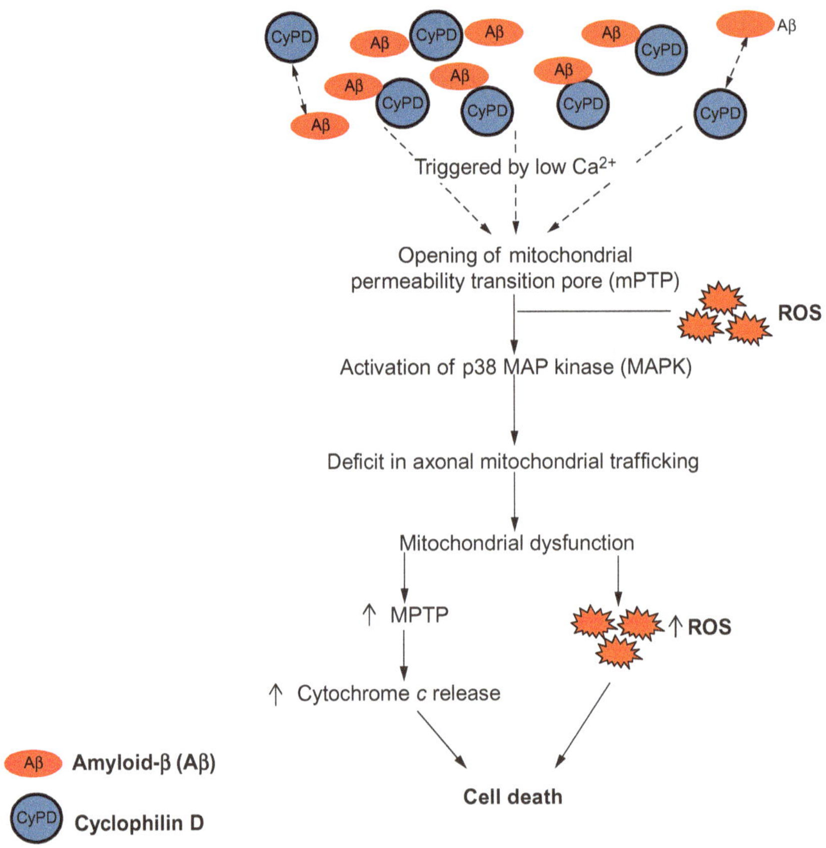

Fig. 1 Effect of Aβ on CypD-involved mPTP formation. Aβ-cyclophilin D interaction mediates impairments in axonal mitochondrial transport due to an increase in the opening of CypD-mediated mitochondrial permeability transition pore (mPTP). This leads to the disruption of Ca^{2+} balance and increases the production/accumulation of reactive oxygen species (ROS). Elevation of Ca^{2+} and oxidative stress activates the downstream p38 MAP kinase signaling pathway, thus contributing to mitochondrial dysfunction.

of stroke, AD, and diabetes.[23,65,87,98–102] Furthermore, blockade of mPTP by genetic depletion or pharmacological inhibition of CypD rescues axonal mitochondrial trafficking and protects synapses from Aβ toxicity. The potential mechanisms underlying the protective effect of CypD deficiency on axonal mitochondrial trafficking include the reduction of Aβ-induced calcium perturbation, the suppression of axonal ROS accumulation, and the activation of the downstream P38/MAPK signaling pathway.

The protein kinase A/cAMP regulatory element-binding (PKA/CREB) signaling pathway, a crucial regulator of synaptic plasticity and learning memory, is adversely affected by an Aβ-rich environment, leading to dendritic spine architecture changes in an AD mouse model.[102] Aβ reduces phosphorylation of PKA, thus disrupting PKA/CREB signal transduction and causing synaptic and cognitive dysfunction.[100,102] Notably, neurons lacking CypD reverse Aβ-induced synaptic dysfunction and are protected against Aβ-induced alterations in PKA and CREB phosphorylation. These results indicate the involvement of CypD in Aβ-induced abnormalities in signal transduction including PKA/CREB signaling. Sustained CypD-induced neuronal/synaptic mitochondrial stress is a potential mechanism underlying synaptic failure in the pathogenesis of AD.

Recently, Wang *et al.* demonstrated that CypD expression levels were significantly elevated in the hippocampi of streptozotocin-induced diabetic mice.[56] The CypD expression levels are further elevated in Aβ-enriched diabetic brain compared to nondiabetic mAPP mice.[56] These results suggest that CypD expression is increased in diabetes mellitus and further enhanced in an Aβ-rich environment. Increased levels of CypD in mitochondria trigger/enhance the mPTP opening, leading to colloidal osmotic swelling of the mitochondrial matrix, dissipation of the inner membrane potential, generation of ROS, and release of many proapoptogenic proteins and procaspases.[99] Hence, blockade of CypD may be a potential therapeutic strategy for preventing and halting synaptic and mitochondrial pathology in AD. Specifically, the development of small-molecule CypD inhibitors could hold therapeutic potential for the treatment of neurodegenerative diseases including AD and diabetes.[103,104]

5. EFFECT OF NEURONAL PreP ACTIVITY AND RAGE SIGNALING ON MITOCHONDRIAL DYSFUNCTION

PreP is a mitochondrial peptidasome that is localized in the mammalian mitochondrial matrix.[105] It is the key for maintenance of mitochondrial health and integrity. PreP proteolytic activity is significantly reduced in AD-affected brain mitochondria and transgenic AD mouse models[106] and is negatively correlated to mitochondrial Aβ accumulation. Du *et al.* demonstrated that increased expression and activity of neuronal PreP significantly reduced mitochondrial Aβ load and the production of proinflammatory mediators, improved mitochondrial function and synaptic plasticity, and attenuated cognitive decline in AD mice.[28] Furthermore, PreP proteolytic activity is required for degradation and clearance of mitochondrial Aβ. Mitochondrial Aβ accumulation may interfere with normal mitophagy and release of mitochondria-derived damage-associated molecular patterns from the injured neurons, leading to increased production of TNF-α, IL-1β, and MCP1, the cytokines known to be involved in the inflammatory process of AD.[107] Thus, dysfunctional or damaged mitochondria can produce excessive inflammation and tissue damage possibly via overproduction of cytokines and ROS.

RAGE-dependent signal transduction via Aβ-RAGE interaction plays an important role in mitochondrial dysfunction. RAGE serves as an important cell-surface receptor mediating chemotactic and inflammatory reaction to Aβ and other proinflammatory ligands.[69,108–113] RAGE signaling in neurons and microglia is known to promote induction of proinflammatory mediators, including cytokines and chemokines, and activation of microglia by increased expression of microglial markers (CD4 and CD11).[107,110] Additionally, overexpression of neuronal PreP in mAPP mice not only reduces Aβ accumulation in the brain but also remarkably suppresses RAGE expression as compared with mAPP mice,[69] suggesting a possible connection between mitochondrial defects and RAGE signaling relevant to the activation of transcription and the proinflammatory response.[28,69,107,109,110,112] Further investigation is required to elucidate the role of RAGE in mitochondrial dysfunction relevant to the pathogenesis of AD and diabetes.

6. EFFECTS OF METHIONINE SULFOXIDE REDUCTASE ON Aβ SOLUBILITY AND MITOCHONDRIAL FUNCTION

Accumulation of oxidized proteins, especially Aβ, is thought to be one of the common causes of AD. Induced ROS generation is one of the earliest consequences of toxic insults mediated by soluble Aβ oligomers.[81] Mitochondria are particularly sensitive to ROS, and reduced metabolic activity resulting from oxidative damage to vital mitochondrial components has been demonstrated in AD.[10]

Methionine (Met) is highly susceptible to oxidation in vivo, particularly under conditions of oxidative stress. The sulfoxide form comprises 10%–50% of Aβ in amyloid plaques of AD brain.[114] Oxidation of Met to Met(O) is reversible and the reverse reaction is catalyzed in vivo by the methionine sulfoxide reductase (Msr) system, composed of peptide-methionine (*S*)-S-oxide reductase (MsrA) and peptide-methionine (*R*)-S-oxide reductase (MsrB), which, respectively, reduce the *S* and *R* enantiomers of the sulfoxide group. These enzymes provide both an efficient repair mechanism for oxidative damage to Met residues and general protection against oxidative stress by scavenging ROS through the recycling of Met.

Studies from primary hippocampal and cortical neurons show increased total Msr activity, ascribed to increased activity in both MsrA and MsrB, in conjunction with protection against cell death induced by the sulfoxide forms of Aβ40 or Aβ42. Exposure of wild-type and *MsrA* knockout mouse cortical neurons to Aβ42 and Met(O)-Aβ demonstrated that lack of MsrA abolishes the protective effect induced by Met(O)-Aβ.[115] Furthermore, lack of MsrA promotes a shift from aggregated forms of Aβ toward soluble oligomers. Given that soluble oligomer Aβ are thought to be more toxic to neurons and synapses than aggregated Aβ forms,[115] enhancing MsrA activity by regulating transcription may have therapeutic applications. Alterations in MsrA expression levels and Aβ structure during normal aging might be a cofactor in AD-related mitochondrial malfunction.[115]

7. IMPACT OF MITOCHONDRIAL DYNAMICS IN MCI AND AD

Mitochondria are highly dynamic organelles that undergo continuous fission and fusion, which are regulated by the GTPase hydrolysis activity mitochondrial fission proteins (DLP1 and Fis1) and mitochondrial fusion protein [mitofusin 1 and 2 (Mfn1 and 2) and optic atrophy (Opa1)]. Mitochondrial dynamics are important for the proper distribution of mitochondria within cells, which is particularly critical for morphologically complex cells such as neurons.[116] Alterations in mitochondrial dynamics significantly impact almost all aspects of mitochondrial function including energy metabolism, calcium buffering, ROS generation, and apoptosis regulation.[117,118] Unbalanced fusion and fission lead, respectively, to mitochondrial elongation and excessive mitochondrial fragmentation, both of which impair the function of mitochondria. It has been shown that exchange of mitochondrial contents is important for mitochondrial function as well as organelle distribution in neurons. Mitochondrial fusion, in particular that mediated by Mfn2, is required for proper development and maintenance of the cerebellum.[119] Mutations in the Mfn2 gene cause neurodegenerative diseases, such as Charcot–Marie–Tooth type 2A, and mutations in OPA1 cause dominantly inherited optic atrophy. Increasing evidence implicates altered mitochondrial trafficking and fusion–fission dynamics in aging-related AD, PD, HD, and amyotrophic lateral sclerosis.

7.1 Effect of Mfn2 on Mitochondrial Function

Mitofusins Mfn1 and Mfn2 are outer membrane GTPases that mediate outer mitochondrial membrane fusion. Mfn2 expression is crucial for maintaining the morphology and operation of the mitochondrial network and mitochondrial metabolism. Recent studies demonstrate that markedly reduced mitochondrial mass and transport may contribute to neuronal loss due to the specific loss of Mfn2 but not Mfn1.[120] Du *et al.* examined the role of Mfn2 in the human-induced pluripotent stem cells (hiPSCs) differentiation system and reported that knockdown of Mfn2 results in mitochondrial dysfunctions and defects in neurogenesis and synapse formation.[119] By contrast, Mfn2 overexpression in neural progenitor cells directs differentiation and maturation into neurons with enhanced mitochondrial functions, suggesting that Mfn2 is crucial for mitochondrial development, and thereby essential to hiPSCs differentiation. Importantly, this also provides a novel neurophysiologic model of mitochondrial development in neurogenesis, which enhances our understanding of the involvement of dysfunctional mitochondria in aging and neurodegenerative diseases.[119] Under pathological conditions, Mfn2 expression levels are increased such as mild cognitive impairment (MCI)-derived mitochondria, leading to aberrant mitochondrial fusion and fission event evidenced by abnormal mitochondrial morphology and function.

7.2 Oxidative Stress and MCI- and AD-Related Mitochondrial Dynamics

MCI is characterized by a decline in cognitive abilities that is noticeable yet not severe enough to completely disrupt an individual's daily activity. MCI is generally considered to be a transitional phase between normal aging and early dementing disorders, especially AD.[121]

In cybrid model, MCI-induced mitochondrial defects manifest as alterations in mitochondrial dynamics, function, and morphology. These dysfunctional MCI cybrid mitochondria exhibit impaired fission/fusion events, impaired mitochondrial respiratory chain enzyme activity, decreased membrane potential, increased mitochondrial and intracellular ROS, and impairment in energy metabolism with decreased ATP levels when compared to non-MCI cybrid mitochondria. Given that mitochondrial Mfn2 is involved in mitochondrial fusion,[119] increased mitochondrial Mfn2 levels in MCI cybrids suggest that altered Mfn2 expression likely contributes to enhanced mitochondrial fusion. Accordingly, changes in MCI mitochondrial morphology display as elongated mitochondria. Interestingly, suppression of Mfn2 overexpression by inhibiting oxidative stress-mediated activation of extracellular signal-regulated kinases (ERK) reverses abnormalities in mitochondrial structure and function.[122] Thus, generation of Mfn2 antagonist may hold potential for prevention and treatment at the early stage of AD.[123]

In contrast to MCI-derived mitochondria, AD mitochondria exhibit fragmentation as shown by overabundant fission, elongate, and aggregated mitochondria, compared to cybrid cells containing mitochondria from normal age-matched subjects with the relatively normal cognitive function. DLP1, which plays a key role in balancing mitochondrial dynamics by regulating mitochondrial fission, was significantly increased in AD mitochondria.[123] Additionally, the abnormal interaction of DLP1 with hyperphosphorylated tau was found in AD neurons.[124] Interaction of DLP1 with glycogen synthase kinase-3 (GSK3β) mediates changes in mitochondrial morphology and dynamics.[125–127] Mitochondrial dynamics modulates the induction of proinflammatory mediators in microglial cells.[128,129] ROS-induced activation of the mitogen-activated protein (MAP) kinase family appears to play a key role in mediating cellular responses to multiple stresses. ERK signaling is involved in mitochondrial function and neuronal stress.[123,130] Taken together, this suggests that oxidative stress-induced activation of MAP kinase via upregulation of DLP1 or Mfn2 expression contributes to mitochondrial dysfunction and abnormal mitochondrial dynamics[122,123] by disrupting the balance of mitochondrial fission and fusion and promoting translocation of DLP1 to mitochondria, leading to mitochondrial fragmentation in AD. Most importantly, suppression of ERK signaling and inhibition of mitochondrial fission or fusion pathways rescues

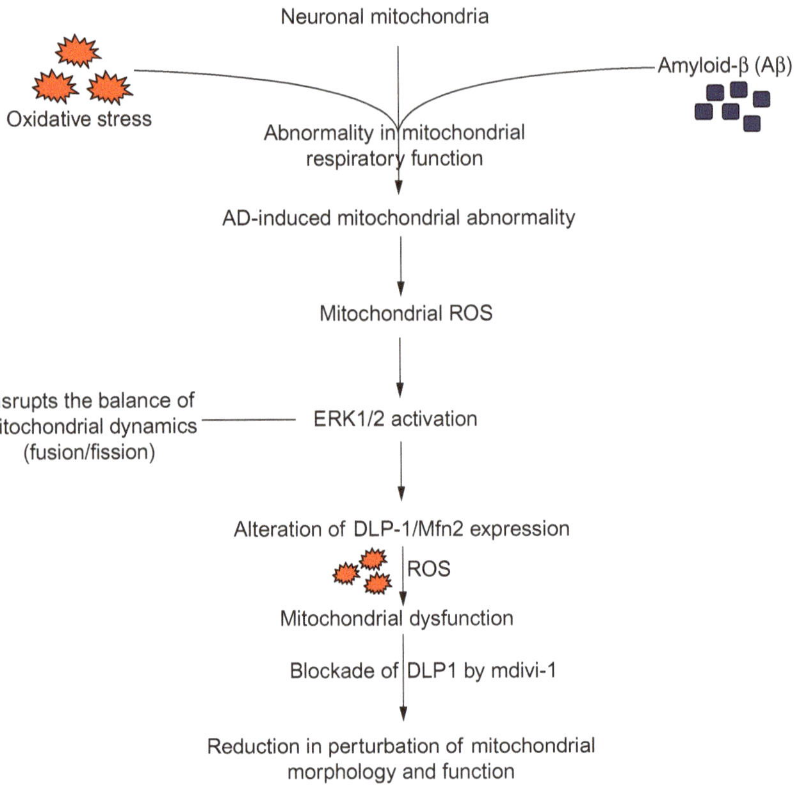

Fig. 2 Effect of AD on mitochondrial dynamics. AD-induced mitochondrial respiratory function abnormality orchestrates ROS generation and accumulation and subsequently activates ERK signal transduction. Activation of ERK signaling disrupts mitochondrial dynamics and results in altered DLP1 and Mfn2 expression, which eventually leads to mitochondrial dysfunction. Inhibition of DLP1 or Mfn2 expression attenuates AD- or MCI-derived mitochondrial and neuronal dysfunction (Mdivi-1, an inhibitor for DLP1).

defective mitochondrial morphology and function induced by AD or MCI[123] (Fig. 2). Antioxidant treatment attenuates AD mitochondrial defects, leading to improvements in axonal mitochondrial transport and mitochondrial bioenergy and function.[4,91]

8. DRP1-MEDIATED MITOCHONDRIAL ABNORMALITIES IN DIABETES

Mitochondria are dynamic organelles that undergo continuous fission and fusion. Fission events are regulated by dynamin-related protein (Drp1), while fusion events are regulated by the large dynamin-related GTPases known as Mfn1 and Mfn2 as well as optic atrophy 1 (OPA1).[131] Alterations in mitochondrial dynamics affect mitochondrial numbers and shape, respiratory enzyme activity, and ATP production. Imbalance between mitochondrial fission and fusion in diabetes results predominantly from upregulation of Drp1, which induces mitochondrial dysfunction (impaired respiration and ATP production) in a variety of cell types, including dorsal root ganglion neurons and β cells.[41] Mitochondrial dysfunction has been implicated in the development of insulin resistance in skeletal muscle cells and hyperglycemia.[132]

A novel and pivotal role of mitochondrial dysfunction in diabetes-induced synaptic impairment involves a GSK3b/Drp1-dependent connection between mitochondrial dysfunction in diabetic neurons and synaptic dysfunction including decline in long-term potentiation. These findings are consistent with diabetic neuropathy as shown by increased Drp1 expression and mitochondrial fission in dorsal root ganglion neurons of 6-month-old type II diabetes (db/db) mice.[2] In contrast to the greater numbers of mitochondria in dorsal root ganglion neurons, hippocampal neurons in 5- to 6-month-old db/db mice displayed smaller numbers of mitochondria, such a decrease was not seen in mice younger than 3 months. Between 3 and 6 months of age, complex I enzyme activity significantly declined by 15%–35% and ATP content was significantly altered. Pharmacologic or genetic inactivation of Drp1 prevented changes in mitochondrial morphology and function in db/db mouse hippocampus or human neuronal cells under hyperglycemic conditions, indicating the role of Drp1 in diabetes-induced mitochondrial dysfunction.[2] Furthermore, genetic activation of GSK3β without high glucose treatment can also promote mitochondrial fragmentation, while inactivation of GSK3β prevents high glucose-induced mitochondrial dysfunction. Taken together, these data suggest that GSK3β likely acts as an upstream signaling mechanism for Drp1 upregulation in diabetes-induced mitochondrial dysfunction.[2]

9. CONCLUSION

Several lines of evidence suggest that age-related AD and diabetes are predominantly associated with mitochondrial dysfunction. Mitochondrial defects result in increased ROS generation, abnormal protein–protein interactions, and decreased mitochondrial ATP production. Overproduction of ROS and mPTP formation with attendant compromised mitochondrial function contribute importantly to neuronal perturbation. Several other

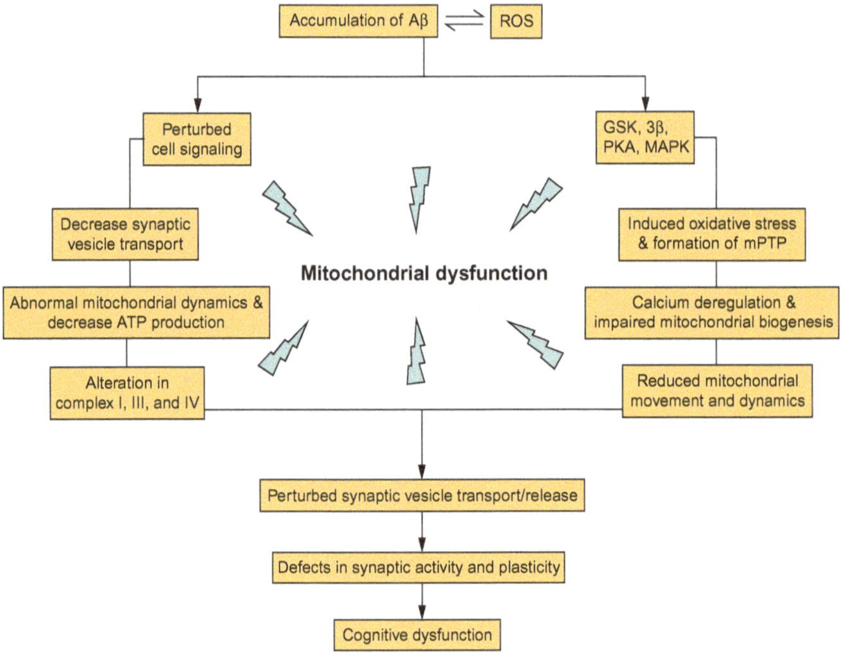

Fig. 3 The cellular factors and related pathways contribute to Aβ-mediated mitochondrial defects and synaptic damage. Aβ accumulation perturbs mitochondrial transport and dynamics, cell signaling, synaptic mitochondrial structure and function, leading to decreased energy metabolism/ATP production, deregulation of calcium homeostasis, perturbed cell signaling cascades, altered key enzymes associated with mitochondrial respiratory chain, induced oxidative stress, and, eventually, synaptic injury and cognitive decline.

factors including intracellular Ca^{2+}, Aβ, and CypD also play an important role in mPTP formation, leading to mitochondrial dysfunction. In addition, disruption of mitochondrial dynamics by altered mitochondrial fusion and fission events contributes to mitochondrial and synaptic injury and cognitive decline relevant to the pathogenesis of AD and diabetes (Fig. 3). Thus, inhibition of mPTP opening by blocking CypD and regulation of mitochondrial dynamics are rational targets for potential therapeutic strategies for AD and diabetes.

REFERENCES

1. Du H, Guo L, Yan SS. Synaptic mitochondrial pathology in Alzheimer's disease. *Antioxid Redox Signal*. 2012;16(12):1467–1475.
2. Huang S, Wang Y, Gan X, et al. Drp1-mediated mitochondrial abnormalities link to synaptic injury in diabetes model. *Diabetes*. 2015;64(5):1728–1742.

3. Du H, ShiDu Yan S. Unlocking the door to neuronal woes in Alzheimer's disease: Abeta and mitochondrial permeability transition pore. *Pharmaceuticals (Basel)*. 2010;3(6):1936–1948.
4. Yu Q, Fang D, Swerdlow RH, et al. Antioxidants rescue mitochondrial transport in differentiated Alzheimer's disease trans-mitochondrial cybrid cells. *J Alzheimers Dis*. 2016;54(2):679–690.
5. Du H, Guo L, Yan S, Sosunov AA. Early deficits in synaptic mitochondria in an Alzheimer's disease mouse model. *Proc Natl Acad Sci USA*. 2010;107(43):18670–18675.
6. Sheng B, Gong K, Niu Y, Liu L. Inhibition of gamma-secretase activity reduces Abeta production, reduces oxidative stress, increases mitochondrial activity and leads to reduced vulnerability to apoptosis: implications for the treatment of Alzheimer's disease. *Free Radic Biol Med*. 2009;46(10):1362–1375.
7. Yao J, Irwin RW, Zhao L, et al. Mitochondrial bioenergetic deficit precedes Alzheimer's pathology in female mouse model of Alzheimer's disease. *Proc Natl Acad Sci USA*. 2009;106(34):14670–14675.
8. Caspersen C, Wang N, Yao J, et al. Mitochondrial Abeta: a potential focal point for neuronal metabolic dysfunction in Alzheimer's disease. *FASEB J*. 2005;19(14): 2040–2041.
9. Manczak M, Anekonda TS, Henson E, et al. Mitochondria are a direct site of A beta accumulation in Alzheimer's disease neurons: implications for free radical generation and oxidative damage in disease progression. *Hum Mol Genet*. 2006;15(9):1437–1449.
10. Takuma K, Yao J, Huang J, et al. ABAD enhances Abeta-induced cell stress via mitochondrial dysfunction. *FASEB J*. 2005;19(6):597–598.
11. Lustbader JW, Cirilli M, Lin C, et al. ABAD directly links Abeta to mitochondrial toxicity in Alzheimer's disease. *Science*. 2004;304(5669):448–452.
12. Reddy PH. Amyloid precursor protein-mediated free radicals and oxidative damage: implications for the development and progression of Alzheimer's disease. *J Neurochem*. 2006;96(1):1–13.
13. Wang X, Su B, Siedlak SL, et al. Amyloid-beta overproduction causes abnormal mitochondrial dynamics via differential modulation of mitochondrial fission/fusion proteins. *Proc Natl Acad Sci USA*. 2008;105(49):19318–19323.
14. Shukkur EA, Shimohata A, Akagi T, et al. Mitochondrial dysfunction and tau hyperphosphorylation in Ts1Cje, a mouse model for Down syndrome. *Hum Mol Genet*. 2006;15(18):2752–2762.
15. Hirai K, Aliev G, Nunomura A, et al. Mitochondrial abnormalities in Alzheimer's disease. *J Neurosci*. 2001;21(9):3017–3023.
16. Crouch PJ, Blake R, Duce JA, et al. Copper-dependent inhibition of human cytochrome c oxidase by a dimeric conformer of amyloid-beta1-42. *J Neurosci*. 2005;25(3):672–679.
17. Devi L, Prabhu BM, Galati DF, et al. Accumulation of amyloid precursor protein in the mitochondrial import channels of human Alzheimer's disease brain is associated with mitochondrial dysfunction. *J Neurosci*. 2006;26(35):9057–9068.
18. Gillardon F, Rist W, Kussmaul L, et al. Proteomic and functional alterations in brain mitochondria from Tg2576 mice occur before amyloid plaque deposition. *Proteomics*. 2007;7(4):605–616.
19. Mungarro-Menchaca X, Ferrera P, Morán J, Arias C. beta-Amyloid peptide induces ultrastructural changes in synaptosomes and potentiates mitochondrial dysfunction in the presence of ryanodine. *J Neurosci Res*. 2002;68(1):89–96.
20. Casley CS, Canevari L, Land JM, et al. Beta-amyloid inhibits integrated mitochondrial respiration and key enzyme activities. *J Neurochem*. 2002;80(1):91–100.
21. Lin MT, Beal MF. Alzheimer's APP mangles mitochondria. *Nat Med*. 2006;12(11): 1241–1243.

22. Reddy PH, Beal MF. Amyloid beta, mitochondrial dysfunction and synaptic damage: implications for cognitive decline in aging and Alzheimer's disease. *Trends Mol Med.* 2008;14(2):45–53.
23. Du H, Guo L, Fang F, et al. Cyclophilin D deficiency attenuates mitochondrial and neuronal perturbation and ameliorates learning and memory in Alzheimer's disease. *Nat Med.* 2008;14(10):1097–1105.
24. Cardoso SM, Santana I, Swerdlow RH, Oliveira CR. Mitochondria dysfunction of Alzheimer's disease cybrids enhances Abeta toxicity. *J Neurochem.* 2004;89(6): 1417–1426.
25. Cardoso SM, et al. Functional mitochondria are required for amyloid beta-mediated neurotoxicity. *FASEB J.* 2001;15(8):1439–1441.
26. Mao P, Manczak M, Calkins MJ, et al. Mitochondria-targeted catalase reduces abnormal APP processing, amyloid beta production and BACE1 in a mouse model of Alzheimer's disease: implications for neuroprotection and lifespan extension. *Hum Mol Genet.* 2012;21(13):2973–2990.
27. Reddy PH, Tripathi R, Troung Q, et al. Abnormal mitochondrial dynamics and synaptic degeneration as early events in Alzheimer's disease: implications to mitochondria-targeted antioxidant therapeutics. *Biochim Biophys Acta.* 2012;1822(5): 639–649.
28. Fang D, Wang Y, Zhang Z, et al. Increased neuronal PreP activity reduces Abeta accumulation, attenuates neuroinflammation and improves mitochondrial and synaptic function in Alzheimer disease's mouse model. *Hum Mol Genet.* 2015;24(18): 5198–5210.
29. Schmidt RE, Parvin CA, Green KG. Synaptic ultrastructural alterations anticipate the development of neuroaxonal dystrophy in sympathetic ganglia of aged and diabetic mice. *J Neuropathol Exp Neurol.* 2008;67(12):1166–1186.
30. Tay SS, Wong WC. Ultrastructural changes in the gracile nucleus of alloxan-induced diabetic rats. *Acta Anat (Basel).* 1990;139(4):367–373.
31. Greenwood CE, Winocur G. High-fat diets, insulin resistance and declining cognitive function. *Neurobiol Aging.* 2005;26(1):42–45.
32. Desrocher M, Rovet J. Neurocognitive correlates of type 1 diabetes mellitus in childhood. *Child Neuropsychol.* 2004;10(1):36–52.
33. Biessels GJ, Kamal A, Ramakers GM, et al. Place learning and hippocampal synaptic plasticity in streptozotocin-induced diabetic rats. *Diabetes.* 1996;45(9):1259–1266.
34. Biessels GJ, Kamal A, Urban IJ, et al. Water maze learning and hippocampal synaptic plasticity in streptozotocin-diabetic rats: effects of insulin treatment. *Brain Res.* 1998;800(1):125–135.
35. Li XL, Aou S, Oomura Y, et al. Impairment of long-term potentiation and spatial memory in leptin receptor-deficient rodents. *Neuroscience.* 2002;113(3):607–615.
36. Stranahan AM, Arumugam TV, Cutler RG, et al. Diabetes impairs hippocampal function through glucocorticoid-mediated effects on new and mature neurons. *Nat Neurosci.* 2008;11(3):309–317.
37. Zorzano A, Liesa M, Palacin M. Role of mitochondrial dynamics proteins in the pathophysiology of obesity and type 2 diabetes. *Int J Biochem Cell Biol.* 2009;41(10): 1846–1854.
38. Yoon Y, Galloway CA, Jhun BS, Yu T. Mitochondrial dynamics in diabetes. *Antioxid Redox Signal.* 2011;14(3):439–457.
39. Yaffe K, Lindquist K, Schwartz AV, et al. Advanced glycation end product level, diabetes, and accelerated cognitive aging. *Neurology.* 2011;77(14):1351–1356.
40. Yu T, Sheu SS, Robotham JL, Yoon Y. Mitochondrial fission mediates high glucose-induced cell death through elevated production of reactive oxygen species. *Cardiovasc Res.* 2008;79(2):341–351.

41. Edwards JL, Quattrini A, Lentz SI, et al. Diabetes regulates mitochondrial biogenesis and fission in mouse neurons. *Diabetologia*. 2010;53(1):160–169.
42. Clodfelder-Miller BJ, Zmijewska AA, Johnson GV, et al. Tau is hyperphosphorylated at multiple sites in mouse brain in vivo after streptozotocin-induced insulin deficiency. *Diabetes*. 2006;55(12):3320–3325.
43. Planel E, Tatebayashi Y, Miyasaka T, et al. Insulin dysfunction induces in vivo tau hyperphosphorylation through distinct mechanisms. *J Neurosci*. 2007;27(50):13635–13648.
44. Ke YD, Delerue F, Gladbach A, et al. Experimental diabetes mellitus exacerbates tau pathology in a transgenic mouse model of Alzheimer's disease. *PLoS One*. 2009;4(11): e7917.
45. Li ZG, Zhang W, Sima AA. Alzheimer-like changes in rat models of spontaneous diabetes. *Diabetes*. 2007;56(7):1817–1824.
46. Zhao WQ, Townsend M. Insulin resistance and amyloidogenesis as common molecular foundation for type 2 diabetes and Alzheimer's disease. *Biochim Biophys Acta*. 2009;1792(5):482–496.
47. Palmeira CM, Rolo AP, Berthiaume J, et al. Hyperglycemia decreases mitochondrial function: the regulatory role of mitochondrial biogenesis. *Toxicol Appl Pharmacol*. 2007;225(2):214–220.
48. Makino A, Scott BT, Dillmann WH. Mitochondrial fragmentation and superoxide anion production in coronary endothelial cells from a mouse model of type 1 diabetes. *Diabetologia*. 2010;53(8):1783–1794.
49. Yu T, Robotham JL, Yoon Y. Increased production of reactive oxygen species in hyperglycemic conditions requires dynamic change of mitochondrial morphology. *Proc Natl Acad Sci USA*. 2006;103(8):2653–2658.
50. Galloway CA, Lee H, Nejjar S, et al. Transgenic control of mitochondrial fission induces mitochondrial uncoupling and relieves diabetic oxidative stress. *Diabetes*. 2012;61(8):2093–2104.
51. Rosca MG, Mustata TG, Kinter MT, et al. Glycation of mitochondrial proteins from diabetic rat kidney is associated with excess superoxide formation. *Am J Physiol Renal Physiol*. 2005;289(2):420–430.
52. Toth C, Schmidt AM, Tuor UI, et al. Diabetes, leukoencephalopathy and rage. *Neurobiol Dis*. 2006;23(2):445–461.
53. Takeda S, Sato N, Uchio-Yamada K, et al. Diabetes-accelerated memory dysfunction via cerebrovascular inflammation and Abeta deposition in an Alzheimer mouse model with diabetes. *Proc Natl Acad Sci USA*. 2010;107(15):7036–7041.
54. Jolivalt CG, Hurford R, Lee CA, et al. Type 1 diabetes exaggerates features of Alzheimer's disease in APP transgenic mice. *Exp Neurol*. 2010;223(2):422–431.
55. Burdo JR, Chen Q, Calcutt NA, et al. The pathological interaction between diabetes and presymptomatic Alzheimer's disease. *Neurobiol Aging*. 2009;30(12):1910–1917.
56. Wang Y, Wu L, Li J, et al. Synergistic exacerbation of mitochondrial and synaptic dysfunction and resultant learning and memory deficit in a mouse model of diabetic Alzheimer's disease. *J Alzheimers Dis*. 2015;43(2):451–463.
57. Alonso-Nanclares L, Gonzalez-Soriano J, Rodriguez JR, et al. Gender differences in human cortical synaptic density. *Proc Natl Acad Sci USA*. 2008;105(38):14615–14619.
58. Connors BW, Long MA. Electrical synapses in the mammalian brain. *Annu Rev Neurosci*. 2004;27:393–418.
59. Shepherd GM, Harris KM. Three-dimensional structure and composition of CA3 → CA1 axons in rat hippocampal slices: implications for presynaptic connectivity and compartmentalization. *J Neurosci*. 1998;18(20):8300–8310.
60. David G, Barrett EF. Mitochondrial Ca2+ uptake prevents desynchronization of quantal release and minimizes depletion during repetitive stimulation of mouse motor nerve terminals. *J Physiol*. 2003;548(Pt 2):425–438.

61. Martinez M, Hernández AI, Martínez N, et al. Age-related increase in oxidized proteins in mouse synaptic mitochondria. *Brain Res*. 1996;731(1-2):246–248.
62. Selkoe DJ. Alzheimer's disease is a synaptic failure. *Science*. 2002;298(5594):789–791.
63. Brown MR, Sullivan PG, Geddes JW. Synaptic mitochondria are more susceptible to Ca2+overload than nonsynaptic mitochondria. *J Biol Chem*. 2006;281(17): 11658–11668.
64. Wang L, Guo L, Lu L, et al. Synaptosomal mitochondrial dysfunction in 5xFAD mouse model of Alzheimer's disease. *PLoS One*. 2016;11(3):e0150441.
65. Yan S, Du F, Wu L, et al. F1F0 ATP synthase-cyclophilin D interaction contributes to diabetes-induced synaptic dysfunction and cognitive decline. *Diabetes*. 2016;65(11): 3482–3494.
66. Fernández-Vizarra P, Fernández AP, Castro-Blanco S, et al. Intra- and extracellular Abeta and PHF in clinically evaluated cases of Alzheimer's disease. *Histol Histopathol*. 2004;19(3):823–844.
67. Hansson Petersen CA, Alikhani N, Behbahani H, et al. The amyloid beta-peptide is imported into mitochondria via the TOM import machinery and localized to mitochondrial cristae. *Proc Natl Acad Sci USA*. 2008;105(35):13145–13150.
68. Abramowski D, Rabe S, Upadhaya AR, et al. Transgenic expression of intraneuronal Abeta42 but not Abeta40 leads to cellular Abeta lesions, degeneration, and functional impairment without typical Alzheimer's disease pathology. *J Neurosci*. 2012;32(4): 1273–1283.
69. Takuma K, Fang F, Zhang W, et al. RAGE-mediated signaling contributes to intraneuronal transport of amyloid-beta and neuronal dysfunction. *Proc Natl Acad Sci USA*. 2009;106(47):20021–20026.
70. Teng FY, Tang BL. Widespread gamma-secretase activity in the cell, but do we need it at the mitochondria? *Biochem Biophys Res Commun*. 2005;328(1):1–5.
71. Pavlov PF, Wiehager B, Sakai J, et al. Mitochondrial gamma-secretase participates in the metabolism of mitochondria-associated amyloid precursor protein. *FASEB J*. 2011;25(1):78–88.
72. Behbahani H, Pavlov PF, Wiehager B, et al. Association of Omi/HtrA2 with gamma-secretase in mitochondria. *Neurochem Int*. 2010;7(6):668–675.
73. Chui DH, Dobo E, Makifuchi T, et al. Apoptotic neurons in Alzheimer's disease frequently show intracellular Abeta42 labeling. *J Alzheimers Dis*. 2001;3(2):231–239.
74. Chui DH, Tanahashi H, Ozawa K, et al. Transgenic mice with Alzheimer presenilin 1 mutations show accelerated neurodegeneration without amyloid plaque formation. *Nat Med*. 1999;5(5):560–564.
75. Oddo S, Caccamo A, Shepherd JD, et al. Triple-transgenic model of Alzheimer's disease with plaques and tangles: intracellular Abeta and synaptic dysfunction. *Neuron*. 2003;39(3):409–421.
76. Hsia AY, Masliah E, McConlogue L, et al. Plaque-independent disruption of neural circuits in Alzheimer's disease mouse models. *Proc Natl Acad Sci USA*. 1999;96(6): 3228–3233.
77. Li QX, Maynard C, Cappai R, et al. Intracellular accumulation of detergent-soluble amyloidogenic A beta fragment of Alzheimer's disease precursor protein in the hippocampus of aged transgenic mice. *J Neurochem*. 1999;72(6):2479–2487.
78. Takahashi RH, Milner TA, Li F, et al. Intraneuronal Alzheimer abeta42 accumulates in multivesicular bodies and is associated with synaptic pathology. *Am J Pathol*. 2002;161(5):1869–1879.
79. Reddy PH, McWeeney S, Park BS, et al. Gene expression profiles of transcripts in amyloid precursor protein transgenic mice: up-regulation of mitochondrial metabolism and apoptotic genes is an early cellular change in Alzheimer's disease. *Hum Mol Genet*. 2004;13(12):1225–1240.

80. Reddy PH, Mani G, Park BS, et al. Differential loss of synaptic proteins in Alzheimer's disease: implications for synaptic dysfunction. *J Alzheimers Dis.* 2005;7(2):103–117.
81. Nunomura A, Perry G, Aliev G, et al. Oxidative damage is the earliest event in Alzheimer disease. *J Neuropathol Exp Neurol.* 2001;60(8):759–767.
82. Billings LM, Oddo S, Green KN, et al. Intraneuronal Abeta causes the onset of early Alzheimer's disease-related cognitive deficits in transgenic mice. *Neuron.* 2005;45(5): 675–688.
83. Leissring MA, Farris W, Wu X, et al. Alternative translation initiation generates a novel isoform of insulin-degrading enzyme targeted to mitochondria. *Biochem J.* 2004;383(Pt. 3):439–446.
84. Yan SD, Fu J, Soto C, et al. An intracellular protein that binds amyloid-beta peptide and mediates neurotoxicity in Alzheimer's disease. *Nature.* 1997;389(6652):689–695.
85. Du H, Guo L, Zhang W, et al. Cyclophilin D deficiency improves mitochondrial function and learning/memory in aging Alzheimer disease mouse model. *Neurobiol Aging.* 2009;32(3):398–406.
86. Du H, Yan SS. Mitochondrial permeability transition pore in Alzheimer's disease: cyclophilin D and amyloid beta. *Biochim Biophys Acta.* 2009;1802(1):198–204.
87. Du H, Guo L, Zhang W, et al. Cyclophilin D deficiency improves mitochondrial function and learning/memory in aging Alzheimer disease mouse model. *Neurobiol Aging.* 2011;32(3):398–406.
88. Yao J, Du H, Yan S, et al. Inhibition of Amyloid-{beta} (A{beta}) peptide-binding alcohol dehydrogenase-A{beta} interaction reduces A{beta} accumulation and improves mitochondrial function in a mouse model of Alzheimer's disease. *J Neurosci.* 2011;31(6):2313–2320.
89. Manczak M, Mao P, Calkins MJ, et al. Mitochondria-targeted antioxidants protect against amyloid-beta toxicity in Alzheimer's disease neurons. *J Alzheimers Dis.* 2010;20(2):609–631.
90. Yin X, Manczak M, Reddy PH. Mitochondria-targeted molecules MitoQ and SS31 reduce mutant huntingtin-induced mitochondrial toxicity and synaptic damage in Huntington's disease. *Hum Mol Genet.* 2016;25(9):1739–1753.
91. Fang D, Zhang Z, Li H, et al. Increased electron paramagnetic resonance signal correlates with mitochondrial dysfunction and oxidative stress in an Alzheimer's disease mouse brain. *J Alzheimers Dis.* 2016;51(2):571–580.
92. Du H, Yan SS. Mitochondrial permeability transition pore in Alzheimer's disease: cyclophilin D and amyloid beta. *Biochim Biophys Acta.* 2010;1802(1):198–204.
93. Martin LJ, Semenkow S, Hanaford A, et al. Mitochondrial permeability transition pore regulates Parkinson's disease development in mutant alpha-synuclein transgenic mice. *Neurobiol Aging.* 2014;35(5):1132–1152.
94. Martin LJ, Gertz B, Pan Y, et al. The mitochondrial permeability transition pore in motor neurons: involvement in the pathobiology of ALS mice. *Exp Neurol.* 2009;218(2):333–346.
95. Martin LJ. The mitochondrial permeability transition pore: a molecular target for amyotrophic lateral sclerosis therapy. *Biochim Biophys Acta.* 2010;1802(1):186–197.
96. Dougherty SE, Hollimon JJ, McMeekin LJ, et al. Hyperactivity and cortical disinhibition in mice with restricted expression of mutant huntingtin to parvalbumin-positive cells. *Neurobiol Dis.* 2014;62:160–171.
97. Brustovetsky N, Brustovetsky T, Purl KJ, et al. Increased susceptibility of striatal mitochondria to calcium-induced permeability transition. *J Neurosci.* 2003;23(12): 4858–4867.
98. Wang Y, Wu L, Li J, et al. Synergistic exacerbation of mitochondrial and synaptic dysfunction and resultant learning and memory deficit in a mouse model of diabetic Alzheimer's disease. *J Alzheimers Dis.* 2014;43(2):451–463.

99. Zhang Z, Wang Y, Yan S, et al. NR2B-dependent cyclophilin D translocation suppresses the recovery of synaptic transmission after oxygen-glucose deprivation. *Biochim Biophys Acta*. 2015;1852(10):2225–2234.
100. Du H, Guo L, Wu X, et al. Cyclophilin D deficiency rescues Abeta-impaired PKA/CREB signaling and alleviates synaptic degeneration. *Biochim Biophys Acta*. 2013;1842(12):2517–2527.
101. Guo L, Du H, Yan S, et al. Cyclophilin D deficiency rescues axonal mitochondrial transport in Alzheimer's neurons. *PLoS One*. 2013;8(1):e54914.
102. Du H, Guo L, Wu X, et al. Cyclophilin D deficiency rescues Abeta-impaired PKA/CREB signaling and alleviates synaptic degeneration. *Biochim Biophys Acta*. 2014;1842(12):2517–2527.
103. Valasani KR, Vangavaragu JR, Day VW, et al. Structure based design, synthesis, pharmacophore modeling, virtual screening, and molecular docking studies for identification of novel cyclophilin D inhibitors. *J Chem Inf Model*. 2014;54(3):902–912.
104. Valasani KR, Sun Q, Fang D, et al. Identification of a small molecule cyclophilin D inhibitor for rescuing abeta-mediated mitochondrial dysfunction. *ACS Med Chem Lett*. 2016;7(3):294–299.
105. Falkevall A, Alikhani N, Bhushan S, et al. Degradation of the amyloid beta-protein by the novel mitochondrial peptidasome, PreP. *J Biol Chem*. 2006;281(39):29096–29104.
106. Alikhani N, Guo L, Yan S, et al. Decreased proteolytic activity of the mitochondrial amyloid-beta degrading enzyme, PreP peptidasome, in Alzheimer's disease brain mitochondria. *J Alzheimers Dis*. 2011;27(1):75–87.
107. Fang F, Lue LF, Yan S, et al. RAGE-dependent signaling in microglia contributes to neuroinflammation, Abeta accumulation, and impaired learning/memory in a mouse model of Alzheimer's disease. *FASEB J*. 2010;24(4):1043–1055.
108. Zhang H, Wang Y, Yan S, et al. Genetic deficiency of neuronal RAGE protects against AGE-induced synaptic injury. *Cell Death Dis*. 2014;5:e1288.
109. Origlia N, Bonadonna C, Rosellini A, et al. Microglial receptor for advanced glycation end product-dependent signal pathway drives beta-amyloid-induced synaptic depression and long-term depression impairment in entorhinal cortex. *J Neurosci*. 2010;30(34):11414–11425.
110. Origlia N, Criscuolo C, Arancio O, et al. RAGE inhibition in microglia prevents ischemia-dependent synaptic dysfunction in an amyloid-enriched environment. *J Neurosci*. 2014;34(26):8749–8760.
111. Origlia N, Capsoni S, Cattaneo A, et al. Abeta-dependent inhibition of LTP in different intracortical circuits of the visual cortex: the role of RAGE. *J Alzheimers Dis*. 2009;17(1):59–68.
112. Origlia N, Righi M, Capsoni S, et al. Receptor for advanced glycation end product-dependent activation of p38 mitogen-activated protein kinase contributes to amyloid-beta-mediated cortical synaptic dysfunction. *J Neurosci*. 2008;28(13):3521–3530.
113. Yan SD, Chen X, Fu J, et al. RAGE and amyloid-beta peptide neurotoxicity in Alzheimer's disease. *Nature*. 1996;382(6593):685–691.
114. Näslund J, Schierhorn A, Hellman U, et al. Relative abundance of Alzheimer A beta amyloid peptide variants in Alzheimer disease and normal aging. *Proc Natl Acad Sci USA*. 1994;91(18):8378–8382.
115. Moskovitz J, Du F, Bowman CF, Yan SS. Methionine sulfoxide reductase A affects beta-amyloid solubility and mitochondrial function in a mouse model of Alzheimer's disease. *Am J Physiol Endocrinol Metab*. 2016;310(6):388–393.
116. Frazier AE, Kiu C, Stojanovski D, et al. Mitochondrial morphology and distribution in mammalian cells. *Biol Chem*. 2006;387(12):1551–1558.

117. Wang X, Su B, Lee HG, et al. Impaired balance of mitochondrial fission and fusion in Alzheimer's disease. *J Neurosci*. 2009;29(28):9090–9103.
118. Chan DC. Mitochondria: dynamic organelles in disease, aging, and development. *Cell*. 2006;125(7):1241–1252.
119. Fang D, Yan S, Yu Q, et al. Mfn2 is required for mitochondrial development and synapse formation in human induced pluripotent stem cells/hiPSC derived cortical neurons. *Sci Rep*. 2016;6:31462.
120. Pham AH, Meng S, Chu QN, et al. Loss of Mfn2 results in progressive, retrograde degeneration of dopaminergic neurons in the nigrostriatal circuit. *Hum Mol Genet*. 2012;21(22):4817–4826.
121. Karakaya T, Fußer F, Schröder J, Pantel J. Pharmacological treatment of mild cognitive impairment as a prodromal syndrome of Alzheimer s disease. *Curr Neuropharmacol*. 2013;11(1):102–108.
122. Gan X, Wu L, Huang S, et al. Oxidative stress-mediated activation of extracellular signal-regulated kinase contributes to mild cognitive impairment-related mitochondrial dysfunction. *Free Radic Biol Med*. 2014;75:230–240.
123. Gan X, Huang S, Wu L, et al. Inhibition of ERK-DLP1 signaling and mitochondrial division alleviates mitochondrial dysfunction in Alzheimer's disease cybrid cell. *Biochim Biophys Acta*. 2014;1842(2):220–231.
124. Manczak M, Reddy PH. Abnormal interaction between the mitochondrial fission protein Drp1 and hyperphosphorylated tau in Alzheimer's disease neurons: implications for mitochondrial dysfunction and neuronal damage. *Hum Mol Genet*. 2012;21(11): 2538–2547.
125. Chou CH, Lin CC, Yang MC, et al. GSK3beta-mediated Drp1 phosphorylation induced elongated mitochondrial morphology against oxidative stress. *PLoS One*. 2012;7(11):e49112.
126. Hong YR, Chen CH, Cheng DS, et al. Human dynamin-like protein interacts with the glycogen synthase kinase 3beta. *Biochem Biophys Res Commun*. 1998;249(3):697–703.
127. Chen CH, Hwang SL, Howng SL, et al. Three rat brain alternative splicing dynamin-like protein variants: interaction with the glycogen synthase kinase 3beta and action as a substrate. *Biochem Biophys Res Commun*. 2000;268(3):893–898.
128. Park J, Choi H, Min JS, et al. Mitochondrial dynamics modulate the expression of pro-inflammatory mediators in microglial cells. *J Neurochem*. 2013;127(2):221–232.
129. Motori E, Puyal J, Toni N, et al. Inflammation-induced alteration of astrocyte mitochondrial dynamics requires autophagy for mitochondrial network maintenance. *Cell Metab*. 2013;18(6):844–859.
130. Dong XB, Yang CT, Zheng DD, et al. Inhibition of ROS-activated ERK1/2 pathway contributes to the protection of H2S against chemical hypoxia-induced injury in H9c2 cells. *Mol Cell Biochem*. 2012;362(1-2):149–157.
131. Palmer CS, Osellame LD, Stojanovski D, et al. The regulation of mitochondrial morphology: intricate mechanisms and dynamic machinery. *Cell Signal*. 2011;23(10): 1534–1545.
132. Lowell BB, Shulman GI. Mitochondrial dysfunction and type 2 diabetes. *Science*. 2005;307(5708):384–387.

INDEX

Note: Page numbers followed by "*f*" indicate figures, and "*t*" indicate tables.

N

FSC
www.fsc.org
MIX
Papier | Fördert
gute Waldnutzung
FSC® C083411

Kolibri 360 GmbH
Ferdinand-Jühlke-Straße 7
99095 Erfurt, Deutschland
produktsicherheit@kolibri360.de